Organic Crop Production – Ambitions and Limitations

Holger Kirchmann · Lars Bergström
Editors

Organic Crop Production – Ambitions and Limitations

Springer

Editors
Dr. Holger Kirchmann
Swedish University
of Agricultural Sciences
Dept. of Soil and Environment
P.O. Box 7014
SE-750 07 Uppsala
Sweden

Dr. Lars Bergström
Swedish University
of Agricultural Sciences
Dept. of Soil and Environment
P.O. Box 7014
SE-750 07 Uppsala
Sweden

ISBN 978-1-4020-9315-9 e-ISBN 978-1-4020-9316-6

DOI 10.1007/978-1-4020-9316-6

Library of Congress Control Number: 2008938183

Printed on acid-free paper

9 8 7 6 5 4 3 2 1

springer.com

Preface

Organic agriculture is being promoted against a background of intensive discussions about production methods, food and feed quality and renewable resources, with the overall aim of long-term sustainability. Organic agriculture is a subject that triggers many different responses in people. Some are convinced that it is the way forward, while others question its benefits and the wisdom of its large-scale implementation. Even among the scientific community, different views have developed over recent decades.

Organic agriculture is promoted in a number of popular and scientific books and is often described as being superior, the solution to common agricultural problems and a means of producing better food. Organic agriculture is often viewed as being environmentally sound and superior to conventional agriculture through the exclusion of synthetic fertilisers and pesticides. As a result, any questioning of organic practices is unpopular and criticism is often interpreted as impeding the development of sustainable systems. In addition, scientifically-based information contradicting the claims made for organic agriculture can be difficult to communicate and can be regarded as a step backwards and against political mainstream opinion.

The topic was discussed at a Symposium at the 18th World Congress of Soil Science in Philadelphia in 2006, where benefits and problems relating to organic crop production were presented. Some of the key findings from that symposium are presented in this book, together with other central aspects of organic crop production. The aim of this book is to provide the readers with a clear, scientifically-based overview of a number of relevant subjects relating to organic crop production so that they can form a balanced picture of this food production approach.

We are very thankful to all the contributing authors for providing their in-depth views in the various chapters. We would also like to acknowledge all the anonymous reviewers who helped to improve the quality of the different chapters and Dr Mary McAfee for excellent linguistic advice. Finally, we would like to thank Springer for publishing the book, which we hope will provide a better understanding of true long-term sustainability in future crop production.

Uppsala, Sweden
July 2008

Holger Kirchmann
Lars Bergström

Contents

Contributors

Rune Andersson Department of Food Science, Swedish University of Agricultural Sciences, P.O. Box 7051, SE-75007 Uppsala, Sweden

Olof Andrén Department of Soil and Environment, Swedish University of Agricultural Sciences, P.O. Box 7014, SE-75007 Uppsala, Sweden, olof.andren@mark.slu.se

Helena Aronsson Department of Soil and Environment, Swedish University of Agricultural Sciences, P.O. Box 7014, SE-75007 Uppsala, Sweden

Lars Bergström Department of Soil and Environment, Swedish University of Agricultural Sciences, P.O. Box 7014, SE-75007 Uppsala, Sweden, lars.bergstrom@mark.slu.se

Göte Bertilsson Greengard AB, P.O. Box 552, SE-24495 Dösjebro, Sweden, bertilsson@greengard.se

Ragnar Eltun Bioforsk Arable Crops Division, Norwegian Institute for Agricultural and Environmental Research, Apelsvoll, N-2849 Kapp, Norway

Lars-Olov Eriksson Johannelunds Theological University College, Heidenstamstorg 75, SE-75427 Uppsala, Sweden

Martin Gerzabek Institute of Soil Research, University of Natural Resources and Applied Life Sciences, Peter-Jordan-Straße 82, A-1190 Vienna, Austria

Keith Goulding Cross-Institute Programme for Sustainable Soil Function, Rothamsted Research, Harpenden, Hertfordshire, AL5 2JQ, UK, keith.goulding@bbsrc.ac.uk

Thomas Kätterer Department of Soil and Environment, Swedish University of Agricultural Sciences, P.O. Box 7014, SE-75007 Uppsala, Sweden

Holger Kirchmann Department of Soil and Environment, Swedish University of Agricultural Sciences, P.O. Box 7014, SE-75007 Uppsala, Sweden, holger.kirchmann@mark.slu.se

Audun Korsaeth Bioforsk Arable Crops Division, Norwegian Institute for Agricultural and Environmental Research, Apelsvoll, N-2849 Kapp, Norway, audun.korsaeth@bioforsk.no

Lennart Mattsson Department of Soil and Environment, Swedish University of Agricultural Sciences, P.O. Box 7014, SE-75007 Uppsala, Sweden

Megan H. Ryan School of Plant Biology, University of Western Australia, 35 Stirling Hwy, Crawley, WA, Australia, 6009, megan.ryan@uwa.edu.au

Elizabeth Stockdale School of Agriculture Food and Rural Development, King George VI Building, Newcastle University, Newcastle-upon-Tyne, NE1 7RU, UK

Mark Tibbett School of Earth and Geographical Sciences, Centre for Land Rehabilitation, University of Western Australia, 35 Stirling Hwy, Crawley, WA, Australia, 6009

Gudni Thorvaldsson Agricultural University of Iceland, Keldnaholti, IS-112 Reykjavik, Iceland

Gunnar Torstensson Department of Soil and Environment, Swedish University of Agricultural Sciences, P.O. Box 7014, SE-75007 Uppsala, Sweden

Christine Watson SAC Aberdeen, Ferguson Building, Craibstone Estate, Aberdeen, AB21 9YA, UK

Mikael Winninge Department of Religious Studies, Umeå University, SE-90187 Umeå, Sweden

Carl K. Winter Department of Food Science and Technology, University of California Davis, CA 95616, USA, ckwinter@ucdavis.edu

Chapter 1
Widespread Opinions About Organic Agriculture – Are They Supported by Scientific Evidence?

Lars Bergström, Holger Kirchmann and Gudni Thorvaldsson

Abstract Organic agriculture ostensibly offers a concept of sustainable practices based on environmental responsibility. It is widely believed that organic principles based on natural means and methods are environmentally sound and thus superior to systems based on artificial inputs. This overview summarises the main results on organic agriculture and highlights relevant facts in order to provide scientific information about the potential and limitations of organic agriculture. The topics of food security and safety, environmental quality, system sustainability and energy consumption are addressed. Some of the main conclusions are that organic agriculture has consistently lower yields than conventional production and is thereby a less efficient method of land use; that environmental problems caused by processes such as nutrient leaching are not reduced by conversion to organic crop production; and that soil fertility status and microbial biodiversity are not improved a priori by organic cropping. The energy investment for production of artificial N fertilisers results in a five- to ten-fold energy return in the form of biomass and this highly positive energy balance needs to be fully acknowledged. The future challenge of developing sustainable forms of agriculture to provide sufficient food for a growing world population with minimal environmental disturbance deserves our wholehearted and unbiased attention.

Keywords Carbon sequestration · Energy issues · Food production · Natural toxins · Nutrient leaching · Pesticide residues · Soil fertility

1.1 Introduction

During the past two decades, organic agriculture has often been presented as being superior to conventional production in many respects. This has led to a widespread belief among the general public that organic crop production is better and in an ambition to satisfy this opinion, politicians and legislators have strongly promoted

---

L. Bergström (✉)
Department of Soil and Environment, Swedish University of Agricultural Sciences, P.O. Box 7014, SE-75007 Uppsala, Sweden
e-mail: lars.bergstrom@mark.slu.se

H. Kirchmann, L. Bergström (eds.), *Organic Crop Production – Ambitions and Limitations*, DOI 10.1007/978-1-4020-9316-6_1,

this type of agriculture. For example, in Sweden the political goal has been to increase the area used for organic production to 20% of all arable soils. In addition, organically produced food should constitute 25% of the food used in state schools, hospitals and residential homes for the elderly. The main driving forces have been to generate benefits for the environment and to improve food quality. More recently, impacts on energy consumption, climate change and long-term sustainability have also been in focus.

Buying organically produced food, and thereby supporting all these benefits, creates a feelgood factor for consumers of organically produced food. For many people, organically produced food is always better even if it requires long-distance transport. However, the question is whether these common opinions circulating in society are supported by the scientific evidence.

The principles of organic practices derive from natural philosophies and not natural sciences. A deeper scientifically-based analysis of long-term organic field experiments – which is the main focus of this book – gives a different picture of the benefits generated. However, scientific comparisons of organic and conventional farming are unappealing to society since they provide evidence that man-made inventions, such as artificial fertilisers, often lead to production of crops in larger quantities and of good quality. This is in contrast to the common belief that we should always follow rules determined by nature.

In this introductory chapter, we present some common opinions about organic agriculture and discuss them briefly in the context of the results presented in the other chapters of this book. The following topics relating to organic agriculture are addressed:

- Food security and safety
- Environmental quality
- System sustainability
- Energy consumption.

1.2 Widespread Opinions Versus Scientific Evidence

1.2.1 Food Issues

Since the introduction of organic agriculture in the 1920s, food quality issues have been a particular focus. One of the founders of organic agriculture, Rudolf Steiner (1924), believed that artificial fertilisers would degenerate agricultural produce to such an extent that they would not be fit for human consumption by the end of the 20th century. Concern about levels of yield only entered the agenda much later.

1.2.1.1 Food Security

A general opinion in society is that conversion to organic crop production is followed by little or no yield reduction and that organic crop production is therefore capable of feeding the world. In fact, some researchers claim that the solution for famine in Africa is large-scale organic agriculture (Pretty et al., 2003).

Our conclusion is that organic agriculture cannot feed the world, because there is substantial scientific evidence that crop yields are considerably lower in organic systems. The long-term yield reduction could be as much as 40–50% compared with the corresponding conventional crops. Therefore, to obtain equivalent yields in organic systems, significantly more land would be needed for agricultural crops. However, according to recent assessments, such land is not available in the world. It is worthwhile mentioning that most good agricultural soils are already under cultivation and that additional crop production would have to use soils of low fertility or with a high risk of erosion or other degradation processes when cropped.

A 40% yield reduction in developed countries would require 67% more agricultural land to produce the same amount of crops (Chapter 3; Kirchmann et al., 2008a). This does not take into account future population growth, which will primarily occur in developing countries where the situation regarding crop production is already critical in many cases. In a world perspective, population growth is expected to be 50% within the next couple of decades. Crop production in developing countries is largely limited by the lack of artificial fertilisers, water and crop protection strategies. In those systems, crop yields can only be increased by providing such inputs and methods. In this context, it is worth mentioning that a key conclusion presented at the FAO meeting in Rome 2008 by the Secretary General of the UN was that one of the most important ways out of starvation in developing countries is increased use of artificial fertilisers. Chapter 3 (Kirchmann et al., 2008a) presents several arguments concerning the need for increasing yields in developing countries.

Irrespective of the different situations prevailing in developed and developing countries, there is no doubt that when considering the population growth aspect and applying only organic production methods, land demand for crop production would increase considerably.

1.2.1.2 Food Safety

The current opinion is that organic food is healthier since it does not contain toxins and is free from artificial pesticide residues.

The use of pesticides is strictly regulated in developed countries, including pesticide residue levels in food. In fact, almost all pesticides with documented negative side-effects have been taken off the market and cannot be used for agricultural crops. Nevertheless, the low pesticide residue levels that are still detectable in food must be evaluated from a toxicological and potential health risk perspective, which must include appropriate safety margins.

The risks involved with pesticide levels in food products must be put into perspective and related to the risks of being exposed to other toxic substances in food products. Crop products generally contain natural toxins, compounds that in many cases are more toxic than pesticides. As shown in Chapter 11 (Winter, 2008), there is still insufficient conclusive scientific evidence that any form of production can reduce the levels of natural toxins. Nevertheless, it is important to stress that the levels of natural toxins are commonly several orders of magnitude higher in food products than pesticide residues. It is also worth mentioning that exposure, especially over the long-term, to low levels of toxic substances is difficult to evaluate. For example, lifetime exposure to pesticide residues by drinking 2 litres per day of water with levels at the EU drinking water criterion ($0.1\ \mu g\ L^{-1}$) would be less than the exposure to chemicals through ingesting one of most common medical pills. This stresses the importance of proper assessment when evaluating health risks. Furthermore, a recent scientific opinion is that substances that are toxic at higher concentrations can in fact be good for human health at low levels, since they trigger the immune system. This phenomenon is called the hormesis effect (Trewavas and Stewart, 2003).

Pesticides are used to protect the crop from infestation by fungi, insects and other pests in a similar way to humans taking medicine when they have a health problem. The crop itself also produces substances that prevent plant tissues from pest damage. Several such examples that are of possible concern to human health are mentioned in Chapter 11 (Winter, 2008). When a crop is not protected from diseases by appropriate pesticides, there is risk of the production of natural plant defence compounds (secondary metabolites) being increased.

It is noteworthy that some pesticides such as pyrethroids and copper sulphate are used in organic farming although they carry documented environmental and health risks (Felsot and Racke, 2006).

1.2.2 Environmental Issues

As pointed out in Chapter 2 (Kirchmann et al., 2008b), environmental concern was not addressed by the founders of organic agriculture. It was first during the 1960s that organic methods were presented as a solution to the emerging environmental problems caused by agriculture. For example, in the book ‘Silent Spring’ Carson (1962) provided evidence that the pesticides in use at that time were having a detrimental effect on nature. Since then, environmental issues have been used frequently as the main argument for the superiority of organic agriculture. Issues relating to climatic change and biodiversity are currently also on the list.

1.2.2.1 Nutrient Leaching

A common opinion is that widespread use of artificial nitrogen fertilisers causes water quality disturbance such as eutrophication of lakes and coastal waters. By exclusively using organic manures in organic agriculture, this problem is automatically solved since organic manures are adapted to nature and cause less leaching of nutrients.

The use of artificial fertilisers dramatically intensified during the 1950s and 1960s and resulted in large yield increases in agricultural crops. However, increasing levels of nitrate were simultaneously observed in surface waters and groundwaters and it was logical to couple this to the intensive use of fertilisers. In many cases, fertiliser use during these early days was excessive since fertilisers were relatively inexpensive and much more nutrients were applied than the crop needed.

From the mid-1980s onwards, leaching of nitrogen has levelled out and in many cases decreased due to better fertiliser management and introduction of a number of efficient countermeasures such as no autumn application of nitrogen fertilisers, inclusion of cover crops and reduced tillage practices. If fertilisers are applied at rates matching crop demand, significant leaching of nitrogen normally does not occur (Bergström and Brink, 1986; Lord and Mitchell, 1998).

One of the main reasons cited by advocates of organic farming for the superiority of organic manures is that nutrients are organically bound and delivered in synchrony with crop demand, thus reducing leaching losses. However, as described in Chapter 7 (Bergström et al., 2008), a number of long-term field experiments have shown that leaching losses of N are in fact increased when solely organic manures are used. The reason is that crop demand and delivery of nutrients from organic manures are not synchronised over a whole year (Chapter 5; Kirchmann et al., 2008c). A large amount of nitrogen is released after the cropping season and this nitrogen is very exposed to leaching in cold and humid regions.

Far less conclusive results have been published regarding leaching losses of phosphorus from organic and conventional cropping systems. However, the use of green manure and cover crops has the potential to increase P losses due to release of soluble phosphorus from the biomass during autumn and winter (Miller et al., 1994).

1.2.2.2 Carbon Sequestration

A widespread opinion is that organic farming sequesters more carbon in soil and can thereby reduce CO_2 levels in the atmosphere.

As stressed in Chapter 3 (Kirchmann et al., 2008a) and pointed out above, organic yields are significantly lower than conventional. Consequently, less soil organic matter can be formed from the biomass produced, which means that less carbon is sequestered, as outlined in Chapter 8 (Andrén et al., 2008). Furthermore, lower yields in organic production mean less water uptake by crops and thereby higher moisture content in soil, which speeds up decomposition of soil organic matter. An additional factor speeding up decomposition of soil organic matter is the intensive mechanical weed control that commonly occurs in organic crop production. In fact, if for example all cereals in Sweden would be grown organically, this would cause a substantial loss of soil carbon. The associated CO_2 emission will be equivalent to the yearly amount of CO_2 emitted by 675,000 average Swedish cars.

1.2.2.3 Pesticides

A common opinion is that pesticides kill useful organisms and pollute the environment.

Even though the occurrence of pesticide residues in the environment is not specifically discussed in this book, there is certainly a clear difference between organic and conventional agriculture, which is probably the cause of most concern among people. When pesticides are used, they are likely to be found at low levels in surface waters, groundwater and other environmental compartments, something that has to be taken seriously. However, analytical techniques have improved dramatically during the past couple of decades and today residues can be detected at parts per trillion (ppt) levels. This stresses the importance of conducting relevant risk assessments with appropriate safety margins.

In a similar way to those in food products, pesticide residue levels in the environment are controlled by various regulations. Within the European Union (EU), some are common for all EU countries, whereas others are specific for individual countries. The overall goal is to avoid unacceptable environmental disturbances. However, in a similar way as regards pesticide residues in food, the question is what is an acceptable level. Furthermore, the risks arising from pesticide residues in the environment must be related to possible disturbances by other chemical substances, which are also regulated.

To guarantee minimal negative side-effects in natural ecosystems, pesticides, whether natural or artificial, should have no or low toxicity except towards the target organism. There appears to be great potential to develop pesticides that are effective, reliable and have a low environmental risk. In addition, new and more precise application techniques can reduce the dose substantially. The trend today is to develop pesticides that inhibit specific process mechanisms in the target organism, such as enzyme reactions in photosynthesis, rendering them effective with a minimum of environmental side-effects. This development will likely continue and make the use of pesticides less controversial in the future.

1.2.3 Sustainability Issues

Achieving sustainable agricultural production is one of the major goals in organic agriculture. This is assumed to be possible through use of a set of pre-determined rules and methods mimicking nature, as discussed in Chapter 2 (Kirchmann et al., 2008b). According to these rules, sustainable agricultural production is achieved by maintaining/improving soil fertility, recycling of nutrients and imitating natural processes. However, irrespective of system, agricultural crop production is mainly a man-made single-crop cultivation with little resemblance to natural ecosystems.

There are other forms of agriculture for which sustainability is a key goal, commonly grouped under the term 'sustainable agriculture' (Bergström et al., 2005; Bergström and Goulding, 2005; Kirchmann and Thorvaldsson, 2000). However, these forms of agriculture have established goals to reach long-term sustainability without postulating rules and methods. In this type of agriculture, artificial fertilisers and pesticides are applied when needed.

Two major conditions determine the sustainability of farming systems, namely that plant nutrients removed or lost must be replaced or returned to the system to avoid depletion and that plant availability of nutrients in soil must be maintained (Chapter 5; Kirchmann et al., 2008c).

1.2.3.1 Soil Fertility and Nutrient Use

A widespread opinion is that yields increase over time if organic management practices are used. In contrast, artificial fertilisers, which are seen as unnatural and unnecessary chemicals, reduce yields over the long-term. A common opinion is that natural cycling of nutrients in organic agriculture is a guarantee for maintenance of good soil fertility.

Any agricultural system requires nutrient support and crop protection strategies to survive and maintain high crop yields. In conventional systems this is achieved by recycling manures and adding artificial fertilisers and pesticides, whereas in organic agriculture manures, feedstuffs, bedding materials, food wastes and untreated minerals are applied to compensate for export of nutrients through various products and losses (Chapter 5; Kirchmann et al., 2008c). Although the 'law of nutrient replacement' can also be followed in organic agriculture, addition of nutrients in the form of less soluble materials than nutrients present in soil results in lower plant availability (Chapter 4; Goulding et al., 2008), and in less efficient utilization and lower yields (Chapter 3; Kirchmann et al., 2008b). Regarding the argument that yields are reduced by the use of artificial fertilisers, one need only look at the historical trends in crop yield. From the time when artificial fertilisers were first introduced there has been a steady increase in yields, which to a large extent is attributed to use of the fertilisers.

A complicating issue when comparing organic and conventional management is that field experiments are often placed on fertile soils, for example on a soil with a high organic matter content such as newly converted grassland or soils previously enriched with P and K through decades of inorganic fertiliser additions (Chapter 4; Goulding et al., 2008). This results in smaller relative yield differences between organic and conventional systems due to the fact that more nutrients are released at such sites than at sites with normal soil fertility. Over the long-term, this results in a depletion of nutrients and/or soil organic matter. Few, if any, organic cropping experiments have been carried out on arable soils that have never received any artificial fertilisers. The general belief that soil organic matter increases in organically managed soils is not valid for most arable systems (Chapters 6 and 8; Korsaeth and Eltun, 2008; Andrén et al., 2008). Another complicating issue is that crop rotations are often designed to favour environmental status of organic production. For example, the arable crop rotation in the Apelsvoll experiment (Chapter 6; Korsaeth and Eltun, 2008) had insown clover/grass during two years, which most likely had a decreasing effect on N leaching whereas no clover/grass was grown in the conventional rotation. Furthermore, potatoes were grown more frequently in the conventional rotation which presumably increased N leaching. Such differences will also favour the soil fertility situation in the organic rotation.

1.2.3.2 Life in Soil

A common principle in organic production is to fertilise the soil but not to feed the crop directly. The underlying concept is that life in soil is promoted by organic farming practices, which are the key to sustainable crop production. One example is the increased colonisation of roots with mycorrhizas, which is considered beneficial for nutrient uptake by crops. Artificial fertilisers are assumed to have a negative impact on life in soil.

A rich microflora in soil is positive primarily due to its potential to release nutrients from soil constituents and added organic material. Life in soil, i.e. biological activity and the occurrence of microbes, is increased by the addition of any nutrient-rich organic material. A benefit from high abundance of microbes in soil and their degradation of organic material is stabilisation of the soil structure.

There are indications that organically managed soils can develop a mycorrhizal community with an increased capacity for P uptake by plants (Chapter 10; Ryan and Tibbett, 2008). However, mycorrhizas cannot substitute for fertiliser inputs as phosphorus taken up by the fungi primarily originate from the finite pool of soil phosphorus and its removal in farm products must be matched by inputs of off-farm sources. Indeed, high mycorrhizal colonisation may be considered an indicator of low plant-available P and in fact under certain conditions may reduce plant growth due to consumption of photosynthate from the host plant. Therefore, enhanced mycorrhizal activity does not compensate for low plant availability of P.

It is important to stress that organic manures are added in conventional systems too and that the beneficial effects are also present in such systems. Furthermore, there is no negative effect on life in soil of adding artificial fertilisers at normal rates. As stated above, organically bound nutrients are released in poor synchrony with crop demand and thereby used less efficiently than artificial fertilisers.

1.2.4 Energy Issues

1.2.4.1 Energy Requirement in the Perspective of Fertiliser Production and Land Demand

The general opinion is that production of artificial fertilisers is energy-demanding and means careless use of valuable resources. In organic systems, no energy for production of artificial fertilisers is needed as nitrogen can be supplied through nitrogen fixing crops. It is also common among organic advocates to look at energy consumption per unit food produced, which favours organic crop production.

The Haber-Bosch process, which is used in the fertiliser industry to convert atmospheric nitrogen into ammonia, requires considerable amounts of energy. A rule of thumb is that one litre of oil is consumed for each kilogram of nitrogen produced. This means that about 100 litres of oil are used annually per hectare of cultivated soil for N fertiliser production in order to produce about six tons of cereals. However, the net result is that for agricultural crops, between 5 and 10 times more energy in form of carbohydrates is produced than is consumed in fertiliser manufacture. In other words, with the help of artificial fertilisers, a very positive energy balance is obtained (Chapter 9; Bertilsson et al., 2008). Therefore, expressing energy requirement per unit yield is misleading as the total yield (food per energy input) and the total areal demand for crop production are not considered. Disregarding such conditions would lead to the conclusion that the most energy-efficient system would actually be a manual cultivation system without tractors or horses.

In a global perspective, there is a certain (and increasing) food demand and this food can either be produced as efficiently as possible on the available arable land or can be produced by considerably increasing the area for food production. In Chapter 9 (Bertilsson et al., 2008), the authors point out that the surplus land released due to higher productivity in conventional crop production compared with organic can be used for bio-fuel production and thereby replace fossil fuel energy. Instead of saving energy through low-input organic farming, modern conventional agriculture increases energy productivity by land being made available for bio-energy. Conventional methods thereby allow 2–4 times more food/energy to be produced on the total available area.

It is also important to note that legume N is not a source without a cost. As pointed out in Chapter 3 (Kirchmann et al., 2008a), growth of green manure legumes for N supply is often only possible by not using the land for saleable food crops, which must be considered as a reduction factor in any food production system.

1.3 Incorporating Scientific Evidence into Decisions Made in Society

As shown above, there are a number of widespread public opinions about organic agriculture that are not supported by scientific evidence. This is a major problem since scientific views have traditionally been a major and successful driving force for development in society in terms of technical improvements, medical treatments, democratic structures and agriculture. It is worrying that something so fundamental for life and survival as food production has become an issue highly influenced by a philosophical view on nature (Chapter 2; Kirchmann et al., 2008b), without considering long-term sustainability and sufficient food supply.

In other disciplines such as medicine, all treatments and methods have to be evidence-based to prove that they are safe and efficient. This is a widely accepted basis and has long been shown to be the best way of obtaining good results. This way of thinking should also be fully applied in food production. We need to produce sufficient, nutritious and wholesome food with as little environmental disturbance as possible. This goal can only be reached by modern, scientifically-based agriculture, not excluding certain inputs and methods due to philosophically-based views. Problems caused by agriculture are inevitable but solutions can be found through thorough analysis, wise planning and innovative thinking not biased by predetermined organic methods.

Overlooking the scientific evidence in decision-making has implications for a democratic society. Science itself is not democratic, but it can only flourish and survive in an open democratic society. Practices based on the 'Back-to-Nature' movement undermine the development of food production and ultimately the survival of society.

It is quite obvious that scientific results need to be communicated to politicians and legislators, but the question is how this can be achieved. Political views are a reflection of public opinion and if politicians are to be re-elected they must satisfy the wishes of the electorate. As indicated in this chapter, there is strong public opinion in favour of organic food production. To attract political attention, the general public must first be educated about sustainable food production without dogmatic limitations. Success in this will eventually change the political views in favour of the idea that a science-based approach typically solves more problems than it creates. In terms of food production, the bottom line is whether further yield declines and increased starvation must be tolerated before political decisions are based on scientific results instead of nature-based opinions.

References

Andrén, O., Kätterer, T., and Kirchmann, H., 2008, How will conversion to organic cereal production affect carbon stocks in Swedish agricultural soils? in: *Organic Crop Production – Ambitions and Limitations*, H. Kirchmann and L. Bergström, eds., Springer, Dordrecht, The Netherlands.

Bergström, L. and Brink, N., 1986, Effects of differentiated applications of fertilizer N on leaching losses and distribution of inorganic N in the soil, *Plant Soil* **93**: 333–345.

Bergström, L., Bowman, B.T., and Sims, J.T., 2005, Definition of sustainable and unsustainable issues in nutrient management of modern agriculture, *Soil Use Manage.* **21**: 76–81.

Bergström, L. and Goulding, K.W.T., 2005, Perspectives and challenges in the future use of plant nutrients in tilled and mixed agricultural systems, *Ambio* **34**: 283–287.

Bergström, L., Kirchmann, H., Aronsson, H., Torstensson, G., and Mattsson, L., 2008, Use efficiency and leaching of nutrients in organic and conventional systems in Sweden, in: *Organic Crop Production – Ambitions and Limitations*, H. Kirchmann and L. Bergström, eds., Springer, Dordrecht, The Netherlands.

Bertilsson, G., Kirchmann, H., and Bergström, L., 2008, Energy analysis of conventional and organic agricultural systems, in: *Organic Crop Production – Ambitions and Limitations*, H. Kirchmann and L. Bergström, eds., Springer, Dordrecht, The Netherlands.

Carson, R., 1962, *Silent Spring*, Houghton Mifflin Company, Boston, USA, 378p.

Felsot, A.S. and Racke, K.D., 2006, *Certified Organic and Biologically Derived Pesticides: Environmental, Health, and Efficacy Assessment*, ACS Symposium, Series 947, American Chemical Society, Washington DC, 326p.

Goulding, K., Stockdale, E., and Watson, C., 2008, Plant nutrients in organic farming, in: *Organic Crop Production – Ambitions and Limitations*, H. Kirchmann and L. Bergström, eds., Springer, Dordrecht, The Netherlands.

Kirchmann, H. and Thorvaldsson, G., 2000, Challenging targets for future agriculture, *Eur. J. Agron.* **12**: 145–161.

Kirchmann, H., Bergström, L., Kätterer, T., Andrén, O., and Andersson, R., 2008a, Can organic crop production feed the world? in: *Organic Crop Production – Ambitions and Limitations*, H. Kirchmann and L. Bergström, eds., Springer, Dordrecht, The Netherlands.

Kirchmann, H., Thorvaldsson, G., Bergström, L., Gerzabek, M., Andrén, O., Eriksson, L.-O., and Winninge, M., 2008b, Fundamentals of organic agriculture – Past and present, in: *Organic Crop Production – Ambitions and Limitations*, H. Kirchmann and L. Bergström, eds., Springer, Dordrecht, The Netherlands.

Kirchmann, L., Kätterer, T., and Bergström, L., 2008c, Nutrient supply in organic agriculture – plant availability, sources and recycling, in: *Organic Crop Production – Ambitions and Limitations*, H. Kirchmann and L. Bergström, eds., Springer, Dordrecht, The Netherlands.

Korsaeth, A. and Eltun, R., 2008, Synthesis of the Apelsvoll cropping experiment in Norway – Nutrient balances, use efficiencies and leaching, in: *Organic Crop Production – Ambitions and Limitations*, H. Kirchmann and L. Bergström, eds., Springer, Dordrecht, The Netherlands.

Lord, E.I. and Mitchell, R.D.J., 1998, Effect of nitrogen inputs to cereals on nitrate leaching from sandy soils, *Soil Use Manage.* **14**: 78–83.

Miller, M.H., Beauchamp, E.G., and Lauzon, J.D., 1994, Leaching of nitrogen and phosphorus from the biomass of three cover crop species, *J. Environ. Qual.* **23**: 267–272.

Pretty, J.N., Morison, J.L.L., and Hine, R.E., 2003, Reducing food poverty by increasing agricultural sustainability in developing countries, *Agric. Ecosyst. Environ.* **95**: 217–234.

Ryan, M.H. and Tibbett, M., 2008, The role of arbuscular mycorrhizas in organic farming, in: *Organic Crop Production – Ambitions and Limitations*, H. Kirchmann and L. Bergström, eds., Springer, Dordrecht, The Netherlands.

Steiner, R., 1924, *Geisteswissenschaftliche Grundlagen zum Gedeihen der Landwirtschaft. Rudolf Steiner Nachlassverwaltung*, 5th Printing 1975, Dornach, Switzerland, 256p. (In German).

Trewavas, A.J. and Stewart, D., 2003, Paradoxal effects of chemicals in the diet on health, *Curr. Opin. Plant Biol.* **6**: 185–190.

Winter, C.K., 2008, Organic food production and its influence on naturally occurring toxins, in: *Organic Crop Production – Ambitions and Limitations*, H. Kirchmann and L. Bergström, eds., Springer, Dordrecht, The Netherlands.

Chapter 2
Fundamentals of Organic Agriculture – Past and Present

Holger Kirchmann, Gudni Thorvaldsson, Lars Bergström, Martin Gerzabek, Olof Andrén, Lars-Olov Eriksson and Mikael Winninge

Abstract Organic agriculture can be traced back to the early 20th century, initiated by the Austrian spiritual philosopher Rudolf Steiner. It was later diversified by a number of key people, and more recent versions are guided by principles issued by the International Federation of Organic Agricultural Movements (IFOAM), founded in 1972. Organic practices were built upon the life philosophies and convictions of the founders regarding how to perceive nature. Today, those original views and ideas are considered as history. However, to understand the principles and opinions of modern organic agriculture, such as the exclusion of water-soluble inorganic fertilisers, we analysed the original ideas and arguments of the founders, who shared the common principle of relying on natural processes and methods, seen as a prerequisite for human health. For example, the British agriculturalist Sir Albert Howard, who together with Lady Eve Balfour founded the British Soil Association, claimed that healthy soils are the basis for human health on earth. In their view, healthy soils could only be obtained if the organic matter content was increased or at least maintained. Later, the German physician and microbiologist Hans-Peter Rusch together with the Swiss biologists Hans and Maria Müller, focused on applying natural principles in agriculture, driven by the conviction that nature is our master and always superior. Even though these early ideas have been abandoned or modified in modern organic agriculture, the principle of the founders regarding exclusion of synthetic compounds (fertilisers and pesticides) is still the main driver for choosing crops and pest control methods.

Keywords Ethics · Founders · History · Life philosophy · Nature philosophy · Theories · E. Balfour · A. Howard · H.-P. Rusch · R. Steiner

---

H. Kirchmann (✉)
Department of Soil and Environment, Swedish University of Agricultural Sciences, P.O. Box 7014, SE-75007 Uppsala, Sweden
e-mail: holger.kirchmann@mark.slu.se

H. Kirchmann, L. Bergström (eds.), *Organic Crop Production – Ambitions and Limitations*, DOI 10.1007/978-1-4020-9316-6_2,

2.1 Introduction

In this chapter, the thinking and arguments of the founders of organic agriculture are analysed. The origins and characteristics of the different schools of organic agriculture are described and the theories and statements of the founders are discussed and evaluated. Furthermore, to get an in-depth picture of the principles of organic agriculture, it is useful to be familiar with the philosophies of life in which the founders were interested, as these influenced their perception of nature and their views on human activities. Only a few scientifically-based analyses of organic agriculture theories have been performed (Jansson, 1948; Kirchmann, 1994).

We want to emphasise that we acknowledge the sincerity and well-minded intentions of the founders and their followers. Many organic farmers are highly skilled and successful experts. Our analysis is solely focused on the roots of organic agriculture and our perspective is limited to methods developed in Europe. Asian forms of organic agriculture such as natural farming according to Buddhism by the Japanese Masanobu Fukuoka (1978, 1989, 1991) or Zen macrobiotic farming based on the diet of George Oshawa (Oshawa and Dufty, 2002) are not considered in this overview.

2.2 Brief History of the Development of Organic Agriculture

The development of organic agriculture dates back to the beginning of the 20th century summarised in Table 2.1. It started as a reaction against industrialization of agriculture and was a response to concerns over the use of mineral fertilisers and pesticides (Merrill, 1983; Conford, 2001). Critics pointed out the unnaturalness of these compounds and regarded their use as a wrong way to produce food. The message was that organic practices have been around for a several thousand years and that maintenance of these practices is a reliable way to achieve healthy food

Table 2.1 Brief overview of the development of European organic agriculture

Movements	Focus
Early 1900s–1960: Reform movement	
1924 Introduction of bio-dynamic farming	Spiritual food production
1946 Foundation of the Soil Association	Health food production
1960–1990: Environmental movement	
1962 Publication of "Silent Spring" by Carson	Against pesticides and pro-environment
1968 Introduction of bio-organic farming	Holistic food production
1972 Foundation of International Federation of Organic Agriculture Movements (IFOAM)	Standardisation, lobbying for world-wide adaption
1980s Definition as "eco-agriculture"	Marketing environmental superiority
Since 1990: Political movement	
Governmental support	Promotion, subsidies, funding of research, etc.

products. One of the forerunners of organic agriculture was the "life reform movement" (Lebensreform Bewegung) in Germany in the 1920s, which acted against urbanisation and industrialisation, idealising vegetarian food, self-sufficiency, natural medicine, allotment gardens, physical outdoor work and all kinds of nature conservation (Vogt, 2001). In 1927/1928, the first "organic" organisation – Arbeitsgemeinschaft Natürlicher Landbau und Siedlung (Community of Natural Farming and Settlement) – was founded with the focus on fruit and vegetable production without artificial fertilisers and pesticides.

The first distinct form of organic agriculture was introduced in 1924 by the Austrian Rudolf Steiner, forming the basis for bio-dynamic farming (Steiner, 1924). Steiner gave a series of lectures entitled "Geisteswissenschaftliche Grundlagen zum Gedeihen der Landwirtschaft" (Spiritual foundations for the renewal of agriculture), with instructions on how to produce organic food supplying spiritual forces to mankind.

The 1940s brought the next wave of organic pioneers, with Lady Eve Balfour (widow of the British Prime Minister Arthur James Balfour) and Sir Albert Howard as prominent figures in the United Kingdom (Howard, 1940, 1947). In 1943, Lady Balfour published a highly influential book called "The Living Soil" in which she pointed out the importance of a healthy soil and the nutritional superiority of organically grown food. In 1946, Balfour and Howard founded the British Soil Association.

In the 1950s, the Swiss couple Hans and Maria Müller developed biological-organic farming methods, encouraged by the bio-dynamic agriculture of Steiner. In 1968, the German physician Hans-Peter Rusch provided the basis for biological organic agriculture in his book entitled "Bodenfruchtbarkeit" (Soil fertility), stressing the recognition of biological wholeness and a holistic view on food production and nature (Rusch, 1978).

In 1972, during an organic agriculture congress in Versailles (France), five organic organisations founded a global organisation called the International Federation of Organic Agriculture Movements (IFOAM), which since then has promoted its worldwide adoption, set standards, drawn up certification procedures etc.

Although some environmental problems as a result of the industrialisation of societies had already been identified, the breakthrough in broad environmental consciousness in the 1960s enabled advocates of organic agriculture to advance their argumentation. Organic agriculture methods were now also presented as a solution to the environmental problems caused by modern agriculture. The book "Silent Spring" by Carson in 1962 was a keystone pointing out the detrimental effects of widespread pesticide use poisoning nature. Later, the "Club of Rome" book "Limits to Growth" by Meadows et al. (1972) focused on population growth and resource depletion, including the environmental consequences of modern agriculture. The exclusion of pesticides and the additional elimination of limited resources such as phosphates and fossil fuels for fertiliser production, respectively, were now used as arguments for the superiority of organic agriculture.

Water pollution by agriculture through nutrient leaching followed by algal blooms was observed during the 1970s (e.g. Ahl and Odén, 1972). Earlier emphasis by organic agriculture organisations on the better quality of organic food and the benefits of organic agriculture for soil (Koepf et al., 1976; Dloughý, 1981) was now complemented by reports pointing out the benefit of this type of agriculture for the environment (e.g. Koepf, 1973). In the early 1980s, eutrophication of lakes and rivers was intensively reported in Europe and nutrient leaching from agriculture was identified as being a main cause. Somewhat later, the advocates of organic agriculture used this opportunity to claim that organic agriculture would be able to reduce N leaching (Granstedt, 1990; Kristensen et al., (1995).

The period between 1980 and 1990 saw a great revival in organic agriculture, initiated by environmental problems caused by modern agricultural practices. Organic agriculture was attributed to be sustainable and environmentally friendly and was redefined as "ecological" agriculture or "eco-agriculture". The image of organic agriculture as a problem-solver attracted much larger groups of "green" supporters, who made a political case for public support.

Since 1990, "green" and other political parties have initiated a number of activities promoting organic agriculture, such as ear-marked research grants, creation of research foundations and funding of university departments of organic agriculture. Furthermore, subsidies for organic production, educational programmes and extension services for organic agriculture were established. In several countries in Europe, organic agriculture has grown in the past 20 years to be a significant sector within agricultural production, whereas in other countries it has remained at a relatively low level. In Austria, for example, 200 farms were managed according to organic principles in 1980 and 18,360 in 2001, the latter accounting for approximately 25% of Austrian arable land (Freyer et al., 2001). In Sweden, a political programme with the aim of increasing organic production to cover 20% of farmland and to encourage the consumption of organically grown food in schools, hospitals, residential care homes etc. has recently been proposed. Today, organic agriculture is a mainstream interest in Western societies, although it has been criticised for not taking into account contradictory evidence regarding some of its claims (Avery, 2000; Tinker, 2000; Trewavas, 2004; Taverne, 2005; Avery, 2006).

2.3 The Schools of Organic Agriculture

2.3.1 Biological Dynamic Agriculture (Rudolf Steiner)

The Austrian Dr Phil Rudolf Steiner (1861–1925), who taught mysticism and esoteric wisdom, created a spiritual system called anthroposophy, a variant of theosophy. He applied his teachings to a wide range of areas in society, e.g. arts and architecture, medicine, religion, pedagogics and also agriculture. Biological dynamic (biodynamic) agriculture builds upon Steiner's lectures during a one-week agricultural course in 1924 in Koberwitz (now Wroclaw), Poland (Steiner, 1924), when he taught a group of followers on considerations of spiritual matters in agriculture.

Steiner wanted to change agriculture and introduced new practices in accordance with his supernatural insights. He gave detailed instructions on non-visible matter, how it acts in soil, crops and animals and how to affect and control the "forces" related to such matter. The text of his lectures provides the core information for current biodynamic farming and can be seen as the basis for the first distinct form of organic agriculture.

Steiner was worried about food quality and the effect of inorganic fertilisers in decreasing crop quality. For example, he taught that agricultural products would degenerate so that they could not be used as food for humans by the end of the century "....die Produkte so degeneriert sein werden, dass sie noch im Laufe dieses Jahrhunderts nicht mehr zur Nahrung der Menschen dienen können" (Steiner, 1924 p. 12). Furthermore, he stated that nobody could know whether mineral fertilisers would lead to a significant degeneration in the quality of agricultural products "Es weiss zum Beispiel kein Mensch heute, dass alle die mineralischen Dungarten gerade diejeningen sind, die zu dieser Degenerierung, von der ich gesprochen habe, zu diesem Schlechterwerden der landwirtschaftlichen Produkte das Wesentliche beitragen" (ibid. p. 20). He claimed that plants are stimulated by wateriness through inorganic fertilisers; they are not stimulated by the living soil "Daher werden Ihnen Pflanzen, welche unter dem Einfluss irgendwelchen mineralischen Düngern stehen, ein solches Wachstum zeigen, das verrät, wie es nur unterstützt wird von angeregter Wässrigkeit, nicht von lebendiger Erdigkeit" (ibid. p. 94). Furthermore, Steiner pointed out that this is a general law "Denn jeder mineralische Dünger bewirkt, dass nach einiger Zeit dasjeninge, was auf den Feldern erzeugt wird, die mit ihm gedüngt werden, an Nährwert verlieren. Das ist ein ganz allgemeines Gesetz" (ibid. p. 176).

However, Steiner did not teach common crop quality criteria such as mineral, protein, carbohydrate or vitamin content or taste. Instead, he instructed on how to manufacture eight different compounds consisting of mixtures of minerals, wild plants and animal organs. Two compounds are aimed at affecting supernatural crop qualities enabling the transfer of "forces" into soil (humus compound) and crops (silica compound). Six compounds are used for the preparation of animal manure (compost compounds) also transferring "forces" via manure into soils and crops. For example, cow manure and powered silica should be placed into cow horns (humus and silica compound) to accumulate "forces". Thereafter, these materials must be highly diluted with water through both clockwise and counter-clockwise spinning and then sprayed on crops and soil. The "forces" accumulated in the cow horns will thereby enable a balanced exchange of terrestrial and cosmic forces in fields. Steiner also stated that sowing or planting of crops should be carried out according to astrological principles.

Steiner's supernatural views on "radiation" and flows of "forces" were not derived from natural science but gained from views and inspiration received during mental exercises. The "forces" Steiner instructed on are unknown to science. However, this is not a proof of their non-existence. On the other hand, there are other strong indications that Steiner's scientific knowledge was limited, as exemplified by the following quotes. Steiner talked about a secret chemistry in organic processes.

For example, he claimed that potassium is transformed into nitrogen and even lime "Ich habe fortwährend davon gesprochen, ... weil nämlich im organischen Process eine geheime Allchemie liegt, die zum Beispiel das Kali, wenn es nur in der richtigen Weise drin arbeitet, wirklich in Stickstoff umsetzt und sogar den Kalk, wenn der richtig arbeitet, wirklich in Stickstoff umsetzt" (Steiner, 1924 p. 136). According to current scientific knowledge, the energy in biological systems is too low to drive nuclear reactions and transmute elements. In addition, the following quote also reveals Steiner's poor knowledge in the field of chemistry, since he believed that silica is transformed into another element of the utmost importance in organisms "Das Silizium wiederum wird umgewandelt im Organismus in einen Stoff, der von ausserordentlicher Wichtigkeit ist, der gegenwärtig unter den chemischen Elementen überhaupt nicht aufgezählt wird" (ibid. p. 137). Even in 1924, it was common scientific knowledge that there is no element transmutation in biological systems.

The following quotes expose Steiner's lack of understanding of science. He lectured on the effect of wild plants that were used for the preparation of his biodynamic compounds. The stinging nettle (*Urtica dioica*) compound makes the soil reasonable "Es ist wirklich etwas wie eine Durchvernünftigung des Bodens, was man durch diesen Zusatz von Urtica dioica wird bewirken können" (ibid. p. 133). Dandelion (*Taraxacum vulgare*) is the intermediary between the homeopathically distributed silica in the cosmos and the silica really needed in the whole area "Der gelbe Löwenzahn, wo er in einer Gegend wächst, ist .... der Vermittler zwischen der im Kosmos fein homoöpathisch verteilten Kieselsäure und demjeningen, was als Kieselsäure eigentlich gebraucht wird über die ganze Gegend hin" (ibid. p. 137).

Steiner looked upon each farm as a closed entity and as a self-sustaining unit. He said that any import to the farm should be seen as a cure for a sick farm "Landwirtschaft ... kann als eine wirklich geschlossene Individualität aufgefasst werden. Was in die Landwirtschaft hereingebracht wird an Düngemitteln und ählichem von auswärts, das müsste in einer ideal gestalteten Landwirtschaft angesehen werden schon als ein Heilmittel für eine erkrankte Landwirtschaft" (ibid. p. 42). The idea of self-sustaining farms is attractive in many ways as it excludes long-distance transport of animal feedstuffs, purchase of fertilisers, import of animals etc. and only presupposes sale of food products. However, in reality this is difficult to achieve. It is well-known that sale of products from a farm means a significant export of nutrients through food products, leaching and other losses, which will result in nutrient depletion in soil over time. It is impossible to maintain soil fertility and high yields over time through an internal recirculation of manure only. On the other hand, Steiner prohibited the return of nutrients present in toilet wastes. A more thorough analysis of biodynamic agriculture has been published earlier (Kirchmann, 1994).

In summary, Steiner stated that behind visible nature there is a supernatural, spiritual world. According to him, organisms have spiritual bodies (e.g. physical, ethereal and astral) interacting with each other in interwoven flows either emitting or absorbing "forces". Spiritual energies are regarded to fill and pervade all things. The specific biodynamic compounds introduced by Steiner should supply soil and plants with "forces" in order to control the absorbance or emanation of "terrestrial and cosmic forces". He wanted to influence life through control of spiritual forces, presupposing their existence in physical matter. There is one central reason why Steiner had an interest in controlling "spiritual powers" in agriculture production. He wanted to show how to produce food enriched with "spiritual powers" that could help mankind to develop spiritually and reach complete intuition. In order to improve karma, overcome evil and finally reach a complete stage of spiritual enlightenment and liberation, humans need to refine their soul and develop spiritually. For a true follower of Steiner, the use of biodynamic compounds is a way to help mankind to reach this goal.

Steiner wrote an additional gospel text to complement the New Testament (Steiner, 1913). He described Christ as being the spirit of the sun (sun logos). He believed that Earth and Sun were unified when Christ was born on Earth. Furthermore, when the blood of Christ dropped onto the earth at Golgotha, Earth actually became the body of Christ. As a consequence, Earth has become holy and nature has received forces for salvation. This may explain why Steiner maintained that only natural means and methods are to be used and why inorganic fertilisers and synthetic pesticides need to be excluded, as only natural products contain curing and saving forces for mankind.

One may conclude that the exclusion of synthetic fertilisers and pesticides in biodynamic production is not motivated by environmental concerns, resource conservation or improvement of biochemical crop quality properties. Steiner did not instruct on how to improve soil fertility or nutrient recycling in society, reduce nutrient leaching from soil, or decrease ammonia volatilisation from composting. He taught how to channel "forces" into food as an essential contribution to the spiritual development of mankind. His ideas about a supernatural world, on which he gave instructions, are unknown to science.

2.3.2 Organic Agriculture (Eve Balfour and Albert Howard)

Lady Eve Balfour (Evelyn Barbara Balfour; 1899–1990), a British farmer and educator, and Sir Albert Howard (1873–1947), a British agriculturalist in India who developed the Indore composting process, were founding figures in the Anglo-Saxon organic agriculture movement, The Soil Association.

The central hypothesis of Lady Balfour was that there is a close relationship between soil fertility and human health and that a decline in soil humus and fertility results in a decline in human health (Balfour, 1943). Similarly, Howard wrote that perfectly healthy soils are the basis for health on earth "The undernourishment of the soil is at the root of all" and health is a "birthright of life" (Howard, 1947 p. 12). Howard regarded soil humus as the most significant of all nature's reserves and the most fundamental component of a life-giving principle (Howard, 1940, 1947). The essential aim of the movement was to increase and maintain organic matter contents in soils, which was regarded as a guarantee of soil health: "Nature's farming is the care devoted to the manufacture of humus. The great law of return ... the great principle underlying nature's farming has been ignored" (Howard, 1947 pp. 31–32).

However, when we consider soil health and soil quality today, the perspective is wider in scope than soil organic matter content alone (Schønning et al., 2002), as organic materials may include organic pollutants and micronutrients (e.g. Doran and Jones, 1996), metal contaminants (e.g. Kirchmann and Andersson, 2001; Schloter et al., 2003) and environmental risks derived from these pollutants (e.g. Swartjes, 1999).

Even though soil organic matter plays a central role in soil fertility, quantitative soil protection (erosion control) and the sustainability of cropping systems, crop growth and crop quality are also affected by other factors such as non-organically bound macro- and micronutrients, acid-base conditions, natural or man-made subsoil compaction, high native contents of non-essential elements etc. that can have a highly significant impact. An improvement in soil organic matter content alone cannot necessarily compensate completely for the impact of these factors on crop production. Thinking solely of the humus status for soil health and disregarding other major crop production factors is not in line with our current understanding.

Howard (1940) regarded the nutrient supply of plants through soluble fertilisers as a fatal error "Artificial fertilisers were born out of the abuse of Liebig's discoveries of the chemical properties of soil. The effects of the physical properties of the soil were by-passed: its physiological life ignored, even denied, the latter a most fatal error. The essential co-partnership between the soil and the life of the creatures, which inhabit it, to which Darwin's genius had early drawn attention, is wholly forgotten" (Howard, 1947 pp. 71–72). Howard stated that there is "a second method by which plants feed themselves. It is a direct connection, a kind of living bridge, between life in soil and the living portion (plants) of the soil" (ibid. p. 22). Howard believed that only plant nutrients made available through this *second method* can feed plants properly (ibid. pp. 22–29). Although phosphorus and other nutrients can be supplied to crops through mycorrhizae, see Chapter 10 of this book (Ryan and Tibbett, 2008), there is no scientific evidence for a second pathway in general. Agricultural crops can only take up dissolved ions (Mengel and Kirkby, 2004), dissolved chelated metal ions (e.g. Ullah and Gerzabek, 1991; Chen et al., 2001) or dissolved amino acids (e.g. Jones and Darrah, 1994; Näsholm et al., 2000) from the soil solution through roots and root hairs.

Balfour (1943) wrote in her preface that "the physiological and spiritual well-being of man has its roots in soil". She also used the term "you are what you eat", referring to the relationship between dietary composition and human physiology. However, organic matter content in soil does not seem to be the primary link between soil health and human health, but rather the shortage of macro- and micronutrients in soil not mainly stored in organic matter leading to hidden hunger (e.g. Welch and Graham, 1999; Rengel et al., 1999; Sanchez and Swaminathan, 2005). Furthermore, Balfour's view about soil health and spiritual health goes beyond approved knowledge.

Balfour made various statements on soil humus, some of which require comment and correction. She argued that large yields caused by inorganic fertilisers reduce the amount of humus in soil. According to our knowledge from soil biology and systems analysis, the opposite is true. Firstly, higher yields result in larger amounts of crop residues, both roots and above-ground plant parts. Thus, higher yields provide more crop residues, and thus more raw materials are available for humus formation. Secondly, roots cannot take up soil organic matter as such, although dissolved organic matter can promote the uptake of cations as chelates (e.g. Ullah and Gerzabek, 1991; Bocanegra et al., 2006). Carbon is taken up by plants as atmospheric carbon dioxide through photosynthesis. Thus, plants do not "eat up" humus. Thirdly, the rate of humus decay is driven by a variety of environmental factors, especially moisture and to a lesser extent temperature (Davidson and Janssens, 2006). High yielding crops require more water and thus reduce soil moisture content more than low yielding crops (e.g. Andrén et al., 1990). Lowering soil moisture content reduces the rate of humus decomposition. So, contrary to Balfour's statement, numerous studies have shown that soil organic matter is increased through increasing yields (e.g. Balesdent et al., 1988; Andrén and Kätterer, 2001).

Another central view in the Balfour-Howard school is that artificial fertilisers speed up the rate at which soil organic matter is exhausted (Balfour, 1943 p. 53). As humus was considered to be the "most significant of all nature's reserves" (Howard, 1947 p. 26), loss of humus means a decrease in soil fertility and must absolutely be avoided. Consequently, inorganic fertilisers need to be banned. But how true is this reasoning? Results from a great number of isotope studies have revealed that ^{15}N-labelled fertilisers are incorporated into soil organic matter through microbial turnover and that decomposition of soil organic matter is not accelerated through addition of inorganic N (e.g. Jansson, 1958; Jansson and Persson, 1982; Bjarnason, 1987). On the contrary, a depressive effect of inorganic N fertiliser on decomposition of soil organic matter and organic materials has been observed (Söderström et al., 1983; Puig-Gimenez and Chase, 1984; Green et al., 1995; Nyamangara et al., 1999). Furthermore, long-term field experiments with inorganic fertilisers have shown that soil organic matter content is maintained through regular applications of soluble nutrients. A possible initial decrease can be traced back to previously high applications of animal manures or organic-matter build-up when the soil was under grass (e.g. Johnston et al., 1989; Kirchmann et al., 1994; Gerzabek et al., 2001). This is in accordance with the studies cited in the paragraph above. Furthermore, ions present in soil solution through dissolution of minerals, exchange reactions with particle surfaces, mineralisation of soil organic matter etc. can be at similar concentrations as after fertiliser application. The view of Balfour and Howard on humus decomposition is not based on scientific evidence but on a misunderstanding of how inorganic fertilisers react in soil.

A further assumption in the Balfour-Howard school is that only composted organic materials should be applied to soil to maintain soil fertility. Balfour argued that addition of straw or green manure to soil has unreservedly damaging consequences on the crop. Although it is possible to increase soil fertility through large additions of compost (e.g. Johnston et al., 1989), it is equally possible to use other non-composted organic materials such as green-manure crops, anaerobically digested sewage sludge, straw in combination with nitrogen or peat to increase soil organic matter content (Kirchmann et al., 1994, 2004). The treatment of the residues, whether they are directly returned as fresh residues to soil or removed and returned as manure, compost etc. may affect the amount of soil organic matter formed (Kirchmann and Bernal, 1997), but it is primarily the amount of residues accessible for humus formation that controls the organic matter content in soil. The statement that all organic materials need to be composted has no scientific support. Furthermore, composting is followed by high losses of carbon in the form of carbon dioxide (e.g. Sommer and Dahl, 1999; Paillat et al., 2005) and nitrogen in the form of ammonia gas (Kirchmann, 1985; Eklind and Kirchmann, 2000; Beck-Fries et al., 2003), respectively, which is a disadvantage as C and N are not conserved in the material.

Balfour and Howard were inspired by the idea that recirculation of organic wastes produced in society back to soil could enable a permanent maintenance of soil fertility and they referred to early Asian societies described by King (1911). King had documented how early Asian cultures recycled source-separated toilet wastes, food wastes, ashes, sediments from ditches and other natural resources to agricultural land after partial composting. This documentation is often taken as an example and proof of the hypothesis that the complete recirculation of nutrients in society enables a sustainable agricultural production. However, the striking point in King's documentation is that the return of large amounts of organic matter to soil in these societies (many without composting) demanded enormous human effort and organisation. Due to an unevenly distributed availability of organic wastes in society, high water contents and thus expensive transportation, the recirculation of organic wastes to arable land is labour-intensive and costly if one wants to achieve an equitable redistribution (Kirchmann et al., 2005). Maximum recirculation of the plant nutrients found in wastes is definitely a very important goal, which might be achievable through extraction of nutrients from organic materials and their return as concentrated inorganic fertilisers (Kirchmann et al., 2005) or in a recycling system based on a very small spatial scale, e.g. within a village or on farm level itself.

Lady Balfour (1943) strongly argued and stressed the importance of food quality for human health. However, she disregarded the principal difference between changes in the diet and choice of organically grown food as the root cause for her observations on human health. In other words, she did not distinguish between changes in diet and quality of organically grown products as the reason for improved health conditions. This is remarkable since she pointed out the important role of whole food for health, i.e., consumption of non-refined flour (whole-wheat etc.), vegetables and fruits. She actually used this type of diet but never considered this in her conclusions. Her focus was on organically grown food only. It is therefore not possible from her studies to draw conclusions on the main reason for the health improvements reported. One may add that the indistinct mixture of dietary composition and food quality is still a common phenomenon when discussing food and health. Comparative analyses show that there are few consistent differences between organically and conventionally produced food (e.g. Ames et al., 1990; Basker, 1992; Woese et al., 1997; Bourn and Prescott, 2002; Ryan et al., 2004). Thus, there is no imperative logic to conclude that organic food products per se improve human health. On the other hand, there is massive evidence that the composition of the diet, i.e. the proportion of fruits, vegetables, saturated fats, refined sugar, fish and so on, is of great importance for human health (e.g. Willet, 1994; Ames, 1998; Trichopoulou and Critselis, 2004), which is also reflected in the dietary recommendations from government organisations.

2.3.3 Biological Organic Agriculture (Hans-Peter Rusch)

This type of organic agriculture was founded by the German physician and microbiologist Dr Hans-Peter Rusch (1906–1977) in collaboration with the Swiss biologists Dr Hans and Maria Müller. In his search for ecologically sensible forms of agriculture, Rusch observed nature and applied nature's principles in agriculture. He defined this as analogical, biological thinking, which was also the subtitle of his book (Rusch, 1978).

Rusch wrote that life is a unity whereby every part is of equal value and given equal rights, simple organisms to the same extent as humans "Das Lebendige ist eine Einheit. Jedes Glied dieser Einheit ist gleichwertig und gleichberechtigt, ob es sich um eine Amöbe oder einen Menschen handelt" (Rusch, 1978 p. 34). He further wrote that in nature there is nothing for its own purpose; it is only the purpose of wholeness "In der Natur ist kein Ding um seiner selbst willen, es ist nur um des Ganzen willen" (ibid. p. 15). Each organism is through the task of symbiosis indivisibly connected into a unit "Die Gemeinschaft alles Lebenden .... ist durch die Pflicht der Symbiose unlösbar zu einer Ganzheit vereinigt" (ibid. p. 15). He combined these two views of nature; firstly that all life on earth has the same inherent value (coequality) and secondly that the living is only correctly viewed in terms of interacting organisms (holism).

In fact, the same perspective on nature constitutes the basis for eco-philosophy (Fox, 1994; Drengson, 1997) introduced by the Norwegian philosopher Arne Naess (Naess, 1989) and earlier used as an explanation for the evolutionary history of humans by the German nature philosopher Ernst Haeckel (Haeckel, 1900).

Rusch criticised natural sciences. He stressed that the outlook for the wholeness of life has been lost due to reductive and highly specialised science "Es gibt noch keine Naturwissenschaft, die den Aufgaben der Ganzheitsforschung gewachsen wäre" (Rusch, 1978 p. 15). Rusch proposed a holistic view of nature incorporating both wanted and unwanted properties and believed that the control of unwanted organisms to help a weakened organism is meaningless in the long-term. The chemical fight against diseases and pests is not only hazardous, it is also primitive and stupid "Wir müssen der Natur dankbar sein, dass sie uns mit Schädlingen und Krankheitserregern ein zuverlässiges Kriterium für fehlende Gesundheiten bereithält .... und biologische Unter- und Minderwertigkeit sofort mit Sicherheit ablesen lässt" (ibid. p. 66); "Der Giftkampf gegen Krankheiten und Schädlinge ist nicht nur gefährlich, er ist primitiv und dumm" (ibid. p. 25).

Rusch (1978) shared the preference for humus with Balfour and Howard, also looking at soil fertility as the basis of all life. However, Rusch modified the focus of Balfour and Howard by emphasising that the process of humus formation is a sign of fertility and not the material as such is most important. He believed that humification is in fact the greatest biological regulation known to nature "Humifizierung ist ein Regulativ, sie ist in der Tat das grösste biologische Regulativ, das die Natur kennt" (ibid. p. 88); humus is a manifestation of the biological achievement "Humus ist Ausdruck der biologischen Leistung" (ibid. p. 91).

Rusch observed litter decomposition, soil layering and humus accumulation in natural ecosystems and he transformed his observations into practical agricultural measures. According to him, normal humus formation is only achieved if one does not disturb the natural soil layering. Any soil tillage should be kept at a minimum to avoid disorder, one only needs to mimic nature "Es ist also grundsätzlich geboten, jede irgendwie entbehrliche Bodenbearbeitung zu vermeiden. . .man muss die Natur nur getreu nachahmen" (ibid. p. 80 and p. 215). His underlying reasoning was to let nature take its course. However, Rusch's statements cannot be corroborated from results of long-term field experiments in which soil disturbance and no-tillage are compared. Although layering typical of undisturbed ecosystems occurs in non-ploughed topsoils with highest organic matter concentrations at the soil surface, the total amount of soil organic matter is not enhanced in untilled soils in all cases (e.g. Antil et al., 2005; Alvarez, 2005).

According to the view that nature shows us how to treat it, Rusch pointed out that organic manures and composts are not suitable for the root zone and must only be used as surface cover "Organische Dünger und Komposte sind nicht wurzeltauglich und dürfen nur als Bedeckung benutzt werden" (ibid. p. 158). Nature does not compost "Die Natur kompostiert nicht" (ibid. p. 166). However, Rusch's view is only valid if nutrient-poor and energy-rich materials (e.g. Jansson, 1958; Kirchmann, 1990) are applied to the root zone and microbes and plant roots compete for nutrients during decomposition of these materials.

Rusch condemned artificial fertilisers as making it impossible to mimic the natural dynamic of nutrient release from soil to plant. He regarded this as the unavoidable mistake of synthetic fertilisers "Es ist vollkommen unmöglich, die natürliche Dosierung der Mineralbewegungen zwischen Boden und Pflanze nachzuahmen, und das ist der unvermeidliche Fehler der künstlichen Düngung" (ibid. p. 73). This argument against the use of synthetic fertilisers is probably the most sophisticated one proposed by the organic movement. They claim that the application of soluble salts to soil does not fulfil the demands of crops and, the most important point, that the supply is not synchronised with the growth of crops.

Although the argument by Rusch may sound reasonable, it is not in accordance with current scientific findings. Even if the supply of nutrients by the soil and their uptake by the plant are in synchrony in natural ecosystems due to the presence of living roots throughout the year, this is not the case in soils of arable systems. In ploughed soils, nutrients released from soil organic matter or organic manures can be lower in spring/summer when crop demand is highest and higher in autumn when there is little demand or no crops are present due to moisture and temperature conditions (e.g. Dahlin et al., 2005). The lower nutrient use efficiency of organic manures than of inorganic fertilisers both in the short- and long-term (Torstensson et al., 2006; Kirchmann et al., 2007) combined with higher leaching losses from organic manures (Bergström and Kirchmann, 1999, 2004) clearly shows the low level of synchronisation between nutrient supply and crop demand of organically bound nutrients.

It needs to be pointed out that adding salts to soil (fertiliser application) is in no way unnatural. Precipitation of soluble salts and ions derived from marine aerosols, nitrate from thunderstorms can add considerable amounts of nutrients to soil in a similar way as fertilisers. Furthermore, additions of animal urine or slurries, even in organic agriculture, means a supply of soluble salts of the same order of magnitude as fertiliser application, see Chapter 5 of this book (Kirchmann et al., 2008b). From a scientific point of view, soluble salts added either through urine or synthetic fertilisers are identical when present in the soil solution, although the origins of the ions are different. To put it simply, whether salts were produced by animal kidneys or by a specific technical process cannot be claimed as being a fundamental difference for the crop. Finally, healthy crops can be grown in pure nutrient solutions without any soil (e.g. Ingestad and Ågren, 1995).

Rusch argued that losses of nutrients are inevitable and high from artificial fertilisers compared with organic manures and that these losses occur because the organic but not the artificial fertiliser is adapted to the turnover in soil "Ein wertvoller, teurer organischer Dünger kann immer noch billiger sein als der billigste Kunstdünger. Verluste durch Auswaschen und Festlegung der Mineralstoffe müssen in jedem Falle bei einer Kunstdüngung in Kauf genommen werden. Verluste treten dadurch ein, dass nur ein natürlicher, aber nicht ein künstlicher Substanzkreislauf an die wechselnden Lebensbedingungen . . . .angepasst ist" (Rusch, 1978 p. 76). The argument that artificial fertilisers are lost to a larger extent than organic manures needs to be viewed from the other perspective. Leaching losses of N from organic manures are often higher than from inorganic fertilisers (Bergström and Kirchmann, 1999, 2004). Furthermore, leaching losses of N from organic agriculture systems can be significantly larger than those from modern farming systems (Torstensson et al., 2006). The lack of synchrony between nutrient release from organic manures and crop demand is the actual reason for higher losses from organic manures, as pointed out above. Independent of their origin, nutrients go through the same chemical and biological reactions in soil. The turnover of inorganic N fertilisers in soil and their naturalness have been explicitly explained by Jansson (1971).

2.3.4 Modern Organic Agriculture (IFOAM)

In 1972, the International Federation of Organic Agricultural Movements (IFOAM) was founded to represent the common interests of the different schools of organic agriculture but still allow their specific practices. This resulted in a new image of organic agriculture with less emphasis on methods but with a greater focus on aims. Today, the views and ideas of the founders of organic agriculture are regarded as history. It is believed that modern organic agriculture has progressed and bypassed the old schools. But is this so?

The common principle of the founders was to exclude synthetic compounds, such as water-soluble synthetic fertilisers, and use natural means and methods only. In fact, this principle is still a central prerequisite in modern organic agriculture and has not been questioned. It is still the incentive for choice of crops and rotations and weed, pest and insect control. Furthermore, biodynamic farming even now requires the use of compounds according to Steiner's instructions.

A number of additional arguments for not using synthetic fertilisers have developed over the last few decades. For example, inorganic N fertilisers are claimed to cause higher loadings to the environment than organic manures, and their production also requires relatively large amounts of energy. However, recent research has shown that inorganic N fertilisers in fact commonly cause lower N leaching losses than organic manures applied in equal amounts, which is discussed in Chapter 7 of this book (Bergström et al., 2008). The energy argument mentioned above also has to be evaluated in detail. Even though the energy consumption in production of N fertilisers is relatively high, the return in the form of energy build-up in crops is considerably higher, as discussed in Chapter 9 of this book (Bertilsson et al., 2008). In other words, the more recent arguments put forward are also questionable.

IFOAM does not mention the concepts of the founders but accentuates four general principles – health, ecology, fairness and care – as key goals for modern organic agriculture (IFOAM, 2006). Indeed, these principles are excellent and to make them become reality, appropriate methods and tools are required. Below, we analyse and comment on the four IFOAM principles and discuss whether organic agriculture is a suitable way to achieve them.

2.3.4.1 Principle 1 – Health

> To sustain and enhance the health of soil, plant, animal, humans and planet as one and indivisible.
>
> (IFOAM, 2006)

According to this, there is a health chain from soils that produce healthy crops, fostering health of animals and humans etc. Originally, this way of thinking was typical for Balfour (1943) and Howard (1940), saying that a *living soil* in particular is a necessary condition for healthy plant growth and for humans. However, soil health (A) does not necessarily provide a guarantee of crop health (B) or animal or human health (C, D) and planet health (E). There is simply no imperative logic that A leads to B and finally to E, although we would like to believe so. Even if crops greatly benefit from fertile and healthy soil, soil conditions are not the only determining factor for crop health. Other factors can be of greater importance, such as weather and climate, plant protection against non-soil borne diseases through NPK fertilisers (Reuveni and Reuveni, 1998), damage through animals and pests, formation of natural toxins in crops etc. Similarly, healthy crops do not automatically guarantee good health of the consuming organisms. For instance, the micronutrient requirements of animals or humans can be much larger than the requirements and uptake by crops (McDowell, 2003). In simple terms, the nutritional composition of a healthy crop may not be adapted to the consuming organism. The "chain" conclusions that perfect conditions in soil finally lead to a healthy planet are highly questionable.

2.3.4.2 Principle 2 – Ecology

> To base organic farming on living ecological systems and cycles, work with them, emulate them and help sustain them.
>
> (IFOAM, 2006)

In the full IFOAM text it is explained that "production is based on ecological processes and recycling. Organic farming should fit the cycles and ecological balance in nature". In other words, ecological systems and cycles should serve as a prototype for organic agriculture. This view is similar to that proposed by Rusch, who wanted to practise agricultural methods following processes observed in nature. However, both organic and conventional agricultural systems are man-made and not naturally occurring. In fact, the cultivation of natural ecosystems such as forests, wetlands, grassland etc. into agricultural land is a drastic conversion. Agriculture means that crops are sown and harvested, weeds are controlled, soils are tilled, and animal manures are collected and applied. Furthermore, the same ecological processes and cycles exist and take place in organic and conventional agricultural systems.

Ecological processes and cycles are proposed to serve as a model providing guidelines for how to treat nature. However, the purpose of agriculture is not to emulate ecological processes but to use and take care of nature for the purpose of food production. Ecological processes simply follow or react to any prevailing conditions, independent of the cause. For example, application of manure to soil increases microbial activity and nitrogen processes in soil to levels much higher than those occurring in undisturbed nature. Our task is to protect the soil from erosion and pollution, to maintain its fertility by application of necessary nutrients, and to manage agro-ecosystems so that nutrient losses are minimised. If we do that, soil processes will continue to work according to these conditions.

Many unnatural measures can be found within organic agriculture. Various industrial wastes (e.g. slag, vinasse, meat and bone meal from abattoir offal) that are not naturally occurring are applied to soil. On the other hand, recycling of toilet wastes to organically managed soils is not allowed, see Chapter 5 of this book (Kirchmann et al., 2008b). Man-made crop varieties and not wild types are grown. Machinery is powered by fossil fuels and animal or man power is very seldom used.

2.3.4.3 Principle 3 – Fairness

> Organic agriculture should build on relationships that ensure fairness with regard to the common environment and life opportunities.
>
> (IFOAM, 2006)

The principle of fairness adds new aims to organic agriculture, not explicitly addressed by the pioneers, such as respect, justice, eradication of poverty, animal welfare, equitable systems for distribution and trade, as well as social costs. The emphasis on these aspects is, without doubt, commendable and definitely wanted within society. Still, the question is whether organic agriculture is the best way to achieve these aims.

Furthermore, supply and quality of food is addressed "Organic farming should contribute to a sufficient supply of good quality food and other products". Again, sufficient supply of high quality food and other products is a general aim for all agriculture. However, organic agriculture produces much less food per area than conventional agriculture and thus requires more land to be used for cropping (see Chapter 3 of this book; Kirchmann et al., 2008a). Organic products can also be affected by pests, which lower the quality. On the other hand, growth of the human population presupposes that much more food has to be produced in the future. Lal (2006) estimated that it is necessary to increase world-wide average cereal yields from 2.64 Mg ha^{-1} (in the year 2000) to at least 4.30 Mg ha^{-1} (by 2050).

Another topic addressed is animal welfare –"animals should be provided with the conditions and opportunities of life that accord with their physiology, natural behaviour and well-being". We are convinced that humans are obliged to show kindness and respect to livestock, as well as being morally responsible for their health and well-being. However, "natural behaviour" is not always wanted. Humans have kept livestock for many years, resulting in a selection of animals with behaviours that differ from the wild species. Natural behaviour cannot be the only guideline for livestock management because even domesticated animals can do harm by victimisation, fighting and cannibalism. It is important to keep animals in such a way that the special requirements of each species are fulfilled and that destructive forms of behaviour can be prevented.

2.3.4.4 Principle 4 – Care

> Organic agriculture should be managed in a precautionary and responsible manner to protect the health and well-being of current and future generations and the environment.

Caring for the environment is a basic principle necessary for its sustainability in order to provide humans with wellbeing, food and other essentials. In most societies, there is consensus to care for the environment and the responsibility of humans towards nature is clear – to respect, to utilise and to care.

It is the explanation of how to care for the environment that is remarkable in the IFOAM document, which states that "Science is necessary to ensure that organic agriculture is healthy, safe and ecologically sound. However, scientific knowledge alone is not sufficient. Practical experience, accumulated wisdom and traditional and indigenous knowledge offer valid solutions, tested by time". The problem with this explanation is that any kind of tradition including occult practices etc. are regarded as being of similar value to scientific results. For example, use of biodynamic compounds is explicitly accepted as a valid solution.

To make it very clear, our criticism is not based on a negative attitude towards accumulated wisdom or traditional and indigenous knowledge as such – knowledge gained this way can be very valuable – but on the fact that this knowledge can be both useful and of disadvantage for the sustainability of agro-ecosystems. It may also be a barrier to other well-founded practices.

Table 2.2 Summary of characteristics of the schools of organic agriculture

Founders and organisation	Philosophy and view on nature	Reasons for exclusion of synthetic fertilisers and pesticides
R Steiner (1861–1925); Biological dynamic farming	Anthroposophy; "Forces" in nature provide salvation	Artificial materials disturb the "flow of forces" in nature and destroy the "spiritual quality" of crops
A Howard (1873–1947); E Balfour (1899–1990); Soil Association	Nature romanticism; Undisturbed nature embodies harmony. Humus guarantees soil fertility providing health. Health is a birthright	Humus is the most significant of all nature's reserves. Inorganic fertilisers speed up humus decay
H-P Rusch (1906–1977); Biological organic farming	Eco-philosophy; Nature is a perfect unit with parity between all forms of life	Inorganic fertilisers are not adapted to the demand of crops. Diseases and pests are natural destruction processes
International Federation of Organic Agriculture Movements (since 1972); (IFOAM)	Environmentalism; Nature is the master	Organic practices are superior and therefore self-evident

2.4 Ethics in Organic Agriculture

The previous analysis shows that organic practices are originally based on certain philosophical views on nature (summarised in Table 2.2) and that there is a lack of agreement between these practices and the scientific evidence. In this section, we comment on this conflict. As indicated earlier, founders and followers of organic agriculture prefer a holistic rather than a reductive view, organic rather than mechanistic studies and in some cases intuition/feeling rather than logical reasoning. Behind these positions one can trace valuations of nature, which we regard to be the roots of organic agriculture. In the following we characterise these valuations and discuss shortcomings.

2.4.1 Idealisation of Nature and Cooperation with Nature

The literature on organic agriculture describes and positions nature as being ideal and the functioning of nature is the prototype to be emulated. Letting nature renew and restore itself and using and adapting to natural cycles is seen as a model. Processes and functions occurring in nature are regarded as being superlative and naturalness is seen as a prerequisite for sound food production (Verhoog et al., 2003). Ecological wisdom, yet not well-defined, is assumed as the guiding authority guaranteeing sustainability. Technical innovations are generally deemed inferior to natural means and methods. Nature is simply assumed to know best and occasionally is even referred to as being "good" (Vilkka, 1997). The idealisation of nature is one of the fundamental principles from which organic practices can be deduced.

Another underlying principle of organic agriculture is to cooperate with nature rather than to dominate and control. Nature is seen as a partner, where all organisms contribute to the health of the whole. In this relationship, all forms of life are often classed as equal in terms of their intrinsic value. For example, the life of a plant is classed as being as valuable as the life of a human being. By stressing parity among all living organisms (biocentrism), the importance and interdependence of life are given much greater emphasis. Humans can perceive themselves as part of nature and not as a separate entity apart from nature. In other words, through rejection of an anthropocentric view, man and nature can re-establish unity. Applying these ethics, humans can build fair relationships with other organisms and take action to counteract abuses of nature.

In fact, the fundamental view of nature as being ideal and complete motivates many academics to argue for organic practices and may also direct consumer choices in preferring organic food. The crucial question arising is whether the two fundamental views outlined above are a reasonable and sound basis. What type of system is nature? How should humans view themselves in relation to nature? What ethical principles would be in accordance with our experience of nature? Sound ethical principles influence our thinking, decisions and actions in our relationship with nature and thus require special attention.

2.4.2 The Dualistic Character of Nature

We are concerned that the idealisation of nature does not deal with the harsh side of nature in a satisfactory way and is thus not considering nature as a whole. It represents a one-sided view, that nature is fantastic, magnificent, beautiful, admirable, etc. with a proper fit and functioning of species, and particular orders and interrelation of ecosystems that seem to function well without human intervention. All this may lead to the conclusion that nature is perfect. However, natural disasters (e.g. hurricanes, earthquakes and tsunamis) illustrate a self-destructive and non-predictable side of nature. Long-lasting ice ages, land loss, continental movements and meteorite collisions further exemplify desolation and species extinction by natural causes. Actually none of these destructive forces is mainly the result of human activity, but they can all destroy the majority of organisms, including humans. In other words, natural disasters can greatly exceed the environmental damage caused by humans.

Furthermore, nature is dangerous and not at peaceful harmony. Wild animals suffer from predator attacks, malnutrition, parasites, diseases etc., showing the lack of "goodness" in nature. In fact, the natural behaviour of animals has no morality and demonstrates what in human terms would be called cruelty, e.g. the strong bullying the weak and the survival of the fittest. In domestic livestock, traits are continuously being selected that are different from those found in the wild. Over the course of civilisation, humans have devoted great effort to developing means and methods to eliminate the dangers and unwanted properties of nature in order to minimise negative impacts on plants, animals and humans.

It can be concluded that nature from a human perspective has a dualistic character with opposing qualities: beauty and order through its life processes on the one hand and chaos, cruelty and desolation on the other. When this dualistic character of nature is ignored and excused, it is difficult to deal with nature in a competent and efficient way. For example, in an idealised view, the common occurrence of diseases and suffering in nature is either denied or seen as nature's way of regulating itself. However, the suffering found in nature is unacceptable as a model for humans. On the contrary, one of the founders of organic agriculture claimed that perfect health is a birthright (see Table 2.2), but there is absolutely no evidence that disease and suffering would be absent if humans were to revert to nature.

2.4.3 Human Stewardship

Equally important to recognition of the dualistic qualities of nature is understanding of the human relationship to nature. Humans living on planet Earth are dependent on nature for survival and must take care to sustain nature in its wholeness. In fact, how humans position themselves towards other organisms determines how nature is viewed and treated (Table 2.3). Should humans act as cooperating partners or managers of nature?

According to the modern school of organic agriculture, a cooperative relationship between humans and nature is proposed as a fundamental principle. This mainly refers to a biocentric relationship, which means that human life is not classified higher than life of other organisms, although this is not always clearly mentioned. However, this is an untenable position. Humans have the ability to recognise all other organisms and can at least partly comprehend planet Earth. Furthermore, human knowledge enables us to improve poor natural conditions, for example supplying nutrients where natural deficiency is limiting growth. We can work out means and programmes to save species from eradication etc., but we also have a totally

Table 2.3 Ethical positions determine human attitudes towards nature

Ethical fundament	Perception of nature	Relation to nature
Theocentric		
Humans believe that God exists to whom they are accountable for	Nature is included in the Fall of Man affected by sin – not perfect	Use for benefit and joy entrusting man stewardship
Anthropocentric		
Humans are above nature and believe in no higher authority to be accountable for	Nature is dualistic characterized by both wanted and unwanted properties	Use for benefit and joy including stewardship and responsibility
Biocentric		
Humans are in parity with all living and believe in unity	Nature is an ideal system comprising of a perfect wholeness	Humans need to learn from nature how to imitate it

unique ability to destroy everything that grows, crawls or runs. All these abilities automatically place humans in a position of leadership on Earth.

Ethics based on cooperation or biocentrism aim to prohibit humans from playing a dominant role in order to establish biological synergy and harmony between humans and nature. According to the motto "let nature do the job it knows best", the less nature is affected by man-made innovations, the better. The biocentric position sets boundaries to human creativity and limits human activity to the exclusive use of naturally occurring compounds. As a consequence, development of artificial products through science is principally rejected, which is also in line with organic practices. The possibility of humans accomplishing new feats is all but eliminated.

Taking philosophies to logical endpoints can reveal their weaknesses. According to the ethical valuation that all forms of life, including viruses and bacteria, are of equal intrinsic value, disease-causing organisms would not be combated. This means that ethics based on cooperation with nature or biocentrism neglect the issue of human survival through the commitment to conserve the biotic community in total. It is obvious that this position is not pro-human and will ultimately be destructive for human societies. We are convinced that both human needs and environmental stewardship need to be considered in the search for sustainable forms of management, but with a pro-human perspective. Humans must show more respect for human than other forms of life.

As the dualistic character of nature involves desirable and undesirable properties, the difficulty is not only in conserving nature, but also in finding solutions that minimise the negative effects of nature on humans. Humans are the only organisms on Earth that have the capacity, overview and knowledge to use, control and care for nature. Humans have created food production systems that have improved our standard of living and will create new food systems in the future. On the other hand, humans can completely destroy ecosystems and eradicate species and therefore they have the obligation to take full responsibility for ensuring that nature can be preserved and new ecosystems created, e.g. urban or agro-ecosystems. Although human domination can be used for best management or can be misused, avoiding taking a leadership role is not a viable option. The supremacy over other organisms calls humans to be stewards on Earth with a moral commitment firstly towards humans but also towards the environment.

2.5 Conclusions

The European founders of organic agriculture were concerned about deterioration in product quality and a decrease in soil fertility. Their common view was that if industrial applications became a model for agricultural progress and development, this would lead to serious negative consequences for nature and mankind. They were convinced that use of natural means and methods in agriculture are intrinsically better than others. Their mission was to convince others to base food production on the exclusion of modern means of production and to show that industrialization of agriculture was the wrong direction.

All founders mistrusted and disliked science as a valuable tool to explore humans and nature, including agriculture. They condemned the reductive character of science as misleading or degraded science as being of limited value. Actually, none of the organic agriculture theories are based on scientific hypotheses or scientific evidence. Instead, strong views about nature and how to treat and deal with it, derived from philosophies about life are the origin of organic agriculture.

Biological Dynamic Agriculture grew out of Anthroposophy, the Soil Association is based on Nature Romanticism, Biological Organic Agriculture has its roots in Eco-philosophy, and modern organic agriculture is based on Environmentalism.

The analysis of the organic agriculture schools reveals that they are filled with flaws and errors. There is no consensus about agronomic practices among the founders, e.g. how to treat animal manures, how to use organic manures, how to till the soil, how to deal with pests etc. The concept of naturalness, excluding synthetic fertilisers and synthetic pesticides, is simply seen as a guarantee for the superiority of organic production. The analysis of modern organic agriculture reveals that the fundamental ideas of the founders are not explicitly mentioned anymore but are still alive. The four principles of modern organic agriculture uphold the way of thinking of the founders but also emphasising desirable aims common for all types of agriculture.

Two fundamental valuations of nature were identified to be roots of organic agriculture. One principle is to regard nature to be an ideal system comprising a perfect wholeness. Nature's wisdom is seen to be the master, whereby natural processes and functions serve as a model and standard to be emulated. The other principle is to base human relationships towards nature on cooperation in order to achieve biological synergy and harmony. However, both principles are insufficient not taking into account the dualistic character of nature and human needs in a satisfactory way.

It is our responsibility as scientists to use the best knowledge and values available in the search for sustainable forms of agriculture. It remains to be further discussed whether organic agriculture methods can provide a sound contribution to future food production systems, as organic principles exclude other potentially superior solutions.

References

Ahl, T., and Odén, S., 1972, River discharge of total nitrogen, total phosphorus and organic matter into the Baltic Sea from Sweden, *Ambio* **1**: 51–56.

Alvarez, R., 2005, A review of nitrogen fertilizer and conservation tillage effects on soil organic carbon storage, *Soil Use Manage.* **21**: 38–52.

Ames, B.N., Profet, M., and Gold, L.S., 1990, Dietary pesticides (99.99% all natural), *Proc. Nat. Acad. Sci. USA* **87**: 7777–7781.

Ames, B.N., 1998, Micronutrients prevent cancer and delay aging, *Toxicol. Lett.* **102/103**: 5–18.

Andrén, O., Lindberg, T., Paustian, K., and Rosswall, T. (eds.), 1990, Ecology of arable land – organisms, carbon and nitrogen cycling, *Ecol. Bull.* (Copenhagen) **40**, 221p.

Andrén, O., and Kätterer, T., 2001, Basic principles for soil carbon sequestration and calculating dynamic country-level balances including future scenarios, in: *Assessment Methods for Soil Carbon*, R. Lal et al., eds., Lewis Publishers, UK, pp. 495–511.

Antil, R.S., Gerzabek, M.H., Haberhauer, G., and Eder, G., 2005, Long-term effects of cropped vs. fallow and fertilizer amendments on soil organic matter – I. Organic carbon, *J. Plant Nutr. Soil Sci.* **168**: 108–116.

Avery, D.T., 2000, *Saving the Planet with Pesticides and Plastic: The Triumph of High-Yielding Farming*, Hudson Institute, Indianapolis, Indiana, USA, 432p.

Avery, A., 2006, *The Truth about Organic Foods*, Henderson Communications, Chesterfield, MO, USA, 230p.

Balesdent, J., Wagner, G.H., and Mariotti, A., 1988, Soil organic matter turnover in long-term field experiments as revealed by carbon-13 natural abundance, *Soil Sci. Soc. Am. J.* **52**: 118–124.

Balfour, E.A., 1943, *The Living Soil*, Faber and Faber Ltd, London, UK, 276p.

Basker, D., 1992, Comparison of taste quality between organically and conventionally grown fruits and vegetables, *Am. J. Altern. Agric.* **7**: 129–136.

Beck-Fries, B., Smårs, S., Jönsson, H., Eklind, Y., and Kirchmann, H., 2003, Composting organic household wastes at different oxygen levels: gaining an understanding of the emission dynamics, *Compost Sci. Util.* **11**: 41–50.

Bergström, L., and Kirchmann, H., 1999, Leaching of total-N from ^{15}N labeled poultry manure and inorganic fertilizer, *J. Environ. Qual.* **28**: 1283–1290.

Bergström, L., and Kirchmann, H., 2004, Leaching and crop uptake of nitrogen from nitrogen-15-labeled green manures and ammonium nitrate, *J. Environ. Qual.* **33**: 1786–1792.

Bergström, L., Kirchmann, H., Aronsson, H., Torstensson, G., and Mattsson, L., 2008, Use efficiency and leaching of nutrients in organic and conventional systems in Sweden, in: *Organic Crop Production – Ambitions and Limitations*, H. Kirchmann and L. Bergström, eds., Springer, Dordrecht, The Netherlands.

Bertilsson, G., Kirchmann, H., and Bergström, L., 2008, Energy analysis of conventional and organic agricultural systems, in: *Organic Crop Production – Ambitions and Limitations*, H. Kirchmann and L. Bergström, eds., Springer, Dordrecht, The Netherlands.

Bjarnason, S., 1987, Immobilization and remineralization of ammonium and nitrate after addition of different energy sources to soil, *Plant Soil* **97**: 381–389.

Bocanegra, M.P., Lobartini, J.C., and Orioli, G.A., 2006, Plant uptake of iron chelated by humic acids of different molecular weights, *Commun. Soil Sci. Plant Anal.* **37**: 239–248.

Bourn, D., and Prescott, J., 2002, A comparison of the nutritional value, sensory qualities and food safety of organically and conventionally produced foods, *Crit. Rev. Food Sci. Nutr.* **42**: 1–34.

Carson, R., 1962, *Silent Spring*, Houghton Mifflin Company, Boston, USA, 378p.

Chen, Y., Magen, H., and Clapp, C.E., 2001, Plant growth stimulation by humic substances and their complexes with iron, International Fertilizer Society, Proceedings No. 470, York, 14p.

Conford, P., 2001, *The Origins of the Organic Movement*, Floris Books, 15 Harrison Gardens, Edinburgh, UK, 287p.

Dahlin, S., Kirchmann, H., Kätterer, T., Gunnarsson, S., and Bergström, L., 2005, Possibilities for improving nitrogen use from organic materials in agricultural cropping systems, *Ambio* **34**: 288–295.

Davidson, E.A, and Janssens, I.A., 2006, Temperature sensitivity of soil carbon decomposition and feedbacks to climate change, *Nature* **440**: 165–173.

Dloughý, J., 1981, Alternativa odlingsformer – växtprodukters kvalitet vid konventionell och biodynamisk odling. Swedish University of Agricultural Sciences, Department of Plant Husbandy. Dissertation. Report No. 91. Uppsala, Sweden (In Swedish).

Doran, J.W., and Jones, A.J., 1996, *Methods for Assessing Soil Quality*, Soil Sci. Soc. Am. Special Publication No. 49, Madison, Wisconsin, USA, 410p.

Drengson, A., 1997, Ecophilosophy, ecosophy and the deep ecology movement: An overview, *Trumpeter* **14**: 110–111.

Eklind, Y., and Kirchmann, H., 2000, Composting and storage of organic household waste with different litter amendments II. Nitrogen turnover, *Bioresour. Technol.* **74**: 125–133.

Fox, W., 1994, Ecophilosophy and science, *Environmentalist* **14**: 207–213.

Freyer, B., Eder, M., Schneeberger, W., Darnhofer, I., Kirner, L., Lindenthal, T., and Zollitsch, W., 2001, Der biologische Landbau in Österreich – Entwicklungen und Perspektiven, *Agrarwirtschaft* **50**: 400–409 (In German).

Fukuoka, M., 1978, *The One-Straw Revolution: An Introduction to Natural Farming*, Rodale Press, Emmaus, PA, USA, 181p.

Fukuoka, M., 1989, *The Road Back to Nature. Regaining the Paradise Lost*, 2nd Printing. Japan Publications, Inc. Tokyo, Japan, 377p.

Fukuoka, M., 1991, *The Natural Way of Farming. The Theory and Practice of Green Philosophy*, 3rd Printing. Japan Publications, Inc., New York, USA, 284p.

Gerzabek, M.H., Haberbauer, G., and Kirchmann, H., 2001, Soil organic matter pools and carbon-13 natural abundances in particle size fractions of a long-term field experiment receiving organic amendments, *Soil Sci. Soc. Am. J.* **65**: 352–358.

Granstedt, A., 1990, Fallstudier av kväveförsörjning i alternativ odling. Swedish University of Agricultural Sciences. Research Committee for Alternative Agriculture. Dissertation. Report No. 4. Uppsala, Sweden (In Swedish).

Green, C.J., Blackmer, A.M., and Horton, R., 1995, Nitrogen effects on conservation of carbon during corn residue decomposition in soil, *Soil Sci. Soc. Am. J.* **59**: 453–459.

Haeckel, E.H.P.A., 1900, *The Riddle of the Universe: At the Close of the Nineteenth Century*, Harper and Brothers, New York, USA, 390p.

Howard, A., 1940, *My Agricultural Testament*, Oxford University Press, Oxford, UK, 253p.

Howard, A., 1947, *The Soil and Health. A Study of Organic Agriculture*, The Devin-Adair Company, New York, USA, 307p.

IFOAM, 2006, *The Four Principles of Organic Farming*, The International Federation of Organic Agriculture Movements, www.IFOAM.org, Bonn, Germany, assessed 5/6-2006.

Ingestad, T., and Ågren, G.I., 1995, Plant nutrition and growth: Basic principles, *Plant Soil* **168/169**: 15–20.

Jansson, S.L., 1948, Reformtendenser inom jordbruket, *Kungl. Lantbruksakademiens Tidskr.* **87**: 129–160 (In Swedish).

Jansson, S.L., 1958, Tracer studies on nitrogen transformations in soil with special attention to mineralization-immobilization relationships, *Ann. R. Agric. Coll. Sweden* **24**: 101–361.

Jansson, S.L., 1971, Naturalness of commercial fertilizers, An ecological treatise, *Acta Agric. Fennica* **123**: 173–185.

Jansson, S.L., and Persson, J., 1982, Mineralization and immobilzation of nitrogen in soil, in: *Nitrogen in Agricultural Soils*, F.J. Stevenson ed., Agronomy Monograph No. 27, Madison, WI, USA, pp. 229–252.

Johnston, A.E., McGrath, S.P., Poulton, P.R., and Lane, W., 1989, Accumulation and loss of nitrogen from manure, sludge and compost: Long-term experiments at Rothamsted and Woburn, in: *Nitrogen in Organic Wastes Applied to Soil*, J.A. Hansen and K. Henriksen, eds., Academic Press Ltd, London, U.K., pp 126–139.

Jones, D.L., and Darrah, P.R., 1994, Amino acid influx at the soil-root interface of Zea mays L. and its implications in the rhizosphere, *Plant Soil* **163**: 1–12.

King, F.H., 1911, *Farmers of Forty Centuries or Permanent Agriculture in China, Korea and Japan*, The MacMillan Company, Madison, WI, USA, 441p.

Kirchmann, H., 1985, Losses, plant uptake and utilisation of manure nitrogen during a production cycle, *Acta Agric. Scand.*, Supplementum 24, Dissertation, 77p.

Kirchmann, H., 1990, Nitrogen interactions and crop uptake from fresh and composted ^{15}N-labelled poultry manure, *J. Soil Sci.* **41**: 379–385.

Kirchmann, H., 1994, Biological dynamic farming – an occult form of alternative agriculture, *J. Agric. Environ. Ethics* **7**: 173–187.

Kirchmann, H., Persson, J., and Carlgren, C., 1994, The long-term soil organic matter experiment at Ultuna, 1956–1991. Swedish University of Agricultural Sciences, Department of Soil Sciences, Reports and Dissertations 17. Uppsala, Sweden, 55p.

Kirchmann, H., and Bernal, M.P., 1997, Organic waste treatment and C stabilization efficiency, *Soil Biol. Biochem.* **29**: 1747–1753.
Kirchmann, H., and Andersson, R., 2001, The Swedish system for quality assessment of agricultural soils, *Environ. Monit. Assess.* **72**: 129–139.
Kirchmann, H., Haberhauer, G., Kandeler, E., Sessitsch, A., and Gerzabek, H., 2004, Effects of level and quality of organic matter input on carbon storage and biological activity in soil: Synthesis of a long-term experiment, *Global Biogeochem. Cycles* **18**: 38–46.
Kirchmann, H., Nyamangara, J., and Cohen, Y., 2005, Recycling municipal wastes in the future: From organic to inorganic forms? *Soil Use Manage.* **21**: 152–159.
Kirchmann, H., Bergström, L., Kätterer, T., Mattsson, L., and Gesslein, S., 2007, Comparison of long-term organic and conventional crop-livestock systems on a previously nutrient depleted soil in Sweden, *Agron. J.* **99**: 960–972.
Kirchmann, H., Bergström, L., Kätterer, T., Andrén, O., and Andersson, R., 2008a, Can organic crop production feed the world? in: *Organic Crop Production – Ambitions and Limitations*, H. Kirchmann and L. Bergström, eds., Springer, Dordrecht, The Netherlands.
Kirchmann, L., Kätterer, T., and Bergström, L., 2008b, Nutrient supply in organic agriculture – plant-availability, sources and recycling, in: *Organic Crop Production – Ambitions and Limitations*, H. Kirchmann and L. Bergström, eds., Springer, Dordrecht, The Netherlands.
Koepf, H.H., 1973, Organic management reduces leaching of nitrate, *Biodynamics* **108**: 20–30.
Koepf, H.H., Pettersson, B.D., and Schaumann, W., 1976, *Biologische Landwirtschaft*, Verlag Eugen Ulmer, Stuttgart, Germany, 303 p (In German).
Kristensen, L., Stopes, C., Kølster, P., and Granstedt, A., 1995, Nitrogen leaching in ecological agriculture: Summary and recommendations, *Biol. Agric. Hortic.* **11**: 331–340.
Lal, R., 2006, Soil science in the era of hydrogen economy and 10 billion people, in: *The Future of Soil Science*, A.E. Hartemink ed., IUSS International Union of Soil Science, Wageningen, The Netherlands, pp.76–79.
McDowell, L.R., 2003, *Minerals in Animal and Human Nutrition*, Elsevier, Amsterdam, The Netherlands, 644p.
Meadows, D.H., Meadows, D.L., Randers, J., and Behrens, W.W., 1972, *Limits to Growth. A Report of the Club of Rome's Project on Predicament of Mankind*, Universe Books, New York, 205p.
Mengel, K., and Kirkby, E.A., 2004, *Principles of Plant Nutrition*, 5th Edition, Springer Verlag, Doordrecht, The Netherlands, 849p.
Merrill, M.C., 1983, Eco-agriculture: A review of its history and philosophy, *Biol. Agric. Hortic.* **1**: 181–210.
Naess, A., 1989, *Ecology, Community and Lifestyle: Outline of an Ecosophy*, Cambridge University Press, New York, 223p.
Näsholm, T., Huss-Danell, K., and Högberg, P., 2000, Uptake of organic nitrogen in the field by four agriculturally important plant species, *Ecology* **81**: 1155–1161.
Nyamangara, J., Piha, M.I., and Kirchmann, H., 1999, Interactions of aerobically decomposed cattle manure and nitrogen fertilizer applied to soil, *Nutr. Cycl. Agroecosys.* **54**: 183–188.
Oshawa, G., and Dufty, W., 2002, *You are All Sanpuka*, Citadel Press, USA, 224p.
Paillat, J.M., Robin, P., Hassouna, M., and Leterme, P., 2005, Predicting ammonia and carbon dioxide emissions from carbon and nitrogen biodegradability during animal waste composting, *Atmos. Environ.* **39**: 6833–6842.
Puig-Gimenez, M.H., and Chase, F.E., 1984, Laboratory studies of factors affecting microbial degradation of wheat straw residues in soil, *Can. J. Soil Sci.* **64**: 9–19.
Rengel, Z., Batten, G.D., and Crowley, D.E., 1999, Agronomic approaches for improving the micronutrient density in edible portions of field crops, *Field Crops Res.* **60**: 27–40.
Reuveni, R., and Reuveni, M., 1998, Foliar-fertilizer therapy – a concept in integrated pest management, *Crop Prot.* **17**: 111–118.
Rusch, H.P., 1978, *Bodenfruchtbarkeit. Eine Studie biologischen Denkens*, 3rd Printing. Haug Verlag, Heidelberg, Germany, 243 p (In German).

Ryan, M.H., Derrick, J.W., and Dann, P.R., 2004, Grain mineral concentrations and yield of wheat grown under organic and conventional management, *J. Sci. Food Agric.* **84**: 207–216.

Ryan, M.H., and Tibbett, M., 2008, The role of arbuscular mycorrhizas in organic farming, in: *Organic Crop Production – Ambitions and Limitations*, H. Kirchmann and L. Bergström, eds., Springer, Dordrecht, The Netherlands.

Sanchez, P.A., and Swaminathan, M.S., 2005, Hunger in Africa: The link between unhealthy people and unhealthy soils, *Lancet* **365**: 442–444.

Schloter, M., Dilly, O., and Munch, J.C., 2003, Indicators for evaluating soil quality, *Agric. Ecosys. Environ.* **98**: 255–262.

Schønning, P., Elmholt, S., Munkholm, L.J., and Debosz, K., 2002, Soil quality aspects of humid sandy loams as influenced by organic and conventional long-term management, *Agric. Ecosys. Environ.* **88**: 195–214.

Sommer, S.G., and Dahl, P., 1999, Nutrient and carbon balance during the composting of deep litter, *J. Agric. Engin. Res.* **74**: 145–153.

Söderström, B., Bååth, E., and Lundgren, B., 1983, Decrease in soil microbial activity and biomasses owing to nitrogen amendments, *Can. J. Microbiol.* **29**: 1500–1506.

Steiner, R., 1913, *The Fifth Gospel: From the Akashic Record*, Reprinted 2007, Rudolf Steiner Press, East Sussex, England, 262p.

Steiner, R., 1924, *Geisteswissenschaftliche Grundlagen zum Gedeihen der Landwirtschaft (Spiritual Foundations for the Renewal of Agriculture)*, Rudolf Steiner Nachlassverwaltung, 5th Printing 1975, Dornach, Switzerland, 256 p (In German).

Swartjes, F.A., 1999, Risk-based assessment of soil and groundwater quality in the Netherlands: Standards and remediation urgency, *Risk Anal.* **19**: 1235–1249.

Taverne, D., 2005, *The March of the Unreason. Science, Democracy and the New Fundamentalism*, Oxford University Press, Oxford, UK, 310p.

Tinker, P.B., 2000, *Shades of Green – A Review of UK Farming Systems*, Royal Agricultural Society of England. Page Bros Ltd, Norwich, UK, 100p.

Torstensson, G., Aronsson, H., and Bergström, L., 2006, Nutrient use efficiencies and leaching of organic and conventional cropping systems in Sweden, *Agron. J.* **98**: 603–615.

Trewavas, A., 2004, A critical assessment of organic farming-and-food assertions with particular respect to the UK and the potential environmental benefits of no-till agriculture, *Crop Prot.* **23**: 757–781.

Trichopoulou, A., and Critselis, E., 2004, Mediterranean diet and longevity, *Eur. J. Cancer Prev.* **13**: 453–456.

Ullah, S.M., and Gerzabek, M.H., 1991, Influence of fulvic and humic acids on Cu-toxicity and V-toxicity to *Zea-Mays* (L.), *Bodenkultur* **42**: 123–134.

Verhoog, H., Matze, M., Lammertz van Bueren, E., and Baars, T., 2003, The role of the concept of the natural (naturalness) in organic farming, *J. Agric. Environ. Ethics* **16**: 29–49.

Vilkka, L., 1997, *The Intrinsic Value of Nature*, Rodopi, Amsterdam/Atlanta, GA, USA, 168p.

Vogt, G., 2001, *Geschichte des ökologischen Landbaus im deutschsprachigen Raum*, www.orgprints.org/1110/. Assessed 27 Sept. 2006.

Welch, R.M., and Graham, R.D., 1999, A new paradigm for world agriculture: Meeting human needs. Productive, sustainable, nutritious, *Field Crops Res.* **60**: 1–10.

Willet, W.C., 1994, Diet and health: What should we eat? *Science* **264**: 532–537.

Woese, K., Lange, D., Boess, C., and Bögl, K.W., 1997, A comparison of organically and conventionally grown foods – results of a review of the relevant literature, *J. Sci. Food Agric.* **74**: 281–293.

Chapter 3
Can Organic Crop Production Feed the World?

Holger Kirchmann, Lars Bergström, Thomas Kätterer, Olof Andrén and Rune Andersson

Abstract Agriculture provides the most essential service to mankind, as production of crops in sufficient amounts is necessary for food security and livelihood. This chapter examines the question of whether organic agriculture can produce enough food to meet future demand. This question relates to a moral imperative and any evaluation must therefore be based on objective scientific facts excluding ideological bias, political correctness, economic incentives or environmental opinions. The chapter begins by defining the conditions necessary for a stringent evaluation of crop yields and explains potential pitfalls. Yield data from national statistics, organic and conventional long-term experiments and comparative studies are then compiled and evaluated, followed by a discussion of the main factors behind low-yielding production. In a global perspective, the scientific literature shows that organic yields are between 25 and 50% lower than conventional yields, depending on whether the organic system has access to animal manure. The amount of manure available on organic farms is usually not sufficient to produce similar crop yields as in conventional systems and therefore green manures are commonly used. However, organic crop yields reported for rotations with green manure require correction for years without crop export from the field, which reduces average yield over the crop rotation. When organic yields are similar to those in conventional production, nutrient input through manure is usually higher than nutrient addition in conventional agriculture, but such high inputs are usually only possible through transfer of large amounts of manure from conventional to organic production. The main factors limiting organic yields are lower nutrient availability, poorer weed control and limited possibilities to improve the nutrient status of infertile soils. It is thus very likely that the rules that actually define organic agriculture, i.e. exclusive use of manures and untreated minerals, greatly limit the potential to increase yields. Our analysis of some yield-related statements repeatedly used by advocates of organic agriculture reached the following conclusions: Organic manure is a severely limited resource, unavailable in quantities sufficient for sustaining high crop yields; legumes are not a free and

---

H. Kirchmann (✉)
Department of Soil and Environment, Swedish University of Agricultural Sciences, P.O. Box 7014, SE-75007 Uppsala, Sweden
e-mail: holger.kirchmann@mark.slu.se

H. Kirchmann, L. Bergström (eds.), *Organic Crop Production – Ambitions and Limitations*, DOI 10.1007/978-1-4020-9316-6_3,

environmentally sound N source that can replace inorganic fertilisers throughout; and low native soil fertility cannot be overcome with local inputs and untreated minerals alone.

Agricultural methods severely limiting crop yields are counter-productive. Lower organic yields require compensation through expansion of cropland – the alternative is famine. Combining expected population growth and projected land demand reveals that low-yielding agriculture is an unrealistic option for production of sufficient crops in the future. In addition, accelerated conversion of natural ecosystems into cropland would cause significant loss of natural habitats. Further improvement of conventional agriculture based on innovations, enhanced efficiency and improved agronomic practices seems to be the only way to produce sufficient food supply for a growing world population while minimising the negative environmental impact.

Keywords Area demand · Conventional yields · Cropland expansion · Habitat loss · Nutrient input · Organic yields · Population growth · Weeds

3.1 Introduction

Providing healthy food for everyone is probably the most important survival issue for mankind in the future. We are currently producing a slight excess of food in relation to consumption (Alexandratos, 1999; Dyson, 1999). However, the demand for food, feed and fibres will greatly increase during coming decades (Evans, 1998; FAO, 2007) driven by a growing population, which is getting wealthier (Bruinsma, 2003; GeoHive, 2007). The global human population has doubled over the last 40 years, to around 6.5 billion people in 2006, and food plus feed production has tripled during the same period (FAO, 2007). By 2030, the global population may reach 8–9 billion, of which 6.8 billion may live in developing countries (Bruinsma, 2003; GeoHive, 2007). As the projected increase will mainly take place in developing countries, Africa would need to increase food production by 300%, Latin America by 80%, Asia by 70%, but even North America by 30%. Assuming that the additional population consumes only vegetarian food, a minimum of 50% more crops will need to be produced by 2030 to ensure sufficient food supply. As a satisfactory diet has been defined to consist of 40 g animal protein per person and day (Gilland, 2002) and taking into account that diets throughout the world are changing with the rise in income towards more meat and dairy products irrespective of culture, there will be a need to actually increase food plus feed production by 60–70%. For example, in developing countries, meat consumption amounted to 71 g per person and day in 1997–1999 and is projected to further increase to 100 g per person and day in 2030 (Bruinsma, 2003). In developed countries, meat consumption of 180 g per person and day is projected for 2030. Since the largest proportion of the projected increase is expected to come from pork, poultry and aquaculture, meeting future demand will depend on achievable increases in cereal yields (Bradford, 1999). A doubling of cereal yields may be necessary by 2030.

Global food production increased by 70% from 1970 to 1995, largely due to the application of modern technologies in developing countries, where food production increased by 90%. However, global food production must grow to the same extent in the coming three decades, as pointed out above, to meet human demand (Bruinsma, 2003; Cassman et al., 2003; Eickhout et al., 2006). Two principal possibilities for achieving this increase have been identified: intensifying agricultural production on existing cropland or ploughing up natural land into cropland, i.e. clearing pastures and rangelands, cutting forests and woodland areas, etc. Some experts have a positive view that food production can be greatly increased if high-yielding production is widely applied (Bruinsma, 2003), and the expansion of arable land in the world is expected to only slightly increase from 1400 Mha in 2006 (FAO, 2007) to 1600 Mha in 2030 (Bouwman et al., 2005). In 2025, the world's farmers will be expected to produce an average world cereal yield of about 4 metric tons per hectare (Dyson, 1999) if conditions are optimised.

There are recent claims that sufficient food can be produced by organic agriculture, expressed in terms such as 'organic agriculture can feed the world' (e.g. Woodward, 1995; Vasilikiotis, 2000; Leu, 2004; Tudge, 2005; Badgley and Perfecto, 2007). The following three arguments have been put forward: (i) Lower production of most crops can be compensated for by increased production of legumes, in particular of grain legumes, while a change to a diet based mainly on vegetables and legumes will provide enough food for all (Woodward, 1995). (ii) Realities in developing countries must be taken into account: 'Increased food supply does not automatically mean increased food security for all. Poor and hungry people need low-cost and readily available technologies and practices to increase food production' (Pretty et al., 2003). (iii) 'Organic agriculture can get the food to the people who need it and is therefore the quickest, most efficient, most cost-effective and fairest way to feed the world' (Leu, 2004). These arguments confuse the original scientific question with other realities interacting with food sufficiency, such as change in dietary composition, poverty, finance, markets, distribution system, etc. However, the basic scientific question remains and requires a stringent review and evaluation of the production potential of organic and conventional systems.

A fundamental question is whether organic yields can be increased radically or whether more natural ecosystems have to be converted into cropland. The following four observations indicate that intensification rather than area expansion is necessary: (1) Agricultural land is steadily decreasing as it is being taken over for urban or industrial use (Blum et al., 2004); (2) global warming may reduce the potential for higher yields in large parts of the world (Parry et al., 2005); (3) significant areas of farmland may be used for fuel production, competing with food production (Nonhebel, 2005); and (4) cropland simply cannot be expanded, due to shortage of suitable land. On the other hand, current yield increases appear to be falling below the projected rate of increase in demand for cereals (Cassman et al., 2002), challenging scientists to do their best to increase crop productivity per unit area (Evans, 1998).

Food production is coupled to a moral imperative, as sufficient food supply is a cornerstone of human welfare. Development of agricultural practices ensuring food sufficiency is a basic human requirement, a prerequisite for satisfactory social conditions and a necessity for civilisations to flourish. Lack of food, on the other hand, is a tragedy leading not only to suffering and loss of life but also to inhuman behaviour, political instability and war (Borlaug, 1970). In fact, eradication of famine and malnutrition has been identified as the most important task on Earth (UN Millennium Project, 2005). Thus, when discussing different forms of crop production, it is of the utmost importance to examine without prejudice the forms of agriculture that can contribute to food sufficiency and security, at present and in the future. Separation of facts and wishful thinking is absolutely necessary and only an unbiased review of the scientific literature can provide objective answers to the questions put forward below. A strong belief and enthusiasm for certain solutions cannot be allowed to hamper the search for objectivity.

The overall aim of this chapter was to examine a morally important aspect of organic agriculture. This was achieved by examining the following questions:

- Can sufficient crop production be obtained through conversion to and/or introduction of organic production?
- Can future food demand be covered by organic agriculture?
- Is it possible to significantly increase organic yields in the future?

3.2 Defining Conditions Necessary for a Stringent Comparison of Crop Yields from Organic and Conventional Systems

Evaluating crop yields from organic and conventional production seems straightforward but there are restrictions and difficulties to be considered. The conditions outlined below are necessary for stringent scientific comparisons based on robust quantitative thinking.

3.2.1 Evaluate Comparable Systems

Yield examinations require that only systems of the same type are compared. Comparing yields from pure crop production systems with those from mixed crop-animal systems, or biofuel-crop systems, is incorrect. The main reason is that each system is characterised by specific crops and a level of production typical for these crops that is not necessarily related to organic/conventional methods. For example, systems with forage and milk production have a higher production level (tonnes of dry plant matter per hectare) than cereal production without animals and manure application. Furthermore, to avoid misinterpretation of yields, crops grown within the same type of system should also be similar.

3.2.2 Choose Long-Term Studies

A critical aspect of a relevant comparison of crop yields from organic and conventional systems is the time span of the comparison. Short-term comparative studies

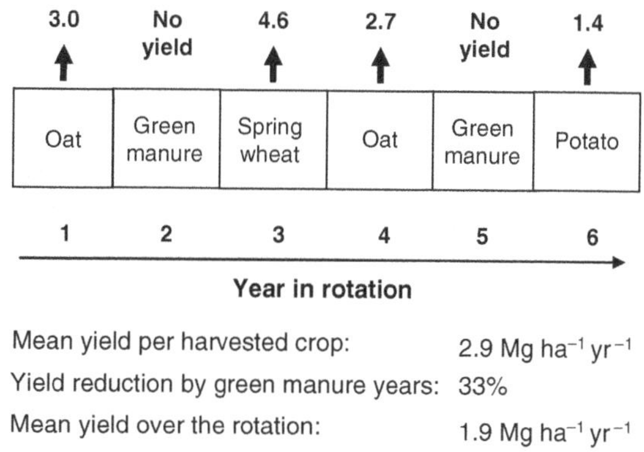

Fig. 3.1 Note the difference between crop yield each year and average yield per 6-year rotation period. Years with non-food crops, e.g. green manure or fallow, reduce mean yield by an equivalent percentage. Data were taken from Torstensson et al. (2006)

can lead to biased conclusions for several reasons. If a conventional system is converted into an organic system, previous soil management practices will affect crop growth in the organic system. A reduced weed population (including seed bank), elevated soil P and K fertility levels, a high organic matter content and amount of recent plant residues, etc. may initially result in higher yields in the organic system than those found after a decade. In other words, only long-term studies with minimal residual effects are really useful.

Using yield data from a single harvest is not valid for a proper evaluation. As yield data vary between years due to weather conditions, fertility management during the previous year, damage through pests, weeds, etc., a single harvest estimate is not representative. More importantly, yields of actual crops may not be a complete measure of the total productivity of a system. When non-food crops (green manure) form part of the rotation or when land lies fallow for a year or so, the year without harvest means a loss of production. Tables of single crop yields may not include these 'lost' years. In other words, total crop output over a whole crop rotation period of several years is the most relevant variable when comparing or discussing crop production in different systems. This has to be taken into account in a scientific comparison of production capacity (Fig. 3.1).

3.2.3 Exclude Yield Data from Organic Systems with High Applications of Manure

A common hidden assumption is that organic manures, composts, etc. are not limited by any means, and that sufficient organic manure is accessible to all farms and can be applied freely. In other words, data from experiments using large quantities of manure are used as proof that it is possible to produce high yields through organic management. However, the amount of nutrients that can be applied through organic manures is actually quite low. For example, only 58 kg N ha^{-1} yr^{-1} (Kirchmann et al., 2005) would potentially be available in Sweden if manure were to be equally distributed on all arable land in the country. Only 50–70% of this amount is available when losses through ammonia volatilisation during storage and handling are considered (Kirchmann and Lundvall, 1998). This quantity is far less than the amount of N applied in the organic studies examined in this review.

Thus, high crop yields are not proof of the productivity of an organic system as long as it uses large amounts of manure transferred from other systems and not produced by the farming system itself. The high yields are actually only proof of the well-known fact that manure can be used as a fertiliser to increase yields. A realistic assessment of the production capacity of organic systems is only possible if any major nutrient transfer from conventional systems is excluded, see Chapter 5 of this book (Kirchmann et al., 2008). If all farming systems were to be organic, it would be impossible to rely on nutrient transfer from conventional production and the amounts of manure applied would be equivalent to the production level of the system. For example, Chen and Wan (2005) showed that the amount of nutrients supplied through organic manures in China is far below the amount required to produce sufficient food for its people.

3.2.4 Consider Whether Differences in the Management of Systems Other than Those Originating from Organic Regulations is a Cause of Bias

A number of management options are not regulated by organic farming regulations, such as use of crop residues, soil tillage, use of catch crops, etc. Differences in management can have a great impact on yields, but these practices are not dependent on an organic/conventional approach and can be managed in the same way in both systems. For example, incorporation of crop residues in organic systems but their removal and sale in conventional systems can affect soil organic matter levels, and ultimately also yields. Using catch crops in one system but not in another can also have a considerable impact on yields (Torstensson et al., 2006).

On the other hand, differences in manure handling can greatly affect the amount of N available for spreading and the release of N in soil. For example, composting of manure, which is a prerequisite in biodynamic agriculture, results in high ammonia losses (e.g. Kirchmann, 1985), while anaerobic storage of slurry limits N losses (Kirchmann and Lundvall, 1998). Thus, systems applying liquid manure (slurry) return more N than those using composted manure. Consequently, slurry with its higher content of plant-available N results in higher yields than composted manure (e.g. Hadas et al., 1996; Svensson et al., 2004).

If differences in management between systems are due to reasons other than organic farming regulations, a bias is added to the comparison. The consequences of any such differences need at least to be discussed and considered before any conclusions are drawn.

3.3 Comparing Organic and Conventional Crop Yields

3.3.1 National Crop Yield Statistics

A search in agricultural statistical databases of EU countries, the USA, Canada and Australia revealed that information on organic crop yields is very scarce. No crop yield data were found, but information on the number of organic farms, the extent of farmland under organic cultivation and, in a few countries, data on milk production are available. We found that during 2007, only Sweden and Finland provided statistics on organic crop yields, which were significantly lower than conventional yields.

Official Swedish statistics (SCB, 2006) reveal that yields of organically grown crops are 20–60% lower than those of conventionally grown crops. Yields of organically grown legumes (peas and beans) and grass/clover leys are, on average, 20% lower (Fig. 3.2), whereas yields of cereals are 46% lower and yields of potatoes as much as 60% lower than in conventional production. National statistics for Finland (Statistics Finland, 2007; Finnish Food Safety Authority, 2006) show a similar picture. Yields of organically produced cereals are 41% lower and yields of potatoes 55% lower (Fig. 3.2). The statistical data represent average figures combining pure cropping systems without animal husbandry and mixed crop-animal systems using animal manure. This means that yields of organic farms using animal manure are probably somewhat underestimated, while the converse is true for organic farms without animals, which have even lower yields. In line with the discussion above, it should be borne in mind that statistical yield data represent single crops in a rotation and do not consider years in the rotation when non-food crops (green manure) are

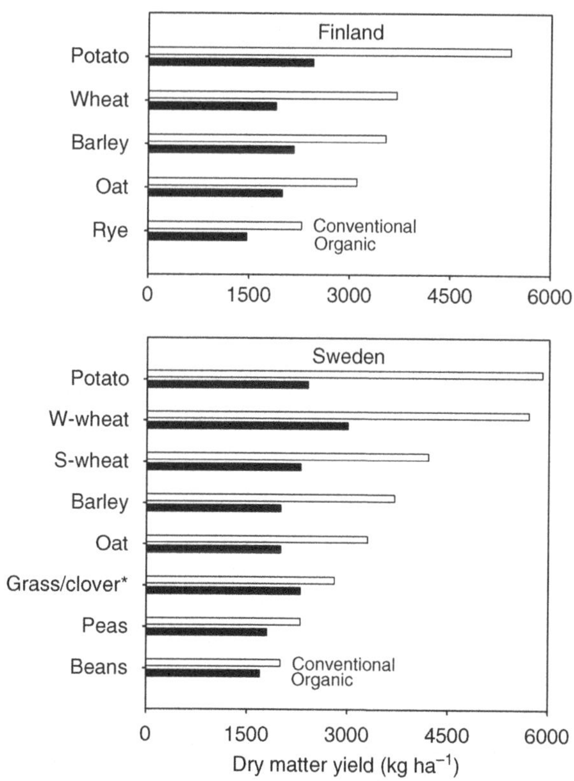

Fig. 3.2 Official national yield data for organically and conventionally grown crops in Finland and Sweden in the year 2005. Figures were derived from Statistics Finland (2007), Finnish Food Safety Authority (2006) and SCB (2006). *Only the first of two or three cuts is represented by the data

grown or when the fields are under fallow. Thus, total crop output per time unit cannot easily be derived from national statistics.

3.3.2 Cropping System Studies in the USA

During the compilation of comparative long-term field studies, we found that some experiments from the USA showed similar yields for organic and conventional systems (Table 3.1). This motivated a separate examination of these studies. For example, there were no differences in yields of soybean in several studies (Sanchez et al., 2004; Pimentel et al., 2005; Smith et al., 2007). Furthermore, some organic maize and oat yields (Porter et al., 2003; Pimentel et al., 2005) were also reported to be similar to those in conventional systems (see Table 3.1). In general, even average yields were reported to be little affected by organic and conventional management, being 13% lower for organic systems that combine crops with animals and 20% for organic systems without animals.

A more thorough examination of the organic systems in the USA revealed that the total amount of nutrients applied was as high as or even higher than that in comparative conventional systems (Teasdale et al., 2000, 2007; Porter et al., 2003; Denison et al., 2004; Sanchez et al., 2004; Pimentel et al., 2005). Similar nutrient application rates have also been reported in other publications dealing with organic cropping systems in the USA (Lockeretz et al., 1980; Liebhardt et al., 1989; Clark et al., 1999). Instead of the inorganic fertilisers that are used in conventional systems, organic farmers in the USA purchase manure, compost, food waste, etc., to satisfy crop nutrient demand and improve soil fertility. However, the critical point is where this manure and compost originate from, i.e. whether the amount of manure or compost applied is sustained by the organic systems or whether it mainly originates from off-farm, non-organic production. In the case of the USA studies, supply of nutrients to organic production was higher than removal, showing that nutrients were purchased, to a large extent from conventional systems as pointed out above. One can therefore conclude that high organic yields can only be achieved if there is an excess of manure/compost, or if other products can be transferred from conventional to organic production. As long as conventional production is the dominant form, this is possible. However, the results are not representative of conditions where modern conventional agriculture is scarce, such as in Africa or in areas completely converted to organic farming.

3.3.3 Cropping System Studies in Europe and Australia

Compilation of a number of long-term field experiments in Europe and Australia (Table 3.2) revealed that yield differences between organic and conventional systems were much larger than those reported from the USA. On average, organic systems in Europe and Australia that combine crops with animals had 25% lower yields and organic systems without animal husbandry had 47% lower yields

Table 3.1 Comparison of yields and N inputs in organic and conventional cropping systems. Data from long-term studies in the USA

Farming system, experiment and crop	Mean yield ($Mg\ ha^{-1}$)		Yield difference (%)	N input[a] ($kg\ ha^{-1}\ yr^{-1}$)		Yield/N input ($kg\ kg^{-1}\ N$)		References
	Con.	Org.		Con.	Org.	Con.	Org.	
Cropping systems with animal manure or compost								
California: Davis, LTRAS (9 year)								Denison et al. (2004)
Maize	11.5	7.6	−66	235	373	49	20	
Tomato	59[b]	66[b]	+11	160	214	369[b]	308[b]	
Maryland: Beltsville, SADP (9 year)								Teasdale et al. (2000, 2007)
Corn	5.5	4.9	−11	159	120	46	41	
Wheat	3.8	2.9	−24	100	130	38	22	
Michigan: Kellogg Biological Station, LFL (8 year)								Sanchez et al. (2004)
Corn	8.6	7.4	−14	140	104	61	71	
Soybean	2.3	2.4	+4	0	0	–[c]	–[c]	
Wheat	3.2	2.7	−15	65	104	49	26	
Minnesota: Lamberton site (7 year)								Porter et al. (2003)
Corn	8.7	7.9	−10	62	185	140	43	
Soybean	2.9	2.3	−20	1	31	–[c]	–[c]	
Oat	1.9	1.8	−5	49	92	38	19	
Pennsylvania: Kutztown, Rodale Institute (FST) (22 year)								Pimentel et al. (2005)
Corn	6.5	6.4	−2	87	198	74	32	
Soybean	2.5	2.5	0	0	0	–[c]	–[c]	
Mean value			***−13***	***88***	***120***	***62***	***34***	

Table 3.1 (continued)

Farming system, experiment and crop	Mean yield (Mg ha^{-1})		Yield difference (%)	N input[a] (kg ha^{-1} yr^{-1})		Yield/N input (kg kg^{-1} N)		References
	Con.	Org.		Con.	Org.	Con.	Org.	
Legume-based cropping systems								
California: Davis, LTRAS (9 year)								Denison et al. (2004)
Rain-fed wheat	4.8	4.1	−15	110	0	–[c]	–[c]	
Irrigated wheat	5.6	4.5	−20	165	0	–[c]	–[c]	
Michigan: Kellogg Biological Station, LTER (12 year)								Smith et al. (2007)
Corn	4.5	4.2	−10	123	0	–[c]	–[c]	
Soybean	2.2	2.2	0	0	0	–[c]	–[c]	
Wheat	3.6	2.1	−42	56	0	–[c]	–[c]	
Pennsylvania: Kutztown, Rodale Institute (FST) (22 year)								Drinkwater et al. (1998)
Corn	6.5	6.4[d](5.1)	−2[d](20)	87	140	75	46	
Soybean	2.5	2.2[d](1.8)	−12[d](30)	0	0	–[c]	–[c]	
Mean value			***−20***	***77***	–[c]	–[c]	–[c]	Pimentel et al. (2005)

[a]N input refers to N sources applied (inorganic fertiliser, manures, compost) excluding N fixation.
[b]Figures refer to fresh weight yield and were excluded from the calculation of mean N yield/N input.
[c]Figures were excluded from the calculation as the input of N through fixation is unknown.
[d]Yield figures refer to crops in single years and do not take into account that during 1 year out of the 5 years in the rotation, non-harvested red-clover-alfalfa hay or hairy vetch was grown and used as green manure. Total crop output over the rotation is therefore 20% lower and corrected figures are given in brackets (for explanation see Fig. 3.2).

Table 3.2 Comparison of yields and N inputs in organic and conventional cropping systems. Data from long-term studies in Europe and Australia

Farming system, experiment and crop	Yield (Mg ha^{-1})		Yield difference (%)	N input (kg ha^{-1} yr^{-1})		Yield/N input (kg kg^{-1} N)		References
	Con.	Org.		Con.	Org.	Con.	Org.	
Mixed crop-animal systems								
Norway: Apelsvoll site (8 year)								Korsaeth and Eltun (2000); Eltun et al. (2002)
Barley, oats, wheat	5.0	3.7	−26	100	–[a]	50	–[a]	
Three-year forage	10.7	8.3	−22	210	143	51	58	
Fodder beet	9.0	9.3	+3	140	–[a]	64	–[a]	
				138[b]	105[b]			
Switzerland: DOK trials (24 year)								Spiess et al. (1993); Besson et al. (1999); Mäder et al. (2002)
Winter wheat	4.5	4.1	−10	–[b]	–[b]	–[b]	–[b]	
Three-year forage	14.0	11.5	−18	–[b]	–[b]	–[b]	–[b]	
Potato	48.0	30.0	−38	–[b]	–[b]	–[b]	–[b]	
Sweden: Bjärröd trial (18 year)								Kirchmann et al. (2007)
Winter wheat	6.1	4.2	−31	120	116	51	36	
Barley	3.7	2.1	−43	80	60	46	35	
One-year forage	7.5	6.1	−19	–[a]	–[a]	–[a]	–[a]	
Australia: New South Wales (30 year)								Ryan et al. (2004)
Wheat	5.5	2.9	−48	17	0	–[a]	–[a]	
Mean value			***−25***	***115***	***102***	***52***	***43***	

Table 3.2 (continued)

Farming system, experiment and crop	Yield (Mg ha^{-1})		Yield difference (%)	N input (kg ha^{-1} yr^{-1})		Yield/N input (kg kg^{-1} N)		References
	Con.	*Org.*		Con.	Org.	Con.	Org.	
Pure cropping systems								
Sweden: Mellby trial (6 year)								Torstensson et al. (2006)
Oats	5.8	2.8^c(1.9)	−52^c (67)	97	71	60	27	
Sweden: Lanna trial (6 year)								Aronsson et al. (2007)
Winter wheat	5.9	3.4^c(2.3)	−42^c (61)	134	84	44	27	
Mean value			***−47^c(64)***	***115***	***77***	***52***	***27***	

[a] Figures were excluded from the calculation as the N input by a specific crop is lacking. Furthermore, the amount of N added through a previous N fixing crop is not given.
[b] Only mean N application for the whole rotation.
[c] Yield figures refer to crops in single years and do not take into account that during 2 out of the 6 years in the rotation, non-harvested green manure crops were grown. Total crop output over the rotation is therefore 33% lower and corrected figures are given in brackets (for explanation see Fig. 3.2).

than equivalent conventional systems. Studies of farms under long-term organic management in Australia (Table 3.2) also showed yields of individual crops to be substantially lower than those on conventional neighbouring farms (Kitchen et al., 2003; Ryan et al., 2004). In addition, Australian organic wheat crops reported by Ryan et al. (2004) were preceded by an average of 4.7 years of pasture, compared with 3.3 years for the conventional crops. The general reason for the large deviation between organic and conventional yields in these studies compared with those in the USA seems to be the limited purchase of manures/compost by organic farms in Europe and Australia.

Nutrient flows to fields and farm-gate balances between organic and conventional farms have been examined to determine whether nutrient inputs in European organic systems are lower throughout (Kaffka and Koepf, 1989; Fowler et al., 1993; Nolte and Werner, 1994; Granstedt, 1995; Halberg et al., 1995; Nguyen et al., 1995; Fagerberg et al., 1996; Wieser et al., 1996). These studies clearly show that the mean input of N, a major yield-determining nutrient, was lower throughout in organic systems over a crop rotation period than in conventional systems. This may explain why there is a greater deviation between organic and conventional yields in Europe. The low nutrient inputs to organic systems can be explained by the European approach of viewing organic crop-animal farms as a self-sustaining unit. The general aims for organic agriculture are to mainly rely on recycling of nutrients from within the system and to enhance the biological activity in soil in order to increase mineral weathering and biological N_2-fixation (Watson et al., 2002; International Federation of Organic Agricultural Movements, 2006). Furthermore, according to the European founders of organic agriculture, high yields caused by easily available nutrients are regarded as being detrimental to crop quality (Steiner, 1924; Balfour, 1943; Rusch, 1978). Thus, the application of nutrients of off-farm origin is often kept to a minimum and this view is reflected in the design of European organic long-term experiments (Table 3.2).

One of the comparative studies cited in Table 3.2 was run on a nutrient-depleted soil that had not received any inorganic fertilisers for 40 years prior to the start of the experiment (Kirchmann et al., 2007). Organic yields amounted to only 50% of those achieved in a comparable conventional cropping system over 18 years, despite use of animal manure at amounts larger than those obtained from on-farm production and large additions of rock phosphate and potassium sources approved by the organic farming organisations. This study underlines the importance of the initial soil fertility status at a site used for a comparative study. Very often, the residual soil fertility effect from previous applications of conventional, inorganic P and K fertilisers before the start of the experiment is overlooked. The continuous decline in organic yields at certain sites can in most cases be explained by further depletion of this soil fertility.

In two organic systems without animals (Table 3.2), green manure crops were grown during two of the six years and no food/feed crops were produced (Torstensson et al., 2006; Aronsson et al., 2007). This means that over the whole crop rotation, organic yields are reduced by a further 33% in accordance with the reasoning in Fig. 3.1. This correction of yield levels by considering years with non-food crops resulted in an overall yield reduction of 64% in the organic system compared with conventional yields.

It is doubtful whether there will be an increased nutrient input through transfer of organic manures or composts from conventional to organic farms in Europe, as this would be against the principle of basing organic farming on 'living ecological cycles' relying on on-farm N input and recycling (International Federation of Organic Agricultural Movements, 2006). Although in certain years N input in organic systems can amount to several hundred kilos per hectare from N_2-fixing crops, the total input over a rotation period is still lower than in conventional as years in the rotation without legumes mean little or no N input. On the other hand, organic forage-ruminant systems, i.e. systems with a large proportion of N_2-fixing forage crops in the rotation, seem to be capable of providing a high continuous input of N to soil (Eltun et al., 2002; Posner et al., 2008) similar to that in conventional systems. These results are in agreement with knowledge from the Norfolk rotation, long before organic agriculture became fashionable.

3.3.4 Organic Yields Higher than Conventional?

A review by Badgley et al. (2007) points out that organic agriculture is misjudged concerning crop production and its potential to supply sufficient food. According to their review, only small yield reductions occur through organic agriculture in developed countries, but organic yields are higher than conventional yields in developing countries. This conclusion is supported by a large number of other papers, which may be taken as evidence of its scientific reliability. We re-examined the papers cited by Badgley et al. (2007) to determine whether their conclusions are based on valid assessments.

A number of the studies cited from developed countries are summarised in Table 3.3. However, due to their limited accessibility and often lower scientific credibility, non-peer-reviewed conference papers, institution reports and magazine articles were not considered. The re-examination of papers reporting high organic yields showed that the data were used in a biased way, rendering the conclusions flawed. Firstly, none of the organic studies cited reported higher crop output from organic production than from conventional over a whole rotation, but only for single years. Secondly, when yields were higher during a single year in organic production, this was coupled to one or both of the following conditions: (1) The amount of nutrients applied to the organic system through manure and compost was equal to or even higher than that applied to the conventional system through inorganic fertilisers; (2) non-food crops (legumes) were grown and incorporated in the preceding year to provide the soil with N. Thirdly, on-farm data were compared with mean yield data within a region. Such comparisons have no validity, since the possible factors behind the differences are not given. In summary, the yield data reported were misinterpreted and any calculations based on these data are likely to be erroneous.

The paper by Badgley et al. (2007) also presents comprehensive yield figures from developing countries. However, of the 137 yield figures reported, 69 originate

Table 3.3 Studies showing higher yields in organic systems than in conventional and the reasons for this

Type of study and year	Yield (Mg ha^{-1}) Con.	Organic	Organic yield increase (%)	Remarks	References
Scientifically non-valid comparisons					
Germany: Talhof, (conventional 1922–1929 vs biodynamic performance 1930–1937)					Koepf et al. (1976)
Winter wheat	1.9	2.4	21	Comparing different periods, neglecting general yield improvement over time	
USA: Western corn belt, survey of 363 farms (1974–1978)					Lockeretz et al. (1981)
Corn			−30	Only very few organic corn yields out of 81 measurements were higher than conventional. On average, organic corn yields were 30% lower	
Ohio: Spray Brothers Farm (1981–1985)					National Research Council (1989)
Corn	–	–	25	Farm yields were compared with mean yields from the county	
Soybean	–	–	40		
Amount of nutrients added not supported by the organic systems					
California: Davis, SAFS project (1994–1998)				Higher applications of N, P and K through compost to the organic system showing higher in 2 years out of 4	Poudel et al. (2002)
Corn			20		
Tomato			10		
Iowa: Neely-Kinyon LTAR (1998–2001)				Higher N, P and K additions through compost to the organic systems resulted in higher yields after 3 years. Only N fertiliser to the conventional system	Delate and Cambardella (2004)
Corn	7.1	8.1	15		
Soybean	2.7	3.1	15		

Table 3.3 (continued)

Type of study and year	Yield ($Mg\ ha^{-1}$)		Organic yield increase (%)	Remarks	References
	Con.	Organic			
Canada: Nova Scotia, field plot (1990–1992)					
Cabbage	45.3	46	2	Nutrient additions through compost to the organic plots were twice as high as through NPK fertiliser	Warman and Harvard (1997)
Carrot	26.2	27.9	6		
Yield reduction by non-food (green manure) years in rotation not considered					
South Dakota: field plots (1986–1992)					
Spring wheat	–	–	–18	The set-aside land for green manure years was 25%, which means that a 9% higher single-crop organic yield was –18% over a whole crop rotation	Smolik et al. (1993, 1995)

from the same paper (Pretty and Hine, 2001). A closer inspection revealed that crop yields were based on surveys and there was no possibility to check crop performance variables and the science behind the data. In fact, only six papers for developing countries cited by Badgley et al. (2007) were derived from peer-reviewed journals. In four papers, rice yields in conventional systems were compared with so-called intensified rice production. However, intensified rice production uses mineral fertilisers, although at lower rates, and is not an organic form of agriculture by European standards (e.g. Sheehy et al., 2004; Latif et al., 2005). Our conclusion is therefore that the argument that organic agriculture can produce similar or even higher yields than conventional does not hold given the boundary conditions outlined above.

3.3.5 Trends in Organic Crop Yields

Yield trends over time were analysed in four Swedish comparative studies to determine the potential to increase production through organic and conventional management. The underlying question is whether yields are following the same trends in organic agriculture as in conventional.

In the study by Kirchmann et al. (2007) (Fig. 3.3), the initial 10-year period was characterised by a relatively constant yield difference between the organic and conventional system. Thereafter, yields increased in both systems but the increase was larger in the conventional system than in the organic, despite higher additions of animal manure to the organic system. In two other studies without animal manure (Torstensson et al., 2006; Aronsson et al., 2007), which used green manure for organic production and fertiliser for conventional, the relative yield differences between systems were much larger (see Fig. 3.3). Furthermore, no yield increase was observed in the organic system over the 5–6-year experimental period, whereas conventional yields increased in one experiment and remained constant in the other. In studies without animal manure, there is good reason to assume that organic yields barely increase over the longer term, as residual soil nutrients are depleted at faster rates than in studies with manure application. For instance, in relatively fertile soils, a decade or more may be needed before residual soil nutrients are sufficiently exhausted for a yield reduction to become apparent (Denison et al., 2004).

In another experiment run for 12 years at a fertile site, each crop in the rotation was grown every year and animal manure was applied in relation to the level of nutrient removal by harvested crops (Ivarson and Gunnarsson, 2001) (Fig. 3.4). Differences between organic and conventional yields were smaller at this site, in particular for forage crops. However, there were no indications that organic yields would increase more or decrease less over time than conventional yields.

Based on the four experiments presented above, we conclude that there is no evidence that yields increase more in organic agriculture than in conventional. However, there is evidence that conventional agriculture has a greater capacity for increased yields than organic agriculture.

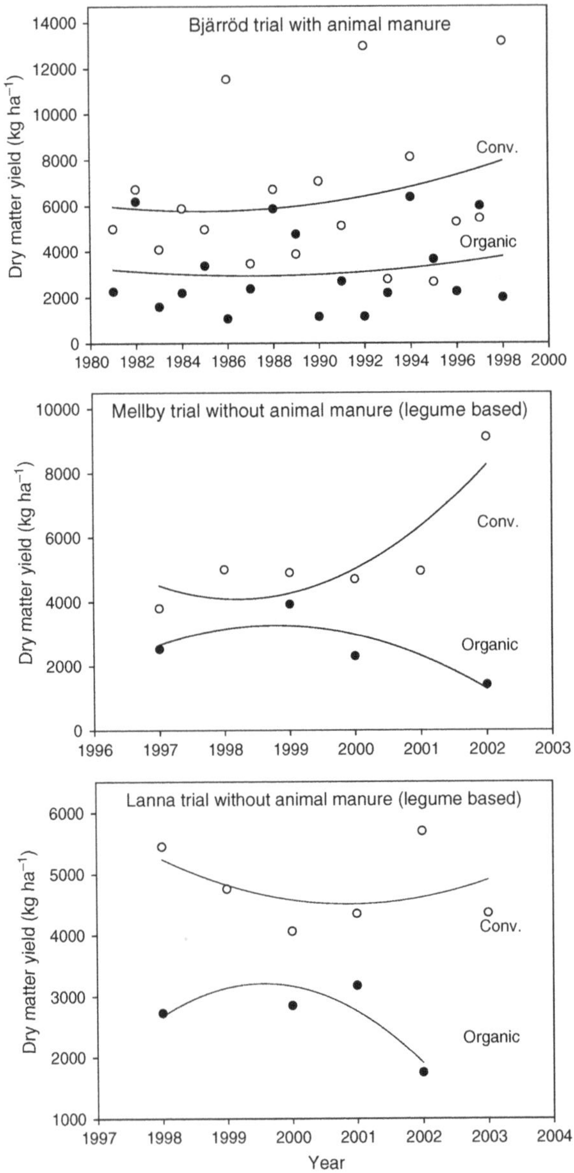

Fig. 3.3 Trends in crop yield in three Swedish long-term studies. Data for the Bjärröd study are from Kirchmann et al. (2007), for the Mellby study from Torstensson et al. (2006) and for the Lanna study from Aronsson et al. (2007). Absent yield data for 1998 and 2001 in the Mellby study and for 2003 in the Lanna study are due to years with green manure crops not being harvested. (*Open circles* = conventional, *filled dots* = organic)

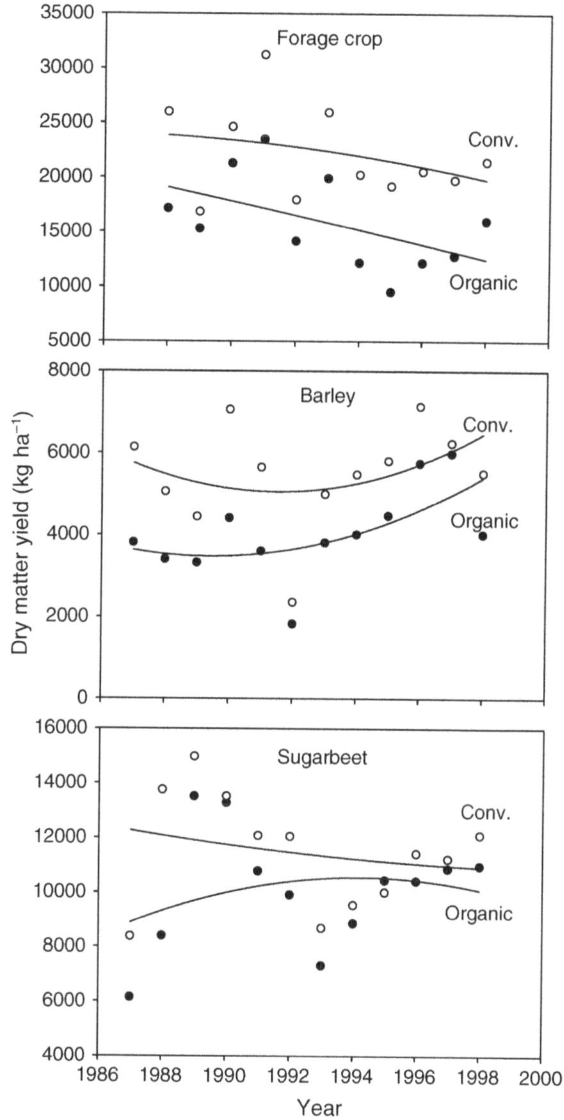

Fig. 3.4 Trends in yield during two 6-year rotations on a fertile site in southern Sweden (Bollerup) of organically and conventionally grown crops to which animal manure was applied in relation to the level of harvested crops. Data from Ivarson and Gunnarsson (2001). (*Open circles* = conventional, *filled dots* = organic)

3.3.6 Food Production at the Global Scale

In summary, this review shows that the reduction in crop yields through large-scale conversion to organic agriculture would, on average, amount to 40%, with a range of variation of 25–50%. A 40% reduction in yield on a global scale is equivalent to the amount of crops required by 2.5 billion people. This estimate is in fact identical to that calculated by Smil (2001), who assessed the role of industrial nitrogen fixation for global food supply. Smil (2001, 2002) concluded that the Haber-Bosch process for industrial fixation of atmospheric nitrogen provides the very means of survival for 40% of humanity and that only half of the current world population could be supported by pre-fertiliser farming, even with a mainly vegetarian diet. The similarity of these estimates confirms the strategic role of fertilisers as a keystone for the well-being and development of mankind.

It is obvious that world-wide adoption of organic agriculture would lead to massive famine and human death. This is something that advocates of organic agriculture are silent about, perhaps because of the severe moral dilemma it poses.

3.4 Factors Limiting Yields in Organic Systems

The yield data presented above lead to the fundamental question of what causes lower yields in organic systems than in conventional. According to conventional agronomic understanding, the following factors can be suggested: (i) Low nutrient input; (ii) low nutrient use efficiency; (iii) high weed abundance; (iv) limited possibilities to improve low native soil fertility in resource-poor areas; and (v) poor control of pests and diseases. Reports detailing the causes of low organic yields are scarce, but in the following section we discuss some of the potential factors.

3.4.1 Low Nutrient Input and Lower Efficiency

A low N input has been identified as one yield-limiting factor for organic systems (Clark et al., 1999; Kirchmann et al., 2007), since organic manures cannot be produced by organic systems themselves in sufficient quantities to deliver enough N for high yields. On the other hand, numerous studies of N_2-fixing crops reveal that grain legumes, green manures and tree legumes have the capacity to fix several hundred kilograms of N during one year (Giller, 2001). So why is the potential of N_2-fixing crops to replace inorganic N fertilisers not fully used? Legumes can indeed improve yields of subsequent crops (e.g. Giller and Cadish, 1995; Sanchez, 2002), improve soil fertility and break pest and disease cycles in a crop rotation, but the total amount of N available to other crops over a whole rotation must be sufficiently high, which means that legumes must be grown every second year at a minimum. There are other limitations associated with the use of legumes. Legume N is not a free source of N. As pointed out earlier, growth of green manure legumes for N supply and soil fertility improvement is often only possible at the expense of not using the land for export of food from the field (Fig. 3.1).

In addition, biologically fixed N and other forms of organic N are not necessarily released in synchrony with the demand of the following crop. The overall lower agronomic effect of N in the organic systems (Tables 3.1 and 3.2) is a clear indication that the N use efficiency of organic inputs (animal manure, green manure, compost, etc.) is lower. The main reason for this is that N from legume residues and animal manures is released even at times when there is no crop growth (Marstorp and Kirchmann, 1991; Dahlin et al., 2005), which carries a risk of significant leaching losses (Bergström and Kirchmann, 1999, 2004) and thus lower N use efficiency.

Using legumes as cash crops in a rotation does not necessarily mean that the amount of N fixed increases the soil N pool. Harvesting grain legumes can remove more N than is fixed. As pointed out by Giller (1998), legumes are not necessarily N providers but can be plunderers. For example, soybeans can remove more N than they add (Toomsan et al., 1995), while peas do not necessarily contribute a net supply of N to soil (Jensen, 1987, 1996). As pointed out earlier, only organic grassland-ruminant systems, i.e. systems with mainly N_2-fixing forage crops in rotation, have the capacity to provide continuous inputs of N to soil similar to those in conventional systems.

Phosphorus has also been suggested to be a limiting nutrient in certain cropping systems. On grazed organic systems in southern Australia, where legume-based annual pastures are rotated with crops, P can become the yield-limiting element (Ryan et al., 2004). Similarly, permanent clover-based pastures of biodynamic farms in Australia show lower production than conventional, which is also ascribed to lower inputs of nutrients, particularly of P (Burkitt et al., 2007a;b). A correct view on this is that over the long-term, less can be taken out of a system if less is put in (Goulding, 2007). However, crop yield also depends on the availability of the nutrients and not only on the quantity added. The same amount of nutrients can be added to organic production as would be supplied by inorganic fertilisers, but in the form of untreated minerals. This was actually done with P and K in the organic treatments in the Swedish long-term study discussed above (Kirchmann et al., 2007) but that study showed that despite very large applications of untreated minerals, the availability of N remained the yield-limiting factor. The same conclusion was drawn by Pang and Letey (2000) and Berry et al. (2002), who found that inadequate N availability and not necessarily nutrient addition was the bottleneck for organic production and stressed that the amount of N available during the period of rapid growth restricts crop productivity in organic systems.

3.4.2 Poor Weed Control

Weed control is a primary concern in all types of agriculture. Weeds compete with the main crops for water, nutrients and light and can thereby significantly reduce yields. However, weeds can also contribute carbon to soil.

As organic agriculture is limited by two options for weed control, physical-mechanical treatment and choice of crop rotation, weed populations are larger in most organically grown crops. For example, Kirchmann et al. (2007) compared weed biomass production over 18 years in an organic and a conventional cropping system and found a slightly increasing trend in the organic system with irregular fluctuations between years. On average, about 1 Mg dry weed biomass per hectare was produced over the 18-year period, with peak values of around 3 Mg dry matter (Fig. 3.5), which was 25 times more than in the conventional system in which pesticides were used. Other comparative studies of organic and conventional production also report an overall higher weed biomass under organic management (Barberi et al., 1998; Poudel et al., 2002; Smith and Gross, 2006). The yield decreasing effect of weeds alone in organic systems can amount to 20–30% according to Posner et al. (2008), who compared effective weed control with ineffective treatment.

Weed management can be improved through a diverse and long rotation period (Teasdale et al., 2004) and especially through growth of perennial forage crops (Sjursen, 2001). Otherwise, the potential for a reduction in the weed seed bank through organic practices is small (e.g. Clark et al., 1999; Porter et al., 2003). Teasdale et al. (2007) showed that weed control in organic no-till systems is barely possible. Weeds with deep root systems cannot be fully disrupted and their biomass can even increase. Peigné et al. (2007) stressed that pressure from weed grasses is much greater in organic conservation tillage than in conventional. Thus overall, we can conclude that effective weed control in organic agriculture is rather limited.

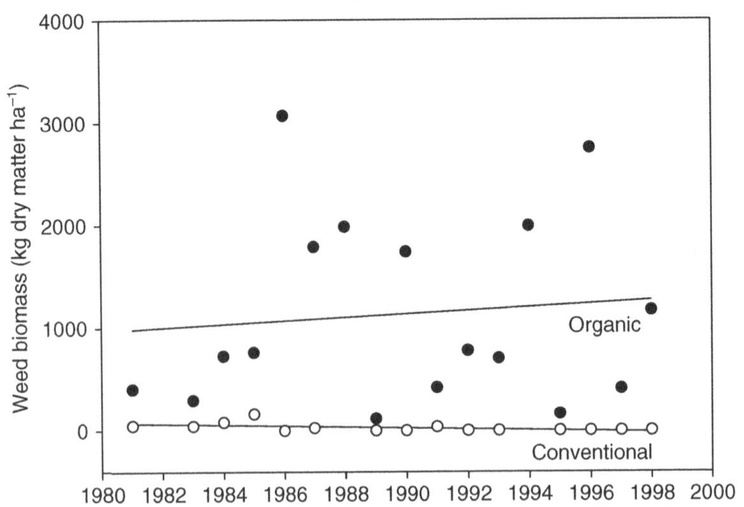

Fig. 3.5 Trends and fluctuations in weed biomass in an organic and conventional cropping system over an 18-year period (data from Kirchmann et al., 2007). Weed data are lacking for 1982, 1988 and 1994 when grass/clover forage was grown. (*Open circles* = conventional, *filled dots* = organic)

3.4.3 Low Native Soil Fertility – The Example of Sub-Saharan Africa

The central question put forward in this section is whether introduction of organic practices in developing countries can increase yields and thereby ensure food supply. Smallholder farmers in developing countries have practised organic methods for thousands of years, as these methods have been the only approach available to manage soil fertility in such systems. The principal question is whether these smallholder farmers can significantly improve food production with locally available resources and improved low cost technologies. In other words, are organic or near-organic practices the way forward? Supporters of organic agriculture point out that the only solution for poor people living under difficult conditions is to apply existing organic methods and improve these practices.

Three of the authors of this chapter have been involved for decades in soil fertility projects in developing countries with the aim of improving agricultural production using local resources and simple technologies. Moreover, all of us have a reasonable understanding of the malnutrition, poverty, lack of infrastructure and other socio-economic weaknesses that are realities in many developing countries. Furthermore, our research work fully supports existing agricultural methods such as erosion control, use of legumes in rotation, application of animal manure, recycling of organic waste, double dig for gardening, etc. However, our experience has forced us to question whether the proposed exclusive use of organic inputs and natural resources to increase crop production makes sense in resource-poor areas. In fact, it is the limited amount of nutrient resources available and/or their inappropriate quality (e.g. Campbell et al., 1998; Palm et al., 2001) that constrain agricultural production in many developing countries. In addition, many soils in these countries are natively poor in plant nutrients and soil depletion is continuing in sub-Saharan Africa (e.g. Smaling and Braun, 1996; Smaling et al., 1997; Mugwira and Nyamangara, 1998). Applications of nutrients to soil through transfer from adjacent areas to agricultural fields by cut-and-carry of organic matter are insufficient, as even these systems are poor in nutrients. The transfer may help to increase the fertility status at a very small scale, for example in domestic gardens (Prudencio, 1993), but at the larger scale the fertility of arable soils cannot be restored by such practices. Crops cannot be supplied with sufficient nutrients through the removal of vegetation from nutrient-depleted, adjacent ecosystems (e.g. Vanlauwe and Giller, 2006).

Here is an example of how crop yields from remote and resource-poor areas employing organic practices can be presented: 'Maize yields increased four to nine times. The organically grown crops produced yields that were 60% higher than crops grown with expensive chemical fertilizers' (Leu, 2004). A yield increase of between 400 and 900% is dramatic but such an enormous increase shows that initial yield levels must have been extremely low, indicating the very difficult conditions for crop production in general. Enhanced production from 250 kg to 1000–2000 kg maize per hectare could represent the actual figures behind the quote. Furthermore, the reader is mistakenly led to believe that chemical fertilisers produce lower yields than organic materials. Higher organic yields than conventional are not proof of the superiority of organic practices. The application of organic material means addition of micronutrients, which are often also lacking in infertile soils. To make the comparison unbiased, the same micronutrients need to be applied with conventional fertilisers. Again, there is no information about the amount of manure applied or what is available for agricultural crops in the region as a whole. It is unclear whether the production increase would be possible for a larger region or just a single field. Viewed over a period of several years, the improvement may not last due to shortage of high-quality organic material. Furthermore, no information is provided on the potential yields from organic resources combined with inorganic fertilisers. A combination of organic material and inorganic nutrient sources has been shown to result in much higher yields than with organic inputs alone (e.g. Murwira and Kirchmann, 1993; Bekunda et al., 1997). In reality, a combination of organic and inorganic nutrient sources is the most successful approach to increase crop yields in resource-poor areas with low fertility soils (Palm et al., 1997; Vanlauwe et al., 2001; TSBF, 2006). The approach of applying exclusively organic products is based on misinformation about the effects of inorganic fertilisers on soils (Vanlauwe and Giller, 2006) and misunderstanding of their environmental impact.

On the other hand, the exclusive use of inorganic fertilisers without applying animal manures and without returning crop residues or other organic materials to the soil can result in a decline in crop yields over time, as shown in a number of long-term field experiments from sub-Saharan Africa (Singh and Balasubramanian, 1979; Swift et al., 1994; Laryea et al., 1995; Pieri, 1995). Advocates of organic agriculture use this type of result to claim that artificial fertilisers damage the soil and decrease soil fertility. A wide-spread view within organic agriculture is that 'more and more synthetic fertilizers are needed to maintain yields. The system error of conventional farming is the independence of natural regulating processes and local resources. The main cause for lower production is found in unutilized or inefficient use of natural resources' (Rundgren, 2002). This is incorrect. As pointed out above, limited supply of natural resources and their poor quality is the main reason for low yields in areas with low soil fertility, not inefficient use of nutrients. The major reason why yields sometimes decline when inorganic fertilisers are used on highly depleted soil is the lack of other essential nutrients not applied with NPK fertilisers. Organic manures and composts usually contain other essential plant nutrients (Ca, Mg, S, Cu, Zn, etc.) in addition to N, P and K. Comparing organic practices with fertiliser application on highly depleted soils is only possible when the fertiliser treatment is not deficient in any other way. In fact, the experiments cited above showed that combining animal manure with inorganic fertilisers led to steadily increasing yields. However, long-term use of artificial N fertilisers such as ammonium sulphate or urea can reduce yields over time due to acidification (Kirchmann et al., 1994). On the other hand, this only occurs if the standard agronomic practice of liming is neglected.

As mentioned above, all efforts to increase yields with locally available resources are positive and the knowledge on how best to use organic and local resources is of the utmost importance. However, there is no scientific reason why conditions cannot be improved through the development of practicable and sustainable management practices utilising the benefits of combined application of organic resources and fertilisers (Palm et al., 1997; Vanlauwe et al., 2001; TSBF, 2006).

Occasionally, erroneous conclusions are drawn based on the fact that hungry and poor people cannot afford to buy food. Therefore, the only option proposed for poor farmers is low-cost organic management (e.g. Vandermeer and Perfecto, 2007). Most hungry and poor people are rural and agriculture is their mainstay. They are hungry because they are not able to produce sufficient food and they are poor because they have nothing to sell. Such disastrous conditions are often caused by a number of factors, such as poor economic and agricultural policy; inadequate investment in infrastructure and rural education; insufficient agricultural services such as research, extension, credit, input supply and marketing; and low investment in rural healthcare. However, the bottom line is that lack of nutrients, poor soil fertility, limited amounts of organic manures, etc. are causing low yields and these causes cannot be overcome by organic methods – the critical shortages will remain. Only introduction of higher yielding technologies producing more food per capita, together with other necessary actions, will improve food security and income for the poor (UN, 2005).

3.5 Placing Organic Yields in Perspective

Yield decreases – not of a few percent but of 25–50% – demand attention. Such drastic yield reductions highlight the fact that sufficient food through organic production cannot be taken for granted. Moreover, foreseeable environmental consequences must be outlined. What would actually happen if organic agriculture were to be introduced all over Western Europe?

3.5.1 Low-Yielding Agriculture Demands Additional Cropland Area

Data in the literature clearly show that organic yields are significantly lower, as is discussed in detail above. In order to produce the same amount of organically grown crops, countries would be forced to convert more land into cropland. Based on yield data from European long-term experiments (Table 3.3) and excluding any major nutrient transfer from conventional agriculture, we assessed the additional cropland required if organic practices were to be introduced (see Fig. 3.6). Conversion to organic cropping systems without animals would require 100% more cropland, since yields of such systems amount to roughly 50% of conventional yields, while organic crop-animal systems would require 33% more land as yields from these amount to about 75% of those in conventional systems. Mean estimates of relative yields for organic cropping and mixed crop-animal systems (derived from Swedish National Statistics; Fig. 3.1) indicate the need to expand agricultural land by approx. 67%. As the additional cropland would be used less efficiently than in

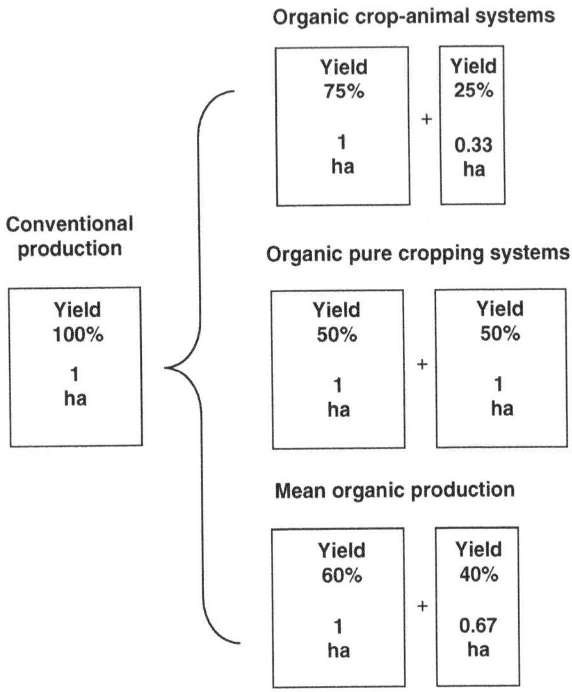

Fig. 3.6 Additional demand for cropland to produce the same amount of crops through organic agriculture as in conventional agriculture. Data for crop-animal and pure cropping systems were taken from Table 3.2 and mean values for both systems from national statistics (Fig. 3.2)

conventional agriculture, the land area needed would be correspondingly larger than the percentage yield decrease.

Calculated values for additional land demand due to conversion to organic agriculture reported by Halberg and Kristensen (1997) for Danish dairy farming show that organic production would require the area used for farming to be extended by 47% in order to maintain yields. Conversely, impressive savings in land area have been achieved through the introduction of modern agricultural practices and the associated increase in yields. Had yields in China and India remained at the level of the 1960s, land area would have needed to be increased two- and threefold, respectively (Quinones et al., 1997). Moreover, in many areas of the world, there is no additional land available for agriculture. For example, China has 7% of the world arable land area and 20–25% of the world population and there is no more agricultural land available (Chen and Wan, 2005).

The need for more farmland to produce the same amount of crops through low-yielding systems instead of high-yielding adds an important boundary condition to the comparison of these systems, as illustrated in Fig. 3.7. Conversion of other ecosystems into cropland means lost production of other raw materials (wood, timber, bio-energy, etc.) from this area and a decline in specific functions and ecosystem services such as biodiversity. These conditions must be considered and must be part of a stringent comparison of agricultural systems. The slogan 'Growing less food per acre leaving less land for nature' (Borlaug and Dowswell, 1994), must find its

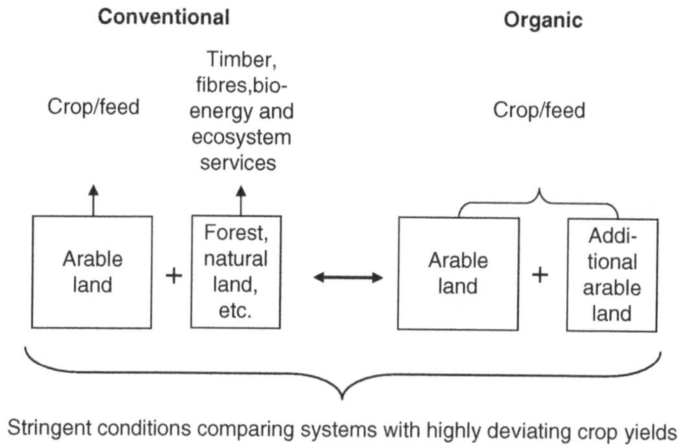

Fig. 3.7 Land demand to produce the same amount of crops must be considered in a scientifically meaningful comparison between different types of agriculture. The additional area required for low-yielding agriculture competes with other land uses, e.g. forests, grazing land, bio-fuel cropping, etc.

way into the conceptual framework for comparing land use, analysis of ecosystem services and computer modelling.

As more cropland is required for low-yielding agriculture, the question arises as to what type of land could be used as cropland to produce sufficient food. Furthermore, population growth and the need for improved human nutrition indicate that more food must be produced in the future. How can we cope with this demand through low-yielding agriculture? Is introduction of low-yielding agriculture a realistic option to meet future needs?

Combining expected population growth and projected land demand indicates that it seems unrealistic to introduce low-yielding agriculture as an option to produce sufficient food in the future. Population growth paired with introduction of low-yielding agriculture would roughly require at least a doubling of global arable land, from 1400 to 2500–3000 Mha. However, land suitable for agriculture is a limited resource and both the best and the second-best land is already in agricultural production. What remains is often only less suitable land, which is characterised by lower soil fertility, the presence of stones and gravel, or high risks for erosion or other rapid degradation when cropped. In most cases, only forests are at hand for conversion, as pointed out by Gregory et al. (2002). Thus, intensification on existing cropland seems to be the main path forward.

One major consequence of a great expansion in cropland would be further loss of natural habitats, as pointed out by e.g. Green et al. (2005), Hole et al. (2005) and Trewavas (2001). Advocates of organic agriculture are silent about how to cope with increasing demand for crops and pay little attention to the necessity for expanding cropland. The consequence of converting natural ecosystems into low-yielding production systems means loss of biodiversity, whereas comparisons of biodiversity in organic and conventional agricultural systems do not include the boundary conditions outlined in Fig. 3.7 (e.g. Mäder et al., 2002; Bengtsson et al., 2005; Gabriel et al., 2006).

3.6 Conclusions

The evaluation of organic yield data by advocates of organic agriculture is flawed in many ways, and different viewpoints are discussed in this chapter. The important points can be summarised as follows:

- Yields of organically grown crops in Europe are in most cases significantly lower than those of conventional crops.
- High organic yields, as reported in certain studies in the USA, are not relevant for comparisons with conventional yields, since they rely on the purchase of large amounts of animal manure.
- Average organic yields from rotations based on green manure are misleading unless years with crops not yielding exportable products are included in the calculations.
- Organic yields are limited by both nutrient shortages and high weed populations, and they are more difficult to increase through on-farm manures and exclusive use of untreated minerals than if the whole toolbox of modern production were allowed.
- Organic agriculture uses cropland less efficiently and requires more cropland to produce the same crop yields. There is good reason to believe that a large-scale conversion to organic agriculture would lead to severe food shortages.
- In order to secure a sufficient food supply in the future, emphasis should be placed on further development of modern but locally adapted forms of production without an ideological bias that *a priori* excludes potential solutions.

References

Alexandratos, N., 1999, World food and agriculture: outlook for the medium and longer term, *Proc. Natl. Acad. Sci. U.S.A.* **96**: 5908–5914.

Aronsson, H., Torstensson, G., and Bergström, L., 2007, Leaching and crop uptake of N, P and K from organic and conventional cropping systems on a clay soil, *Soil Use Manage.* **23**: 71–81.

Badgley, C. and Perfecto, I., 2007, Can organic agriculture feed the world? *Renew. Agric. Food Syst.* **22**: 80–82.

Badgley, C., Moghtader, J., Quintero, E., Zakern, E., Chappell, J., Avilés-Vázquez, K., Samulon, A., and Perfecto, I., 2007, Organic agriculture and the global food supply, *Renew. Agric. Food Syst.* **22**: 86–108.

Balfour, E.A., 1943, *The Living Soil*. Faber and Faber Ltd., London, UK, 276p.

Barberi, P., Cozzani, A., Macchia, M., and Bonari, E., 1998, Size and composition of the weed seedbank under different management systems for continuous maize cropping, *Weed Res.* **38**: 319–334.

Bekunda, M.A., Bationo, A., and Ssali, H., 1997, Soil fertility management in Africa: a review of selected research trials, in: *Replenishing Soil Fertility in Africa*, R.J. Buresh, P.A., Sanchez, and F. Calhourn, eds., SSSA Special Publications No. 51, Madison, WI, pp. 63–79.

Bengtsson, J., Ahnström, J., and Weibull, A-C., 2005, The effect of organic agriculture on biodiversity and abundance: a meta-analysis, *J. Appl. Ecol.* **42**: 261–269.

Bergström, L. and Kirchmann, H., 1999, Leaching of total nitrogen from nitrogen-15-labeled poultry manure and inorganic nitrogen fertilizer, *J. Environ. Qual.* **28**: 1283–1290.

Bergström, L. and Kirchmann, H., 2004, Leaching and crop uptake of nitrogen from nitrogen-15-labeled green manures and ammonium nitrate, *J. Environ. Qual.* **33**: 1786–1792.

Berry, P.M., Sylvester-Bradley, R., Philipps, L., Hatch, D.J., Cuttle, S.P., Rayns, F.W., and Gosling, P., 2002, Is the productivity of organic farms restricted by the supply of available nitrogen? *Soil Use Manage.* **18**: 248–255.

Besson, J., Myre, K., and Niggli, U., 1999, DOK-Versuch: Vergleichende Langzeituntersuchungen in den drei Anbausystemen biologisch dynamisch, organisch-biologisch und konventionell. II. Ertrag der Kulturen: Kartoffeln, 1. und 2. Fruchtfolgeperiode, *Schweiz. Landw. Forsch.* **31**: 127–155 (In German).

Blum, W.E.H., Buesing, J., and Montanella, L., 2004, Research needs in support of the European thematic strategy for soil protection, *Trends Analyt. Chem.* **23**: 680–685.

Borlaug, N.E., 1970, *The Green Revolution, Peace and Humanity* – Nobel Lecture, December 11, 1970, www.agbioworld.org/biotech-info/topics/borlaug/nobel-speech.html, Agbioworld, Tuskegee Institute, AL 36087-0085, USA. Assessed 24 November 2007.

Borlaug, N.E. and Dowswell, C.R., 1994, *Feeding a Human Population that Increasingly Crowds a Fragile Planet*, 15th World Congress of Soil Science, Keynote lecture, Supplement to Transactions, Acapulco, Mexico July 10–16, 1994.

Bouwman, A.F., van der Hoek, K.W., Eickhout, B., and Soenario, I., 2005, Exploring changes in world ruminant production systems, *Agric. Syst.* **84**: 121–153.

Bradford, G.E., 1999, Contributions of animal agriculture to meeting global human food demand, *Livest. Prod. Sci.* **59**: 95–112.

Bruinsma, J., 2003, *World Agriculture towards 2015/2030–An FAO Perspective*. Earthscan, London, UK, 432p.

Burkitt, L.L., Small, D.R., McDonald, J.W., Wales, W.J., and Jenkin, M.L., 2007a, Comparing irrigated biodynamic and conventionally managed dairy farms. 1. Soil and pasture properties, *Aust. J. Exp. Agric.* **47**: 479–488.

Burkitt, L.L., Wales, W.J., McDonald, J.W., Small, D.R., and Jenkin, M.L, 2007b, Comparing irrigated biodynamic and conventionally managed dairy farms. 2. Milk production and composition and animal health, *Aust. J. Exp. Agric.* **47**: 489–494.

Campbell, B., Frost, P., Kirchmann, H., and Swift, M., 1998, A survey of soil fertility management in small-scale farming systems in North Eastern Zimbabwe, *J. Sus. Agric.* **11**(2/3): 19–39.

Cassman, K.G., Döbermann, A.D., and Walters, D.T., 2002, Agroecosystems, N-use efficiency and N management, *Ambio* **31**: 132–140.

Cassman, K.G., Döbermann, A.D., Walters, D.T, and Yang, H, 2003, Meeting cereal demand while protecting natural resources and improving environmental quality, *Ann. Rev. Environ. Resour.* **28**: 10.1–10.44.

Chen, F. and Wan, K., 2005, The impact of organic agriculture on food quantity, food quality and the environment: a Chinese perspective, *Soil Use Manage.* **21**: 73–74.

Clark, M.S., Horwarth, W.R., Shennan, C., Scow, K.M., Lantni, W.T., and Ferris, H., 1999, Nitrogen, weeds, and water as yield-limiting factors in conventional, low-input and organic tomato systems, *Agric. Ecosyst. Environ.* **73**: 257–270.

Dahlin, S., Kirchmann, H., Kätterer, T., Gunnarsson, S., and Bergström, L., 2005, Possibilities for improving nitrogen use from organic materials in agricultural cropping systems, *Ambio* **34**: 288–295.

Delate, K. and Cambardella, C.A., 2004, Agroecosystem performance during transition to certified organic grain production, *Agron. J.* **96**: 1288–1298.

Denison, R.F, Bryant, D.C., and Kearney, T.E., 2004, Crop yields over the first nine years of LTRAS, a long-term comparison of field crop systems in a Mediterranean climate, *Field Crops Res.* **86**: 267–277.
Drinkwater, L.E., Wagoner, P., and Sarrantonio, M., 1998, Legume based cropping systems have reduced carbon, nitrogen losses, *Nature* **396**: 262–265.
Dyson, T., 1999, World food trends and prospects to 2025, *Proc. Natl. Acad. Sci. U.S.A.* **96**: 5929–5936.
Eickhout, B., Bouwman, A.F., and van Zeijts, H., 2006, The role of nitrogen in world food production and environmental sustainability, *Agric. Ecosyst. Environ.* **116**: 4–14.
Eltun, R., Korsaeth, A., and Nordheim, O., 2002, A comparison of environmental, soil fertility, yield, and economical effects in six cropping systems based on an 8-year experiment in Norway, *Agric. Ecosyst. Environ.* **90**: 155–168.
Evans, L.T., 1998, *Feeding the Ten Billions – Plants and Population Growth*, Cambridge University Press, Cambridge, UK, 247p.
Fagerberg, B., Salomon, E., and Jonsson, S., 1996, Comparisons between conventional and ecological farming systems at Öjebyn, *Swedish J. Agric. Res.* **26**: 169–180.
FAO, 2007, Food and Agriculture Organization of the United Nations, Statistical Yearbook 2005/06, Rome, www.fao.org/statistics/yearbook/vol_1_1/site_en.asp?page=resources. Assessed 28 April 2007.
Finnish Food Safety Authority (EVIRA), 2006, Organic farming 2005 – Statistics, http://www.evira.fi/portal/se/vaxtproduktion_och_foder/ekoproduktion/aktuellt_inom_ekovervakningen/ Loimaa, Plant Production Inspection Centre, Finland. Assessed 20 December 2007.
Fowler, S.M, Watson, C.A., and Wilman, D., 1993, N, P and K on organic farms: herbage and cereal production, purchases and sales, *J. Agric. Sci.* **120**: 353–360.
Gabriel, D., Roschewitz, I., Tscharntke, T., and Thies, C., 2006, Beta diversity at different spatial scales: plant communities in organic and conventional agriculture, *Ecol. Appl.* **16**: 2011–2021.
GeoHive, 2007, Global Statistics, World Population Prospects, www.geohive.com/earth/pop_prospects2.aspx. Assessed 27 March 2007.
Giller, K.E. and Cadish, G., 1995, Future benefits from biological nitrogen fixation: an ecological approach to agriculture, *Plant Soil* **174**: 255–277.
Giller, K.E., 1998, Tropical legumes: providers and plunderers of nitrogen, in: *Carbon and Nutrient Dynamics in Natural and Agricultural Tropical Ecosystems*, L. Bergström and H. Kirchmann, eds., CAB International, Wallingford, UK, pp. 33–45.
Giller, K.E., 2001, *Nitrogen Fixation in Tropical Cropping Systems, 2nd edition*, CAB International, Wallingford, UK, 423p.
Gilland, B., 2002, World population and food supply. Can food production keep pace with population growth in the next half-century? *Food Policy* **27**: 47–63.
Goulding, K., 2007, Nutrient management on farms, or 'You get out what you put in', *J. Sci. Food Agric.* **87**: 177–180.
Granstedt, A., 1995, Studies of the flow, supply and losses of nitrogen and other plant nutrients in conventional and ecological agricultural systems in Sweden, *Biol. Agric. Hortic.* **11**: 51–67.
Gregory, P.J., Ingram, J.S.I., Andersson, R., Betts, R.A., Brovkin, V., Chase, T.N., Grace, P.R., Gray, A.J., Hamilton, N., Hardy, T.B., Howden, S.M., Jenkins, A., Meybeck, M., Olsson, M., Ortiz-Monasterio, I., Palm, C.A., Payn, T.W., Rummukainen, M., Schulze, R.E., Thiem, M., Valentin, C., and Wilkinson, M.J., 2002, Environmental consequences of alternative practices for intensifying crop production, *Agric. Ecosyst. Environ.* **88**: 279–290.
Green, R.E., Cornell, S.J., Scharleman, J.P.W., and Balemford, A., 2005, Farming and the fate of wild nature, *Science* **307**: 550–555.
Hadas, A., Kautsky, L., and Portnoy, R., 1996, Mineralization of composted manure and microbial dynamics in soil as affected by long-term nitrogen management, *Soil Biol. Biochem.* **28**: 733–738.
Halberg, N., Kristensen, E.S., and Kristensen, I.S., 1995, Nitrogen turnover on organic and conventional mixed farms, *J. Agric. Environ. Ethics* **8**: 30–51.

Halberg, N., and Kristensen, I.S., 1997, Expected crop yield loss when converting to organic dairy farming in Denmark, *Biol. Agric. Hortic.* **14**: 25–41.
Hole, D.G., Perkins, A.J., Wilson, J.O., Alexander, I.H., Grice, P.V., and Evans, A.D., 2005, Does organic farming benefit biodiversity? *Biol. Conserv.* **122**: 113–130.
International Federation of Organic Agricultural Movements, 2006, The four principles of organic farming, Bonn, Germany, www.IFOAM.org. Assessed 19 March 2007.
Ivarson, J., and Gunnarsson, A., 2001, Försök med konventionella och ekologiska odlingsformer 1987–1998, Meddelande från Södra Jordbruksförsöksdistriktet Nr 53, Swedish University of Agricultural Sciences, Uppsala, Sweden, 165 pp (In Swedish).
Jensen, E.S., 1987, Seasonal patterns of growth and nitrogen fixation in field-grown pea, *Plant Soil* **101**: 29–37.
Jensen, E.S., 1996, Grain yield, symbiotic N_2 fixation and interspecific competition for inorganic N in pea-barley intercrops, *Plant Soil* **182**: 25–38.
Kaffka, S. and Koepf, H., 1989, A case study on the nutrient regime in sustainable farming, *Biol. Agric. Hortic.* **6**: 89–106.
Kirchmann, H., 1985, Losses, plant uptake and utilisation of manure nitrogen during a production cycle, *Acta Agric. Scand.* **24**(Suppl.): 77.
Kirchmann, H., Persson, J., and Carlgren, K., 1994, The Ultuna long-term soil organic matter experiment, 1956–1991, Swedish University of Agricultural Sciences, Department of Soil Sciences, Report 17, Uppsala, Sweden, 55p.
Kirchmann, H. and Lundvall, A., 1998, Treatment of solid animal manures: identification of low NH_3 emission practices, *Nutr. Cycl. Agroecosyst.* **51**: 65–71.
Kirchmann, H., Nyamangara, J., and Cohen, Y., 2005, Recycling municipal wastes in the future: from organic to inorganic forms? *Soil Use Manage.* **21**: 152–159.
Kirchmann, H., Bergström, L., Kätterer, T., Mattsson, L., and Gesslein, S., 2007, Comparison of long-term organic and conventional crop-livestock systems in a previously nutrient- depleted soil in Sweden, *Agron. J.* **99**: 960–972.
Kirchmann, H., Kätterer, T., and Bergström, L., 2008, Nutrient supply in organic agriculture – plant availability, sources and recycling, in: *Organic Crop Production – Ambitions and Limitations*, H. Kirchmann and L. Bergström, eds., Springer, Dordrecht, The Netherlands.
Kitchen, J.L., McDonald, G.K., Shepherd, K.W., Lorimer, M.F., and Graham, R.D., 2003, Comparing wheat grown in South Australian organic and conventional farming systems, I. Growth and grain yield, *Aust. J. Agric. Res.* **54**: 889–901.
Koepf, H.H., Pettersson, B.D., and Schaumann, W., 1976, *Biologische Landwirtschaft*, Verlag Eugen Ulmer, Stuttgart, Germany, 303 pp (In German).
Korsaeth, A. and Eltun, R., 2000, Nitrogen mass balances in conventional, integrated and ecological cropping systems and the relationship between balance calculations and nitrogen runoff in an 8-year field experiment in Norway, *Agric. Ecosyst. Environ.* **79**: 199–214.
Laryea, K.B., Anders, M.M., and Pathak, P., 1995, Long-term experiments on Alfisols and Vertisols in the semiarid tropics, in: *Soil Management: Experimental Basis for Sustainability and Environmental Quality*, R. Lal and B.A. Stewart, eds., CRC Press, Boca Raton, FL, pp. 267–292.
Latif, M.A., Islam, M.R., Ali, M.Y., and Saleque, M.A., 2005, Validation of the system of rice intensification (SRI) in Bangladesh, *Field Crops Res.* **93**: 281–292.
Leu, A., 2004, Organic agriculture can feed the world, *Acres – a Voice for Eco-Agric.* **34**(1): 1–4.
Liebhardt, W.C., Andrews, R.W., Culik, M.N., Harwood, R.R., Janke, R.R., Radke, J.K., and Rieger-Schwartz, S.L., 1989, Crop production during conversion from conventional to low-input methods, *Agron. J.* **81**: 150–159.
Lockeretz, W., Shearer, G., Sweeney, S., Kuepper, G., Wanner, D., and Kohl, D.H., 1980, Maize yields and soil nutrient levels with and without pesticides and standard commercial fertilisers, *Agron. J.* **72**: 65–72.
Lockeretz, W., Shearer, G., and Kohl, D.H., 1981, Organic farming in the corn belt, *Science* **211**: 540–547.
Marstorp, H. and Kirchmann, H., 1991, Carbon and nitrogen mineralization and crop uptake of nitrogen from six green manure legumes decomposing in soil, *Acta Agric. Scand.* **41**: 243–252.

Mäder, P., Fliesbach, A., Dubois, D., Gunst, L., Fried, P., and Niggli, U., 2002, Soil fertility and biodiversity in organic farming, *Science* **296**: 1694–1697.
Murwira, H.K. and Kirchmann, H., 1993, Nitrogen dynamics and maize growth in a Zimbabwean sandy soil under maize fertilisation, *Commun. Soil Sci. Plant Anal.* **24**: 2343–2359.
Mugwira, L.M. and Nyamangara, J., 1998, Organic carbon and plant nutrients in soil under maize in Chinamhora communal area, Zimbabwe, in: *Carbon and Nutrient Dynamics in Natural and Agricultural Tropical Ecosystems*, L. Bergström and H. Kirchmann, eds., CAB International, Wallingford, UK, pp. 15–21.
National Research Council, 1989, *Alternative Agriculture*, National Academy Press, Washington, DC, 448p.
Nolte, C. and Werner, W., 1994, Investigations on the nutrient cycle and its components of a biodynamically-managed farm, *Biol. Agric. Hortic.* **10**: 235–254.
Nonhebel, S., 2005, Renewable energy and food supply: will there be enough land? *Renew. Sus. Energ. Rev.* **9**: 191–201.
Nguyen, M.L., Haynes, R.J., and Goh, K.M., 1995, Nutrient budgets and status in three pairs of conventional and alternative mixed cropping farms in Canterbury, New Zealand, *Agric. Ecosyst. Environ.* **52**: 149–162.
Palm, C.A., Myers, R.J.K., and Nandwa, S.M., 1997, Combined use of organic and inorganic nutrient sources for soil fertility maintenance and replenishment, in: *Replenishing Soil Fertility in Africa*, R.J. Buresh, P.A. Sanchez, and F. Calhourn, eds., SSSA Special Publications No. 51, Madison, WI, pp. 193–217.
Palm, C.A., Gachengo, C.N., Delve, G., Cadish, G., and Giller, K.E., 2001, Organic inputs for soil fertility management in tropical agroecosystems: application of an organic resource database, *Agric. Ecosyst. Environ.* **83**: 27–42.
Pang, X.P. and Letey, J., 2000, Organic farming: challenges of timing nitrogen availability to crop nitrogen requirements, *Soil Sci. Soc. Am. J.* **64**: 247–253.
Parry, M., Rosenzweig, C., and Livermore, M., 2005, Climate change, global food supply and risk of hunger, *Philosophical Trans. R. Soc. B: Biol. Sci.* **360**: 2125–2138.
Peigné, J., Ball, B.C., Roger-Estrade, J., and David, C., 2007, Is conservation tillage suitable for organic farming? A review, *Soil Use Manage.* **23**: 129–144.
Pieri, C., 1995, Long-term soil management experiments in semiarid Francophone Africa, in: *Soil Management: Experimental Basis for Sustainability and Environmental Quality*, R. Lal and B.A. Stewart, eds., CRC Press, Boca Raton, FL, pp. 225–266.
Pimentel, D., Hepperley, P., Hanson, J., Douds, D., and Seidel, R., 2005, Environmental, energetic, and economic comparisons of organic and conventional farming systems, *BioScience* **55**: 673–582.
Porter, P.M., Huggins, D.R., Perillo, C.A., Quirling, S.R., and Crookston, R.K., 2003, Organic and other management strategies with two- and four-year crop rotations in Minnesota, *Agron. J.* **95**: 233–244.
Posner, J.L., Baldock, J.O., and Hedtcke, J.L., 2008, Organic and conventional cropping systems in the Wisconsin integrated cropping systems trials: I. Productivity 1990–2002, *Agron. J.* **100**: 265–260.
Poudel, D.D., Horwarth, W.R., Lanini, W.T., Temple, S.R., and van Bruggen, A.H.C., 2002, Comparison of soil N availability and leaching potential, crop yields and weeds in organic, low-input and conventional farming systems in northern California, *Agric. Ecosyst. Environ.* **90**: 125–137.
Pretty, J.N. and Hine, R.E., 2001, Reducing food poverty with sustainable agriculture: a summary of new evidence, Centre for Environment and Society, Essex University, UK, www.essex.ac.uk/ces/esu/occasionalpapers/SAFE%20FINAL%20-%20Pages1-22.pdf. Assessed 29 December 2007.
Pretty, J.N., Morison, J.L.L., and Hine, R.E., 2003, Reducing food poverty by increasing agricultural sustainability in developing countries, *Agric. Ecosyst. Environ.* **95**: 217–234.
Prudencio, C.Y., 1993, Ring management of soils and crops in the West African semi-arid tropics: the case of the mossi farming systems in Burkina Faso, *Agric. Ecosyst. Environ.* **47**: 237–264.

Quinones, N.A., Borlaug, N.E., and Dowswell, C.R., 1997, A fertilizer-based green revolution for Africa, in: *Replenishing Soil Fertility in Africa*, R.J. Buresh, P.A. Sanchez and F. Calhourn, eds., SSSA Special Publications No. 51, Madison, WI, pp. 81–95.
Rundgren, G., 2002, *Organic Agriculture and Food Security*, IFOAM, Dossier 1, Bonn, Germany, 20p.
Rusch, H.P., 1978, *Bodenfruchtbarkeit. Eine Studie biologischen Denkens*, 3rd Printing. Haug Verlag, Heidelberg, Germany, 243 pp (In German).
Ryan, M.H., Derrick, J.W., and Dann, P.R., 2004, Grain mineral concentrations and yield of wheat grown under organic and conventional management, *J. Sci. Food Agric.* **84**: 207–216.
Sanchez, P.A., 2002, Soil fertility and hunger in Africa, *Science* **295**: 2019–2020.
Sanchez, J.E., Harwood, R.R., Willson, T.C., Kizilkaya, K., Smeenk, J., Parker, E., Paul, E.A., Knezek, B.D., and Robertson, G.P., 2004, Managing soil carbon and nitrogen for productivity and environmental quality, *Agron. J.* **96**: 769–775.
SCB, 2006, *Yearbook of Agricultural Statistics*, Official Statistics of Sweden, SCB, Örebro, Sweden.
Sheehy, J.E., Peng, S., Dobermann, A., Mitchell, P.L., Ferrer, A., Yang, J., Zou, Y., Zhong, X., and Huang, J., 2004, Fantastic yields in the system of rice intensification: fact or fallacy? *Field Crops Res.* **88**: 1–8.
Singh, L. and Balasubramanian, V., 1979, Effects of continuous fertilizer use on a ferruginous soil (Halustalf) in Nigera, *Exp. Agric.* **15**: 257–265.
Sjursen, H., 2001, Change of weed seed bank during the first complete six-course rotation after conversion from conventional to organic farming, *Biol. Agric. Hortic.* **19**: 71–90.
Smaling, E.M.A. and Braun, A.R., 1996, Soil fertility research in sub-Saharan Africa: new dimensions, new challenges, *Commun. Soil Sci. Plant Anal.* **27**: 365–386.
Smaling, E.M.A., Nandwa, S.M., and Jansson, B., 1997, Soil fertility in Africa is at stake, in: *Replenishing Soil Fertility in Africa*, R. J. Buresh, P. A. Sachez and F. Calhourn, eds., SSSA Special Publications No. 51, Madison, WI, pp. 47–62.
Smil, V., 2001, *Enriching the Earth: Fritz Haber, Carl Bosch, and the Transformation of World Food Production*, MIT Press, Cambridge, MA, 338p.
Smil, V., 2002, Nitrogen and food production: proteins for human diets, *Ambio* **31**: 126–131.
Smith, R.G. and Gross, K.L., 2006, Weed community and corn yield variability in diverse management systems, *Weed Sci.* **54**: 106–113.
Smith, R.G., Menalled, F.D., and Robertson, G.P., 2007, Temporal yield variability under conventional and alternative management systems, *Agron. J.* **99**: 1629–1634.
Smolik, J.D., Dobbs, T.L., Rickerl, D.H., Wrage, L.J., Buchenau, G.W., and Machacek, T.A., 1993, Agronomic, economic, and ecological relationships in alternative (organic), conventional, and reduced-till farming systems, Bull. 718, South Dakota Agric. Exp. Stat., Brookings, USA.
Smolik, J.D., Dobbs, T.L., and Rickerl, D.H., 1995, The relative sustainability of alternative, conventional and reduced-till farming systems, *Am. J. Altern. Agric.* **10**: 25–35.
Spiess, E., Stauffer, W., Niggli, U., and Besson, J.M., 1993, DOK-Versuch: Vergleichende Langzeit-Untersuchungen in den drei Anbausystemen biologisch-dynamisch, organisch-biologisch und konventionell. IV. Aufwand und Ertrag: Nährstoffbilanzen, 1. und 2. Fruchtfolgeperiode, *Schweiz. Landw. Forsch.* **32**: 565–579 (In German).
Statistics Finland, 2007, Finland in Figures. Agriculture, Forestry and Fishery, Statistics Finland, Helsinki. http://www.stat.fi/tup/suoluk/suoluk_maatalous_en.html. Assessed 18 December 2007.
Steiner, R., 1924, Geisteswissenschaftliche Grundlagen zum Gedeihen der Landwirtschaft. Steiner Verlag, 5. Auflage 1975. Dornach, Schweiz, 256 pp (In German).
Svensson, K., Odlare, M., and Pell, M., 2004, The fertilizing effect of compost and biogas residues from source separated household waste, *J. Agric. Sci.* **142**: 461–467.
Swift, M.J., Seward, P.D., Frost, P.G.H., Qureshi, J.N., and Muchena, F.N., 1994, Long-term experiments in Africa: developing a database for sustainable land use under global change, in: *Long-term Experiments in Agricultural and Ecological Sciences*, R.A. Leigh and A.E. Jonston, eds., CAB Int. Wallingford, England, pp. 229–251.

Teasdale, J.R., Rosecrance, R.C., Coffman, C.B., Starr, J.L., Paltineanu, I.C., Lu, Y.C., and Watkins, B.K., 2000, Performance of reduced-tillage cropping systems for sustainable grain production in Maryland, *Am. J. Altern. Agric.* **15**: 79–87.

Teasdale, J.R., Mangum, R.W., Radhakrishnan, J., and Cavigelli, M.A., 2004, Weed seedbank dynamics in three organic farming crop rotations, *Agron. J.* **96**: 1429–1435.

Teasdale, J.R., Coffman, B., and Mangum, R.W., 2007, Potential long-term benefits of no-tillage and organic cropping systems for grain production and soil improvement, *Agron. J.* **99**: 1297–1305.

Toomsan, B., McDonagh, J.F., Limpinuntana, V., and Giller, K.E., 1995, Nitrogen fixation by groundnut and soyabean and residual nitrogen benefits to rice farmers' fields in Northeast Thailand, *Plant Soil* **175**: 45–56.

Torstensson, G., Aronsson, H., and Bergström, L., 2006, Nutrient use efficiency and leaching of N, P and K of organic and conventional cropping systems in Sweden, *Agron. J.* **98**: 603–615.

TSBF, CIAT, 2006, Integrated soil fertility management in the tropics, http://www.tsbf.org/pdf/tsbf_ciat_achievements_2002–2005.pdf, Nairobi, Kenya. Assessed 29 December 2007.

Trewavas, A., 2001, The population/biodiversity paradox. Agricultural efficiency to save wilderness, *Plant Physiol.* **125**: 174–179.

Tudge, C., 2005, Can organic farming feed the world? http://www.colintudge.com, Oxford, England. Assessed 29 December 2007.

UN Millennium Project, 2005, *Halving Hunger: It Can Be Done*, P. Sanchez, ed., Task Force on Hunger. Earthscan, UK, 245p.

Warman, P.R. and Harvard, K.A., 1997, Yield vitamin and mineral contents of organically and conventionally grown carrots and cabbage, *Agric. Ecosyst. Environ.* **61**: 155–162.

Watson, C.A., Atkinson, D., Gosling. P., Jackson, L.R., and Rays, F.W., 2002, Managing soil fertility in organic farming systems, *Soil Use Manage. Suppl.* **18**: 239–247.

Wieser, I., Heß, J., and Lindenthal, T., 1996, Nutrient balances on organically managed grassland farms in Upper Austria, *Die Bodenkultur* **47**: 81–88.

Woodward, L., 1995, Can organic farming feed the world? www.population-growth-migration. info/essays/woodwardorganic.html, Elm Research Centre, England. Assessed 29 December 2007.

Vandermeer, J. and Perfecto, I., 2007, The agricultural matrix and a future paradigm for conservation, *Conserv. Biol.* **21**: 274–277.

Vanlauwe, B., Wendt, J., and Diels, J., 2001, Combined application of organic matter and fertilizer, in: *Sustaining Soil Fertility in West-Africa*, G. Tian, F. Ishida and J.D.H. Keatinge, eds., SSSA Special Publications No 58, Madison, WI, pp. 247–280.

Vanlauwe, B. and Giller, K.E., 2006, Popular myths around soil fertility management in sub-Saharan Africa, *Agric. Ecosyst. Environ.* **116**: 34–46.

Vasilikiotis, C., 2000, Can organic farming "Feed the World"? http://nature.berkeley.edu/~christos/espm118/articles/organic_feed_world.pdf, University of California Berkeley, CA. Assessed 29 December 2007.

Chapter 4
Plant Nutrients in Organic Farming

Keith Goulding, Elizabeth Stockdale and Christine Watson

Abstract Effective nutrient management is essential in organic farming systems. Processed soluble fertilisers such as ammonium nitrate, which feed the plant directly and are thought to bypass the natural processes of the soil, are not generally acceptable. Nutrient supply to crop plants is supported through recycling, the management of biologically-related processes such as nitrogen fixation by clover and other legumes, and the limited use of unrefined, slowly-soluble off-farm materials that decompose in the same way as soil minerals or organic matter. The aim is to achieve as far as possible a closed nutrient cycle on the farm and to minimise adverse environmental impact. Effective management of any 'waste' materials such as manures and crop residues is a key to nutrient cycling on organic farms. However, not all organic farms have easy access to manures and recycling is limited by the prohibition of the use of sewage sludge because of current concerns over the introduction of potentially toxic elements, organic pollutants and disease transmission. In addition, the current global market, in which food is transported large distances from the farm, results in a significant export of nutrients. Exported nutrients must be replaced to avoid nutrient depletion of soils. Nutrient budgeting suggests some cause for concern over the sustainability of organic systems because of their dependence on feedstuffs and bedding for inputs of phosphorus (P) and potassium (K), and on the very variable fixation by legumes or imports of manure or compost for nitrogen (N); air pollution and net mineralisation from soil reserves appear to comprise a large part of the N supply on some organic farms. Losses of N from organic systems can also be as large as those from conventional systems and, being dependent on cultivation and the weather, they are even more difficult to control than those from fertilisers applied to conventional farms. There is some evidence of P deficiency in soils under organic production, and replacing K sold off the farm in produce is especially difficult. Organic farming systems may be sustainable and have the potential to deliver significant environmental benefits, but these depend on specific cropping and management practices on each farm. It is important that we study

K. Goulding (✉)
Cross-Institute Programme for Sustainable Soil Function, Rothamsted Research, Harpenden, Hertfordshire, AL5 2JQ, UK
e-mail: keith.goulding@bbsrc.ac.uk

H. Kirchmann, L. Bergström (eds.), *Organic Crop Production – Ambitions and Limitations*, DOI 10.1007/978-1-4020-9316-6_4,

and improve nutrient management on *all* farm systems and in the context of plant, animal and human health in order to develop more sustainable farming systems.

Keywords Organic farming · Plant nutrients · Nitrogen (N) · Potassium (K) · Phosphorus (P) · Nutrient budgets · Nutrient management · Soils

4.1 Introduction

Organic farming systems are very diverse and are found across the world, with a range of crop and animal enterprises often linked together. However, strong unifying principles link this wide range of systems and management practices (IFOAM, 1998, 2006). In many countries organic farming has a clear legislative basis and certification schemes for production and processing. Within the European Union (EU), crop and livestock products sold as organic must be certified as such under EC Regulations 2092/91 and 1804/99. In the EU (including the new member states) there was 5.7 million ha land under organic management (3.4% of the agricultural land area) and 143,000 organic farms at the end of 2003 (Willer and Yussefi, 2005). Organic farming systems fall into similar categories to those of conventional agriculture: mixed, livestock, stockless and horticultural. In the EU15 (i.e. excluding the new member states) around 26% of the organic land area was under arable crops in 2003 (Nic Lampkin, personal communication). In the UK, 89% of fully organic land is under grass, 7.7% arable and 1.2% horticulture (Soil Association, 2004). This is in marked contrast with countries like Denmark where 53% of the organic land was in arable cultivation in 2003 (Nic Lampkin, personal communication).

In his book 'The Control of Soil Fertility', Cooke (1967) began by reminding his readers that 'The inevitable result of farming is always to diminish *natural* fertility because portions of the total supply of plant nutrients ... are removed.' Cooke believed that the use of fertilisers within agriculture was necessary to meet food requirements and regarded such systems as efficient and productive. Organic farmers disagree. Processed soluble fertilisers such as ammonium nitrate, which feed the plant directly are thought to bypass the natural processes of the soil and are not generally acceptable (although some soluble materials are permitted under certain circumstances).

From the early days of research into organic farming systems, holistic approaches were held to be more appropriate for organic systems than reductionist ones, embracing the whole philosophy of organic farming and the idea that the 'health of soil, plant, animal and man is one and indivisible' (Howard, 1943; Woodward, 2002). This has, in part, led to the idea that holistic research is the only acceptable approach. If so, what are the most appropriate ways to research nutrient cycling in organic farming systems, and to what extent do they, and should they, differ from approaches in conventional agriculture? There has been a long-standing and unfruitful conflict between 'reductionist' and 'holistic' science in connection with agricultural and ecological research (for example, see Lockeretz and Anderson, 1993; Rowe,

1997). We need a combination of these approaches if we are to understand not only nutrient cycling processes in soils, but also their role in the efficiency of nutrient use at the whole farm level and beyond. In this context, very useful reviews are available from RASE (2000) and a Supplement to Soil Use and Management (Vol. 18, 2002).

Research into nutrient cycling on organic farms can be classified in several ways other than reductionist and holistic. For example, there is a clear split in research approaches between studies that compare organic and conventional systems (e.g. Mäder et al., 2002) and studies that compare different management systems within organic farming (e.g. Olesen et al., 1999). One of the complicating factors in interpreting results of the former approach is whether these trials truly compare farming systems or simply different rotations. Factorial crop rotation experiments (e.g. Mäder et al., 2002; Watson et al., 1999) and field scale testing of crop rotations (e.g. Cormack, 1999), which allow factorial experiments within them, contribute to different aspects of the understanding of nutrient cycling in crop rotations. As soon as the crops or even varieties within a rotation are changed, the impact of that rotation both in terms of yield and productivity, nutrient supply and environmental impact will change regardless of the production system. However, under given soil and climatic constraints, the most productive choice of crops and varieties in a rotation will differ depending on whether the system is managed conventionally or organically. It is also essential to remember that:

> ...although nutrient management in organically farmed soils is fundamentally different to soils managed conventionally, the underlying processes supporting soil fertility are not.
> (Stockdale et al., 2002)

Effective management of any 'waste' materials such as manures and crop residues is a key to nutrient cycling on organic farms. However, not all organic farms have easy access to manures, and recycling is limited by the prohibition of the use of sewage sludge because of current concerns over the introduction of potentially toxic elements, organic pollutants and disease transmission. In addition, the current global market, in which food is often transported large distances from the farm, results in a significant export of nutrients. These must be replaced otherwise, as Cooke (1967) said, soil is impoverished and the system unsustainable. Johnston (1991) has expressed concern that, on organic farms, phosphorus (P) and potassium (K) may be mined from soil reserves because of the paucity of acceptable sources to replace these nutrients. This dilemma is recognised by the organic farming movement and a list of allowed amendments is published (UKROFS, 1999). Most of these are organic materials such as calcified seaweed, or relatively insoluble mined products such as rock phosphate, and are not suitable in all circumstances. However, in certain circumstances, where plant or soil analysis clearly demonstrates a need and permission is obtained from the certifying organisation, more soluble materials such as potassium sulphate are allowed. Such exceptions can appear contradictory to conventional farmers. This problem is acknowledged by the organic movement, which is seeking to base its list of approved products on a more sound scientific footing (Peter Crofts, UKROFS, personal communication).

Organic farming is often characterised in terms of what it does not do. It is therefore described as a system which produces food without the use of fertilisers, pesticides or pharmaceuticals. More correctly, organic farming should be regarded as a system which is designed in order to minimise the need for external inputs of nutrients, crop protection aids and preventative medicines. This more accurate definition explains the existence of a list of last resort measures, such as those given above, that are available for use when the checks and balances of the normally balanced system have failed (reluctantly and only with appropriate safeguards). These two elements of the philosophy are critical to understanding research needs at a technical and scientific level. The elements in the definitions remind us that all agricultural systems, both organic and those most easily described as conventional, require mechanisms to appropriately regulate a balance in resource capture between the crop and other vegetation, micro-organisms and invertebrates, and the provision of a range of mineral nutrients for crop growth. In a conventional system these are provided largely through inputs of externally generated materials. The research needs of conventional farming systems have therefore dominantly related to the efficient use of external materials and the design of appropriate inputs. In an organic system these requirements are provided primarily through recycling within the total system. Consequently, the research needs of organic farming have been rather different, such that only the most basic of results from many studies, e.g. information on soil chemistry, have been directly of relevance to organic systems.

In this chapter we examine the sustainability of nutrient cycling on organic farms, mostly using a nutrient budget or audit approach, including some consideration of losses to the environment, and focusing on the situation in Europe.

4.2 Nutrient Management on Organic Farms

The main methods of nutrient management in organic farming in Europe are set out in the EU Council Regulation 2092/91 (Table 4.1). Regulations differ between countries. The International Federation of Organic Agricultural Movements (IFOAM, 2006) has information for all participating countries. Though these approaches are only legally enforceable for food sold in Europe, the basic practices described within this regulation apply to a wide range of systems in temperate and Mediterranean climates. The emphasis is on the use of multi-annual rotations and organic materials of plant and animal origin from organic farms. Where these methods cannot provide adequate nutrition a limited range of other organic materials and mineral

Table 4.1 Extract from European Council regulation (EEC) No. 2092/91

2.1 The fertility and the biological activity of the soil must be maintained or increased, in the first instance by:
I a. cultivation of legumes, green manures or deep-rooting plants in an appropriate multi-annual rotation
II b. incorporation of livestock manure from organic livestock production in accordance with the provisions and within the restrictions of part B, point 7.1 of this annex;
III c. incorporation of other organic material, composted or not, from holdings producing according to the rules of this Regulation.

fertilisers can be used, although their use is permitted only where the need can be demonstrated to the certifying body (for example by soil analysis or by presentation of a nutrient budget). Amendments include rock phosphate, various ground rock products that contain K and magnesium (Mg) (see Fortune et al., 2004), and gypsum. Products such as rock phosphate release nutrients over a period of years rather than weeks (Rajan et al., 1996) and thus their use is planned to build fertility in the longer-term. Trace elements may also be applied, with approval, if they are necessary. The use of lime to maintain pH levels is also acceptable.

Research aimed at improving nutrient supply in organic systems therefore generally focuses on improved use of these technologies on farm. However, it is also important that research is able to challenge the organic standards and feed into the process of future standards development. This may involve working beyond the confines of organic farms, and the use of models may be particularly appropriate.

4.2.1 Nitrogen

Crop rotations, including a mixture of leguminous 'fertility building' and cash crops, are the main mechanism for nitrogen (N) supply within organic systems. Organic rotations are divided into phases that increase the level of soil N and phases that deplete it. The N building and depleting phases must be in balance, or show a slight surplus, if long-term fertility is to be maintained (Berry et al., 2002). Organic rotations must include legumes, usually in a ley phase, to provide N in the absence of soluble N fertiliser. The ratio of ley to arable will be determined by a combination of the system (stocked or stockless) and the soil type, being lower on N retentive soils and higher on sandy soils.

In North West Europe, a typical rotation on a mixed organic farm with a three-year grass and clover ley will support two or three years of arable cropping. This may be extended by including a N-fixing cash crop, such as beans, or by including a short period of N-fixing green manure, such as vetch, between cash crops. Cultivation itself leads to an increase in nutrient availability, particularly N, as microbial activity is stimulated and organic matter breakdown occurs (Silgram and Shepherd, 1999). Mechanical weed control, commonly used in organic horticultural systems, can thus provide a mid-season boost to crops by stimulating mineralisation, although at other times additional stimulation of mineralisation may cause losses by leaching or denitrification.

Nitrogen fixation represents a major input of N into organic farming systems. The amount of N fixed by leguminous crops is very variable, being dependent on such factors as climate, soil pH, available N, P and K, age of legume, species, cultivar and strain of symbiotic rhizobium (Ledgard and Steele, 1992). A number of empirical relationships have been proposed for estimating N fixation by legumes (e.g. Haraldsen et al., 2000; Høgh-Jensen et al., 2004). Some of the models have recently been tested against measured values (Topp et al., 2005). Their performance varies but, generally, their ability to predict N fixation is very good (Fig. 4.1).

Fig. 4.1 N fixation measurements in different ages of grass-clover leys by year compared with the model predictions (Topp et al., 2005): 1997, 1-year ley only; 1998, 1- and 2-year leys; 1999, 1-, 2- and 3-year leys; 2000, 1-, 2-, 3- and 4-year leys

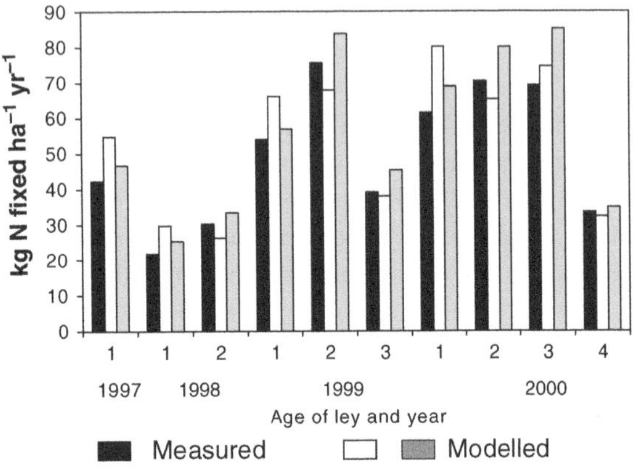

As Fig. 4.1 shows, sometimes as much as 250–300 kg N ha^{-1}, but often as little as 60–70 kg ha^{-1}, are fixed by a 3-year ley. The input of N through biological fixation is associated with seasonal changes in the availability of soil N. Studies at an organic rotation in the North East of Scotland (Watson et al., 1999), comparing flows of N in different ages of a grass-clover ley have identified a significant ($P < 0.001$) annual fluctuation in the soil nitrate pool (Fig. 4.2). This is likely to be a consequence of lower plant uptake during the winter months coupled with decomposition of residues from the clover. There was also a significant ($P < 0.05$) effect of the age of the grass-clover ley, with first and second year leys containing higher concentrations of NO_3 than third and fourth year leys. This was probably

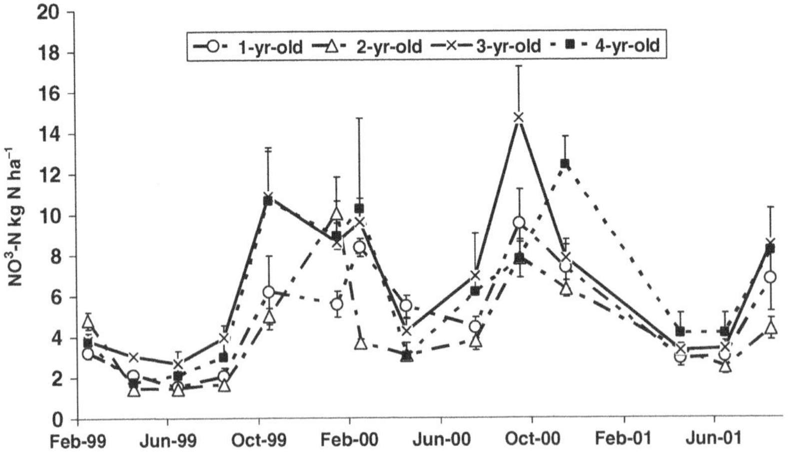

Fig. 4.2 Changes in the amounts of nitrate-N in the 0–15 cm layer of soil at the Tulloch organic rotation in NE Scotland between February 1999 and August 2001 during the growth of leys and following cultivation. Treatments compared are: One year old grass clover leys (1-yr-old), Two year old grass clover leys (2-yr-old), Three year old grass clover leys (3-yr-old), and Four year old grass clover leys (4-yr-old); Watson et al. (1999)

a consequence of the higher rates of N fixation associated with the former. Such information on the input and availability of N is invaluable in helping to design and predict the performance of organic cropping systems.

This seasonal variation in soil mineral N is common to all farm systems but is much more difficult to control in organic systems through cultivation of the ley. Poor matching of soil supply and crop demand for N can lead to losses by leaching or denitrification or both; the transition from ley to arable cropping in an organic rotation is generally associated with the highest loss, with up to 180 kg N ha^{-1} leached in the winter after ploughing (Philipps and Stopes, 1995; Lord et al., 1997). Season, timing and intensity of cultivation have been shown to have a substantial effect on this loss, but when comparable organic and conventional systems are examined over a rotation, losses of N per area are similar or a little smaller in organic systems (Watson et al., 1993; Silgram and Shepherd, 1999). Cultivation of leys in spring, followed immediately by spring cropping, reduced nitrate leaching considerably over a comparable conventional system (Watson et al., 1993). It should be noted, however, that the 'comparable' organic and conventional systems should also be representative of typical organic and conventional systems.

Other management decisions also affect soil mineral N levels and the available of N to a growing crop or its risk of loss. Autumn and winter grazing of leys or other fodder crops can leave large amounts of mineral N at risk of loss when compared with an ungrazed winter cereal (Fig. 4.3), but this is true of all grazed pasture systems. This illustrates a key point: management rather than the system *per se* determines nutrient cycling, crop growth and losses.

In addition to symbiotic N fixation and atmospheric deposition, nutrients may be brought in to an organic system in imported animal feeds, manures, composts and permitted fertilisers. The nature and quantity of imported nutrients will depend on the farming system and the soil type. Manures from non-organic livestock production may be brought onto the holding but there are restrictions (e.g. it must originate from an 'ethical' source, i.e. the animals producing it must be kept un-

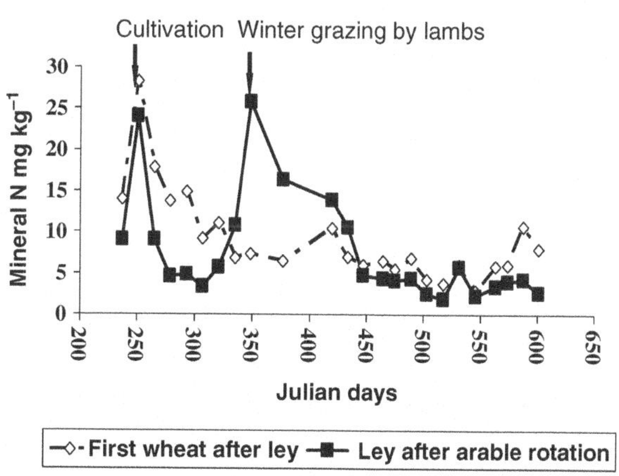

Fig. 4.3 Seasonal changes in mineral N levels in soil as affected by cultivation and grazing patterns

der high welfare standards and not have been fed on a diet containing Genetically Modified Organisms). EU legislation (Council Regulation (EC) No 1804/1999), which came into force in August 2000, requires that a maximum of 170 kg total N ha^{-1} yr^{-1} is applied in manure; where necessary, stocking rate must be reduced to meet this limit. The careful management of animal manure to minimise losses and optimise nutritional benefits is a key feature of stocked organic systems, on which the manure represents a valuable source of N. In this context, it should be noted that composting can lead to large losses of N via ammonia volatilisation.

Stockless systems present a challenge for organic farming. In an exemption to the 'Set-aside' rules in Europe ('Set-aside' is land taken out of production to reduce food surpluses), organic farmers are permitted to use green cover containing more than 5% legumes in the seed mixture (MAFF, 1999). Clearly, such systems are dependent on EU policy.

4.2.2 Phosphorus and Potassium

As noted above, the application of acceptable mineral nutrient sources are permitted only where the need can be demonstrated to the certifying body, e.g. by soil analysis or by presentation of a nutrient budget. The sources should be unrefined and slowly soluble (to avoid water pollution); in the case of P, rock phosphate is permitted. In mixed and animal-based systems, animal feeds and bedding import relatively large amounts of P and K to organic farms (Fowler et al., 1993; Nolte and Werner, 1994). Any necessary additional P and K is applied strategically within the rotation, with one application of rock phosphate expected to supply P for a number of following crops. However, organic farming seeks to optimise the recycling of P and K and to keep imports as small as possible. In animal-based and mixed systems, good manure management is therefore essential. Manure and slurry are used to redistribute nutrients around the farm. However the grazing patterns of livestock, such as 'camping' under trees, next to hedges and at fixed feed troughs, can increase the spatial heterogeneity of P and K returns, which may persist for many years. Phosphorus occurs in organic and inorganic forms in manures (Peperzak et al., 1959; Gerritse and Vriesema, 1984; Sato et al., 2005) and little is lost by leaching; K is found in soluble forms, and large leaching losses of K can therefore occur during manure storage and composting (Fowler et al., 1993).

Potassium is potentially the most difficult major nutrient to manage in organic systems since K sold in produce must be replaced, but there is no obvious sustainable source of K available to organic farmers. Where deficiency can be demonstrated, organic certification bodies will allow the use of materials such as sulphate of potash, MSL-K (volcanic tuff) and Kali vinasse (by-product of the sugar beet industry). There is a need for information on the long- and short-term effects of newly available materials on soil K status, such as the latter two products. However,

Table 4.2 Yield of grass/clover (g pot^{-1}) in a pot experiment (sum of 4 cuts) with a range of organically acceptable K sources applied at 41.5 kg ha^{-1} −/+ additional N (Fortune et al., 2004)

Treatment	−N	+N
Control	7.24	18.85
DKSI	10.46	21.63
Kali vinasse	8.97	17.2
MSL-K	8.29	19.17
Rapemeal	15.69	24.29
Sulphate of potash	9.4	26.35
Sylvinite	8.1	20.54
LSD	4.96	

as already shown by Fortune et al. (2004), yield responses to many of these materials are small, particularly in situations where N is limiting (Table 4.2). In North America, where products like MSL-K and Kali vinasse are not available, potassium sulphate is allowed (from a mined source or from evaporative sources such as the Dead Sea or Great Salt Lake) and even potassium chloride is allowed if 'derived from a mined source and applied in a manner that minimizes chloride accumulation in the soil'.

In the UK and Western Europe, adequate P and K levels in soils have been achieved through many years of applications of fertilisers. A strong hypothesis exists that these are supporting P and K supply to organic crops on converted land. It is possible to take data from several organic farms of different ages that would appear to support the hypothesis that the lack of inputs of P and K is causing the decline in these nutrients in soils now that fertilisers are no longer applied (Fig. 4.4; the data points in Fig. 4.4 are for measured 'Available P' values according to Olsen's method on three farms under organic management for different periods; the line represents a measured decay curve for available P in soils; Johnston et al., 2001). The analyses from the organic farms fit well onto the P decay curve, supporting the hypothesis that available P declines as the duration of the organic rotation increases and P is exported.

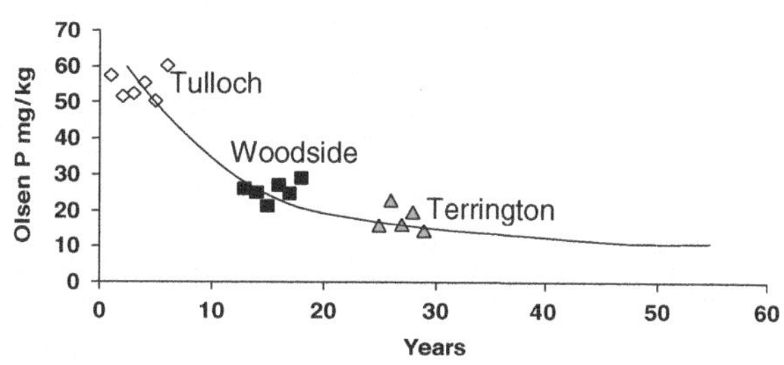

Fig. 4.4 The data points show available P as measured by Olsen's method in soils from three farms in organic rotations for 1 (Tulloch), 15 (Woodside) and 29 (Terrington) years. The line beyond the data points is extrapolated from data of Johnston et al. (2001)

4.3 Sustainability of Nutrient Supplies in Organic Farm Systems

4.3.1 Nutrient Budgets

Nutrient budgets have been compiled around the world, using a variety of scales and methodological approaches (Scoones and Toulmin, 1998; Watson and Atkinson, 1999). They measure or estimate the inputs and outputs of nutrients (usually N, P and K) to a field, farm or system, usually at the 'farm gate'. Farm gate budgets do not usually include the necessarily very detailed measurements of losses such as leaching, denitrification and ammonia volatilisation (but see below), consider each field separately, or measure transfers between fields. Nor do they provide information on soil processes or biological inputs and outputs of nutrients, which are particularly important for N.

Where N fixation is the major external source of N, the balance between N fixing and exploitative arable cropping periods is critical in determining not only productivity but also environmental impact. Figure 4.5 shows that, for a stockless organic rotation, even on what is a very fertile soil, matching N removals in crops with N fixation is difficult.

Goss and Goorahoo (1995) and Halberg et al. (1995) concluded that N budgets were generally positive for organic farms but were probably smaller than for conventional farms (e.g. Watson and Younie, 1995). Leach et al. (2005) calculated a wide range of N surpluses for a series of organic farms from 20 to 120 kg N ha^{-1} yr^{-1}, which is indicative of a wide range of potential N losses.

Looking more widely at N, P and K budgets, the data for the stockless organic system adopted at Terrington farm in the UK show how difficult it can be to balance removals when animals (and their feed and bedding) are not part of the system (Table 4.3). The K deficit on this farm is particularly worrying. Although a detailed

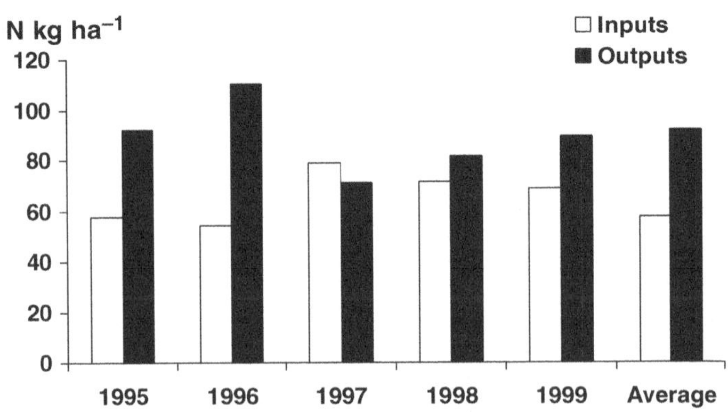

Fig. 4.5 N inputs from biological fixation and N outputs in saleable produce for Terrington stockless organic farm

Table 4.3 N, P and K budgets averaged over the period 1995–1999 on the Terrington stockless organic farm in the UK

	N ($kg\ ha^{-1}\ yr^{-1}$)	P ($kg\ ha^{-1}\ yr^{-1}$)	K ($kg\ ha^{-1}\ yr^{-1}$)
Inputs			
Deposition	30	0.04	3.5
Seed	4.3	1.2	2.6
N fixation	31.9	–	–
Fertiliser	–	9.3	–
Outputs			
Crop offtake	88.8	13.1	45.3
Balance	**–22.6**	**–2.6**	**–39.2**

analysis of the soil might reveal a long-term potential to release K from clay minerals, it would be unwise to rely on this for too long.

Many stockless small-scale organic farms cannot use cover crops to meet their N need, because the land cannot be taken out of production for extended periods of time, and their sole nutrient source is compost. The use of compost in stockless systems can present particular problems because, when added to meet the N requirement of crops, a large surplus of P is usually added at the same time, with accompanying water quality problems.

Where organic systems include animals then nutrient budgets suggest greater sustainability, but with problems remaining for P and especially K (Table 4.4).

Table 4.4 N, P and K budgets ($kg\ ha^{-1}\ yr^{-1}$) averaged over 1993–1998 for two Scottish stocked farm rotations. ('Grazing returns' are for nutrients brought onto the farm in stock; 'Straw for bedding' and 'Silage' are sold off the farm)

	Tulloch			Woodside		
	N	P	K	N	P	K
Inputs						
Deposition	12.0	0.02	2.1	12.0	0.02	2.1
Seed	2.1	0.1	0.7	4.1	1.4	3.2
N fixation	45.0	–	–	35.0	–	–
Manures	56.2	11.9	50.8	57.6	12.2	52.1
Grazing returns	22.1	2.1	29.6	23.0	2.2	30.9
Total	**137.4**	**14.4**	**83.2**	**129.8**	**15.8**	**87.9**
Outputs						
Crop outputs	37.3	11.3	37.3	34.0	10.1	30.8
Straw for bedding	6.5	2.2	19.0	7.4	1.8	15.4
Silage	51.7	13.1	82.0	36.3	7.7	48.0
Liveweight gain	5.5	4.0	8.7	5.8	4.2	9.1
Volatilisation	9.3	–	–	9.9	–	–
Total	**110.4**	**30.6**	**147.0**	**95.4**	**23.7**	**103.3**
Balance	**+27.0**	**–16.2**	**–63.9**	**+34.4**	**–7.9**	**–15.4**

4.3.2 Impact of Nutrient Losses

Data such as those in Table. 4.3 and 4.4 present quite comprehensive budgets including inputs from atmospheric deposition, seed, feed and imported fertilisers, and manures and outputs in saleable produce, but with little information on losses. Measuring full nutrient budgets, including losses and internal flows, is essential for a proper measure of farm sustainability. However, as the complexity of the approach increases there is a need to measure or estimate increasing numbers of variables, and either costs or errors increase. Table 4.5 shows full nitrogen budgets for three organic farms in which an earlier, simple farm gate N budget has been modified by including losses (Goulding et al., 2000). The full budgets, including losses, are very different from the original simple budgets and show that the upland/hill farm (i.e. an extensive mixed farm on poor soil 250 m above sea level) has a small N deficit, the lowland dairy farm is in balance, and the stockless system still has a small N surplus. Clearly nutrient budgets that lack any consideration of losses, especially for N, are likely to provide misleading conclusions on the sustainability of the systems.

Table 4.5 Nitrogen budgets for three organic farms including estimated losses (Goulding et al., 2000)

Farm type	Upland/hill farm	Lowland dairy	Stockless arable
Loss process		N loss (kg ha^{-1} yr^{-1})	
Leaching	5	50	50
Ammonia volatilisation	25	50	0
Denitrification	5	20	20
Total loss	35	120	70
Previous N budget	+18	+122	+96
New N budget	−17	+2	+26

4.4 Discussion

4.4.1 Long-Term Trends

Care must be taken when reaching conclusions about the sustainability of farming from single year farm-gate nutrient budgets. Applications of P and K are often made during the ley phase of an organic rotation to supply the whole multi-year rotation. The weather also has an important effect through its impact on losses. It is likely that there would be a net immobilization of N in soil organic matter in dry years and net mineralisation in wet years. Thus annual deficits and surpluses may be temporary and balanced by corresponding surpluses and deficits in other years. However, where budgets are the averages of 4–5 years (i.e. a rotation), as in Table. 4.3 and 4.4, they can be viewed more confidently and show deficits for N, P and K on the organic farms investigated.

4.4.2 Nutrient Cycling at the Wider Scale

Organic produce is sold into the national and international markets that dominate food production, leading to a net export of plant nutrients from most organic farms. Some organic farms appear to remain sustainable only through the use of feed, bedding and composts brought onto the farm and, in some cases, air pollution (Goulding et al., 2000; Watson et al., 2002). The wider aspects of nutrient cycling urgently need attention, with critical questions such as the sources of nutrients brought onto organic farms: if they originate from conventional farms, are these organic farms sustainable?

At a wider scale there are issues concerning the import of nutrients in fertilising materials over long-distances. The concept of food miles, that is the distance travelled by food products between production and consumption, is now in common use (Paxton, 1994). There is a similar issue in relation to sustainability in terms of 'resource miles' or the distance travelled and energy costs associated with freight of products such as rock phosphate being brought from as far away as Tunisia and Morocco to the UK for use on organic farms. In the long-term, organic farming standards may need to consider more fully the use of locally sourced by-products and waste materials for nutrient supply. The ideal of sustainable organic farms that do not consume non-renewable resources or pollute the environment will remain elusive while the global market economy dominates agriculture.

4.5 Conclusions

The data presented here suggest some cause for concern over the sustainability of organic systems because of their dependence on feedstuffs and bedding for inputs of P and K, and on the very variable fixation by legumes or imports of manure or compost for N. Air pollution and net mineralisation from soil reserves appear to comprise a large part of the N supply on some organic farms, and there is some evidence pointing to P and K depletion in soils.

Although nutrient management in organically managed soils is fundamentally different to soils managed conventionally, the underlying processes supporting soil fertility are not. It is farm management decisions about the timing of cultivation and the application of inputs that will have the largest affect on nutrient availability and the potential for losses, not the farming systems per se, so research on nutrient cycling in both organic and conventional systems requires similar approaches and scientific rigor. In addition, management determines the *risk* of loss to the environment but the weather generally determines the actual loss. Even with the best management possible, the weather can negate even the best practices. With these difficulties and uncertainties, questions need to be asked about whether it is generally easier to manage organic or inorganic nutrient sources? Where is the greatest risk?

Organic farming systems have the potential to deliver significant environmental benefits, but so do other more extensive systems. The effect of changing to organic farming will depend on specific cropping and management practices on each farm. It is important that we study and improve nutrient management on *all* farm systems and in the context of plant, animal and human health in order to develop more sustainable farming systems.

Acknowledgments Rothamsted Research receives grant-in-aid from the UK Biotechnology and Biological Sciences Research Council. Organic farming research at SAC is currently sponsored by SEERAD, Defra and Scottish Natural Heritage. The UK Department for the Environment, Food and Rural Affairs funded some of the research reported here.

References

Berry, P.M., Stockdale, E.A., Sylvester-Bradley, R., Philipps, L., Smith, K.A., Lord, E.I., Watson, C.A., and Fortune, S., 2002, N, P and K budgets for crop rotations on nine organic farms in the UK, *Soil Use Manage.* **18**: 248–255.

Cooke, G.W., 1967, *The Control of Soil Fertility*, Crosby-Lockwood, London, 526p.

Cormack, W.F., 1999, Testing a stockless arable organic rotation on a fertile soil, in: *Designing and Testing Crop Rotations for Organic Farming*, J.E. Olesen, R. Eltun, M.J. Gooding, E.S. Jensen, and U. Kopke, eds., DARCOF, Denmark, pp. 115–124.

Fortune, S., Hollies, J., and Stockdale, E.A, 2004, Effect of different potassium fertilizers suitable for organic farming systems on grass/clover yields and nutrient offtakes and interactions with nitrogen supply, *Soil Use Manage.* **20**: 403–409.

Fowler, S.M., Watson, C.A., and Wilman, D., 1993, N, P and K on organic farms: herbage and cereal production, purchases and sales, *J. Agric. Sci. (Camb.)* **120**: 353–360.

Gerritse, R.G., and Vriesema, R., 1984, Phosphate distribution in animal waste slurries, *J. Agric. Sci. (Camb.)* **102**: 159–161.

Goss, M.J., and Goorahoo, D., 1995, Nitrate contamination of groundwater: measurement and prediction, *Fertil. Res.* **42**: 331–338.

Goulding, K.W.T., Stockdale, E.A., Fortune, S., and Watson, C., 2000, Nutrient cycling on organic forms, *J. R. Agric. Soc. Eng.* **161**: 65–75.

Halberg, N., Kristensen, E.S., and Kristensen, I.S., 1995, Nitrogen turnover on organic and conventional mixed farms, *J. Agric. Environ. Ethic.* **8**: 30–51.

Haraldsen, T.K., Asdal, A., Grasdalen, C., Nesheim, L., and Ugland, T.N., 2000, Nutrient balances and yields during conversion from conventional to organic cropping systems on silt loam and clay soils in Norway, *Biol. Agric. Hortic.* **17**: 229–246.

Høgh-Jensen, H., Loges, R., Jørgensen, F.V., Vinther, F.P., and Jensen, E.S., 2004, An empirical model for quantification of symbiotic nitrogen fixation in grass-clover mixtures, *Agric. Syst.* **82**: 181–194.

Howard, A., 1943, *An Agricultural Testament*, Oxford University Press, Oxford.

IFOAM (International Federation of Organic Agriculture Movements), 1998, *IFOAM Basic Standards for Production and Processing*, IFOAM Publications, Germany.

IFOAM (International Federation of Organic Agriculture Movements), 2006, http://www.ifoam.org/about_ifoam/principles/. Assessed October 2006.

Johnston, A.E., 1991, Potential changes in soil fertility form arable farming systems including organic systems, *Proc. Fertil. Soc. Lond.* **306**: 38.

Johnston, A.E., Goulding, K.W.T., Poulton, P.R., and Chalmers, A.G., 2001, Reducing fertiliser inputs: endangering arable soil fertility? *Proc. Int. Fertil. Soc.* **487**: 44.

Leach, K.A., Allingham, K.D., Conway, J.S., Goulding, K.W.T., and Hatch, D.J., 2005, Nitrogen management for profitable farming with minimal environmental impact: the challenge for mixed farms in the Cotswold Hills, England, *Int. J. Agric. Sust.* **2**: 21–32.

Ledgard, S.F., and Steele, K.W., 1992, Biological nitrogen fixation in mixed legume/grass pastures, *Plant Soil* **141**: 137–153.
Lockeretz, W., and Anderson, M.D., 1993, *Agricultural Research Alternatives*, University of Nebraska Press, Lincoln.
Lord, E., Stopes, C., and Philipps, L., 1997, *Assessment of Relative Nitrate Losses from Organic and Conventional Farming Systems, based on Recent Measurements. Final Report to MAFF*, Contract OFO 141, ADAS, Wolverhampton, UK.
Mäder, P., Fliessbach, A., Dubois, D., Gunst, L., Fried, P., and Niggli, U., 2002, Soil fertility and biodiversity in organic farming, *Science* **296**: 1694–1697.
MAFF (Ministry of Agriculture, Fisheries and Food), 1999, *Review of the Organic Aid Scheme*, MAFF, London, UK.
Nolte, C., and Werner, W., 1994, Investigations on the nutrient cycle and its components of a biodynamically-managed farm, *Biol. Agric. Hortic.* **10**: 235–254.
Olesen, J.E., Rasmussen, I.A., Askegaard, M., and Kristensen, K., 1999, Design of the Danish crop rotation experiment, in: *Designing and Testing Crop Rotations for Organic Farming*, J.E. Olesen, R. Eltun, M.J. Gooding, E.S. Jensen, and U. Kopke, eds., DARCOF, Denmark, pp. 49–62.
Paxton, A., 1994, *The Food Miles Report: The Dangers of Long Distance Food Transport*, SAFE Alliance, London, UK.
Peperzak, P., Caldwell, A.G., Hunziker, R.R., and Black, C.A., 1959, Phosphorus fractions in manures, *Soil Sci.* **87**: 293–302.
Philipps, L., and Stopes, C.E., 1995, Organic rotations and nitrate leaching in the UK, *Biol. Agric. Hortic.* **11**: 123–134.
Rajan, S.S.S., Watkinson, J.H., and Sinclair, A.G., 1996, Phosphate rocks for dirt application to soils, *Adv. Agron.* **57**: 77–159.
RASE (Royal Agricultural Society of England), 2000, *Shades of Green. A Review of UK Farming Systems*, RASE, Stoneleigh, UK.
Rowe, J.S., 1997, From reductionism to holism in ecology and deep ecology, *Ecology* **27**: 147–151.
Sato, S., Solomon, D., Hyland, C., Ketterings, Q.M., and Lehmann, J., 2005, Phosphorus speciation in manure and manure amended soils using XANES spectroscopy, *Environ. Sci. Technol.* **39**: 7485–7491.
Scoones, I., and Toulmin, C., 1998, Soil nutrient balances: what use for policy? *Agric. Ecosyst. Environ.* **71**: 255–267.
Silgram, M., and Shepherd, M.A., 1999, The effect of cultivation on soil nitrogen mineralisation, *Adv. Agron.* **65**: 267–311.
Soil Association, 2004, *Organic Food and Farming Report 2004*, Soil Association, Bristol, UK.
Soil Use and Management, 2002, *Soil Fertility in Organically Managed Soils*, Vol. 18(Suppl.), pp. 238–308.
Stockdale, E.A., Shepherd, M.A., Fortune, S., and Cuttle, S.P., 2002, Soil fertility in organic farming systems – fundamentally different? *Soil Use Manage.* **18**: 301–308.
Topp, C.F.E., Watson, C.A., Rees, R.M., and Sanders, I., 2005, *The Prediction of Biological Nitrogen Fixation*. XX International Grassland Conference: Offered Papers, Proceedings of the IGC, Dublin, Wageningen Academic Publishers, The Netherlands, 877p.
UKROFS, 1999, *Standards for Organic Food Production*, United Kingdom Register of Organic Food Standard, London, UK.
Watson, C.A., and Younie, D., 1995, Nitrogen balances in organically and conventionally managed beef production systems, in: *Grassland into the 21st Century: Challenges and Opportunities*, G.E Pollott, ed., *Proc. Br. Grass. Soc.* **1665**: 197–199.
Watson, C.A., Fowler, S.M., and Wilman, D., 1993, Soil inorganic-N and nitrate leaching on organic farms, *J. Agric. Sci. (Camb.)* **120**: 361–169.
Watson, C.A., and Atkinson, D., 1999, Using nitrogen budgets to indicate nitrogen use efficiency and losses from whole farm systems: a comparison of three methodological approaches, *Nutr. Cycl. Agroecosyst.* **53**: 259–267.

Watson, C.A., Younie, D., and Armstrong, G., 1999, Designing crop rotations for organic farming: importance of the ley-arable balance, in: *Designing and Testing Crop Rotations for Organic Farming*, J.E. Olesen, R. Eltun, M.J Gooding, E.S. Jensen, and U. Kopke, eds., DARCOF, Denmark, pp. 91–98.

Watson, C.A., Bengtsson, H., Løes, A-K., Myrbeck, A., Salomon, E., Schroder, J., and Stockdale, E.A., 2002, A review of farm-scale nutrient budgets for organic farms in temperate regions, *Soil Use Manage*. **18**: 264–273.

Willer, H., and Yussefi, M., 2005, *The World of Organic Agriculture. Statistics and Emerging Trends 2005*, IFOAM, Bonn, Germany.

Woodward, L., 2002, *Science and Research in Organic Farming*, EFRC Pamphlet series, Policy and Research Department, Elm Farm Research Centre, UK.

Chapter 5
Nutrient Supply in Organic Agriculture – Plant Availability, Sources and Recycling

Holger Kirchmann, Thomas Kätterer and Lars Bergström

Abstract This chapter examines the practice of applying nutrients in organic or slowly soluble inorganic form in the belief that plants will obtain balanced nutrition through the actions of soil microbes. The organic principle of only fertilising the soil and not directly feeding the crop with water-soluble nutrients has no support in science. The release of organically bound nutrients in soil through biological activity is not necessarily synchronised with crop demands and occurs even at times when there is no crop growth. Changes in the soil biological community do not overcome this limitation. Despite the ideal of organic agriculture being self-sustaining through cycling of nutrients, in principle only on-farm wastes are recycled and most municipal wastes are excluded due to concerns about pollutants and philosophical views on life (biodynamic agriculture). Nutrient supply in European organic agriculture is mainly covered through purchase of straw, manure and fodder from conventional agriculture and by-products from the food industry. Untreated minerals seem to play a minor role. The fertility of agricultural soils can only be maintained over the long-term if plant nutrients removed are replaced with equivalent amounts and if added sources have a higher solubility than those present in the soil. These conditions are in most cases not fulfilled in organic agriculture. It can thus be concluded that the naturalness of nutrient sources is no guarantee of superior quality and that promotion of organic principles does not improve the supply and recycling of nutrients but excludes other more effective solutions for nutrient use in agricultural systems.

Keywords Nitrogen input · Soil fertility · Crop yield · Waste recycling · Rock phosphate · Nitrate leaching · Nutrient recycling · Organic manures · Fertilisers

---

H. Kirchmann (✉)
Department of Soil and Environment, Swedish University of Agricultural Sciences, Box 7014, SE-750 07 Uppsala, Sweden
e-mail: holger.kirchmann@mark.slu.se

H. Kirchmann, L. Bergström (eds.), *Organic Crop Production – Ambitions and Limitations*, DOI 10.1007/978-1-4020-9316-6_5,

5.1 Introduction

Use of plant nutrients in organic agriculture is viewed differently than in conventional agriculture. Nutrients are supplied in organic forms or as untreated minerals with low solubility in the belief that plants will obtain balanced nutrition through the actions of roots and soil microbes and through weathering of minerals. The premise of organic agriculture is to supply the soil with nutrients but not directly feed the plants with soluble nutrients. Organic agriculture was founded on the dogma, as opposed to the hypothesis, that the use of organically bound nutrients and untreated minerals is superior to the use of artificial fertilisers, which are classified as being unnatural. Therefore, all fertilisation practices are designed and characterised according to this dogma (Watson et al., 2002a). Over time, this has led to the common opinion that the exclusion of soluble inorganic fertilisers contributes towards increased soil fertility and conservation of resources, and is a sound form of nutrient management. However, the most central question is whether the exclusion of artificial fertilisers can be justified scientifically. Is the naturalness of organic manures and untreated minerals associated with superior characteristics and functioning? Another idea central in organic agriculture is to view farms as sustainable units based on nutrient recycling and use of local resources.

In this chapter, we study the literature to determine whether and how the data reject or strengthen the arguments for abolishing soluble inorganic fertilisers in organic crop production. The question of self-sufficiency of farms is also examined. As organic principles have been adopted in many types of agricultural systems with vastly different soils and as social and political pressure increases for an even more widespread adoption of organic agriculture across the globe, the exclusion of inorganic fertilisers requires an in-depth examination.

Central aspects reviewed in this chapter are:

- The basis for exclusion of soluble inorganic fertilisers;
- Problems relating to reliance on organic nutrient sources and untreated minerals;
- The potential to recycle nutrients and thus the sustainability of organic production systems.

5.2 The Organic Principle – To Fertilise the Soil and Not Feed the Crop with Artificial Fertilisers

5.2.1 The Humus Theory and Plant Nutrients

The founders of organic agriculture regarded a living soil and the release of nutrients through soil biological activity as the proper way to supply crop demands. Addition of artificial, water-soluble fertilisers was deemed an unnatural way of plant nutrition. During the search for plant nutrients through history from Aristotle (384–322 BC) to Thaer (1752–1828 AD), the prevailing view was that organic manures such as animal wastes, composts, etc. increase soil fertility and that soil humus is the source of plant growth. Thaer (1837–1839) viewed humus as the residues of animal and plant putrefaction, a black body that formed the source for plant dry matter (Feller et al., 2003). Later, von Liebig (1840) added new, substantiated knowledge to this standpoint by showing that plants take up nutrients in the form of dissolved salts. von Liebig introduced the 'Law of the Minimum', meaning that crops require a minimum of mineral substances for growth. von Liebig (1840) wrote that 'even to great leaders in plant physiology, carbonic acid, ammonia, acids and bases are sounds without meaning, words without sense, terms of an unknown language, which awake no thoughts and no associations'. von Liebig proved that humus is not taken up by crops as such but is a source of plant nutrients released in water-soluble form. He provided the missing knowledge for understanding how humus acts in soil, and thereby corrected and complemented the humus theory with a mechanistic explanation. He showed that independent of origin, a number of essential elements dissolved in the soil solution act as plant nutrients, and this concept forms the basis for modern plant nutrition.

Relating the findings of von Liebig to the organic principle of excluding artificial, water-soluble fertilisers raises some central questions. Is addition of completely water-soluble nutrient sources to plants an unnatural practice? Are artificial fertilisers the only compounds containing water-soluble nutrients or are the organic materials used in organic agriculture also a significant source? Are the amounts of water-soluble nutrients added through artificial and organic fertilisers highly different? Is nutrient addition through artificial fertilisers poorly adjusted to the uptake dynamics of crops?

5.2.2 *Water-Solubility of Nutrients Supplied with Organic Fertilisers and Untreated Minerals*

Artificial fertilisers are excluded in organic agriculture, primarily due their function of directly feeding the crop, which is the result of their high or complete solubility in water. However, a closer examination of natural products used in organic agriculture reveals that these products can also be highly water-soluble (see Table 5.1). For example, urine contains more than 94% of its nitrogen as soluble ammonium/ammonia (e.g. Kirchmann and Pettersson, 1995) and animal slurries contain 50–70% of total N as ammonium (e.g. Bernal et al., 1993). Ulén (1984) found that up to one-third of the P content from clover/grass leys was released upon freezing. The vacuole of each mature plant cell contains most of the plant potassium, phosphate, calcium and magnesium, which are released upon freezing or mechanical destruction of the plant. Ploughing under of green manure crops also provides substantial amounts of soluble minerals to the soil. The release of P and K from fresh, unground barley straw has been shown to amount to 60 and 90%, respectively, of its total content upon cold-water leaching (Christensen, 1985), with inorganic ions being the main form. Moreover, a number of natural minerals approved for organic agriculture are highly water-soluble, e.g. halite (NaCl), kieserite ($MgSO_4 \times 2\ H_2O$), crude potassium salts

Table 5.1 Water-solubility of organic materials, wastes and minerals approved for organic agriculture

Type of material and nutrient	Water-soluble portion (%)	Reference
Organic wastes		
Animal slurry–N	50–70	Bernal et al. (1993)
Animal dung-N (anaerobic storage)	51–75	Kirchmann and Witter (1992)
Urine–N	94	Kirchmann and Pettersson (1995)
Green manures and crop residues		
Clover/grass foliage-P	11–33	Ulén (1984)
White clover foliage-N	36–41	Kirchmann and Bergqvist (1989)
Potato haulm-N	35	Henriksen and Breland (1999)
Barley straw-N	33–58	Reinertsen et al. (1984)
Barley straw-P	60	Christensen (1985)
Barley straw-K	90	Christensen (1985)
Industrial wastes		
Vinasse-K	100	PDA (2008)
Untreated minerals		
Kieserite–Mg and S	100	Härdter et al. (2004)
Kainit-K	100	Ullmann's Agrochemicals (2007)
Halite-Cl	100	Lide (1999)
Copper sulphate-Cu[a]	100	Lide (1999)

[a]Although the EU theoretically banned copper sulphate in 2002, the compound can still be used as a fungicide in organic agriculture as no alternative has been presented.

e.g. kainit ($KMg(ClSO_4)_4 \times 11\ H_2O$), and ash containing e.g. potash (K_2CO_3). In contrast, even artificial fertilisers may not be completely water-soluble, for example di-ammonium phosphate of fertiliser grade quality contains 5–20% water-insoluble phosphates (Gilkes and Mangano, 1983; Nielsson, 1987).

Concerning the amount of soluble nutrients added, application of 30 Mg cattle slurry containing 2.0 kg $NH_4{-}N\ Mg^{-1}$ provides, for example, 60 kg N ha^{-1}, while 40 Mg urine containing 1.7 kg $NH_4{-}N\ Mg^{-1}$ adds 68 kg N ha^{-1}, which is close to the amounts usually applied with inorganic fertiliser. In addition, urine does not add any significant amount of organic material as it is a solution of excreted salts.

Soluble ions are provided to the soil not only through application of agricultural fertilisers, but also as atmospheric deposition of different compounds. For example, deposition of sea salt spray is a significant source of fully soluble sodium, chloride, calcium, etc. Annual fluxes of marine aerosols through deposition can amount to 51 kg of chloride and 25 kg of sodium per hectare in southern Sweden (Hultberg and Grennfeldt, 1992). Ammonia, which is mainly emitted from livestock (ECETOC, 1994) and which is highly water-soluble, is deposited both close to and far from the source and can add tens of kilograms of fully soluble nitrogen to soil (Ferm, 1998). However, the most obvious example of high natural salt concentrations in soil is through high evaporation under arid climatic conditions, where soluble ions from deeper layers are transported through mass flow to the soil surface, where they form saline, saline-sodic or sodic soils in which high salt concentrations actually stress crop growth (Brady and Weil, 2008). The high calcium concentrations in the soil solution of calcareous soils, in excess of crop demand, show that large amounts of soluble ions in the soil solution are in no way an abnormal condition for plants.

5.2.3 Dynamics of Nutrient Release from Organic Matter in Soil

Inorganic ions present in or released from organic fertilisers are identical to ions released from artificial fertilisers. Nutrient uptake by crops is mainly through inorganic ions, with organic nutrient uptake being of minor importance (see Chapter 10 of this book; Ryan and Tibbett, 2008). As plant roots do not distinguish between solutes due to origin, except for possible discrimination against heavier isotopes, ions derived from artificial fertilisers or natural materials are involved in the same processes in the soil and in the crop. In other words, a molecule, for example, ammonium in slurry or in nitrogen fertiliser, undergoes the same reactions in soil since the chemical properties of a molecule are not affected by its origin. However, despite identical characteristics of soluble nutrients derived from organic manures, untreated minerals or artificial fertilisers, a sophisticated argument against artificial fertilisers has been presented by the founder of organic-biological agriculture (Rusch, 1978). Rusch pointed out that the dynamic release of nutrients from soil organic matter and the availability of nutrients over time is the main difference. Artificial fertilisers cause a high initial nutrient concentration in soil solution upon addition, whereas Rusch (1978) assumed that there is synchrony regarding the release of nutrients from soil organic matter and the demand of growing crops. Addition of artificial, soluble salts is regarded as by-passing nutrient release from the soil and is thus considered an unnatural form of supply. This argument sounds convincing but has no basis in science, as revealed below.

Whereas in natural ecosystems, for example in forests, nutrient release from soil organic matter and uptake by trees can take place all year round due to the permanent vegetation and a living root system active throughout the seasons, the situation is very different in arable systems. Annual agricultural crops grow only during a couple of months, during which period they require a high rate of nutrient supply to produce a large biomass. Living roots only exist during this growth period.

Consider an example from a cold-temperate zone in central Sweden, Uppsala. To grow spring barley in this region takes less than four months. A well-grown crop takes up about 150 kg of nitrogen per hectare and about 50% of this amount must be available to the crop during a three-week period (Fig. 5.1). However, the most fertile topsoils in this region deliver less than 1 kg N ha^{-1} per day^{-1} during the cropping period and, on average, only 70 kg N ha^{-1} over the whole year. As can be seen in Fig. 5.1, annual rates of N release also vary greatly between years. Total amounts released were about twice as high in 1961 as in 1976, i.e. 92 and 47 kg N ha^{-1}, respectively, but during both these years about 50% of the mineralisation occurred outside the barley growing season. The calculations of daily rates of nitrogen release are based on a climate index (Andrén et al., 2007) using mean climatic data for the

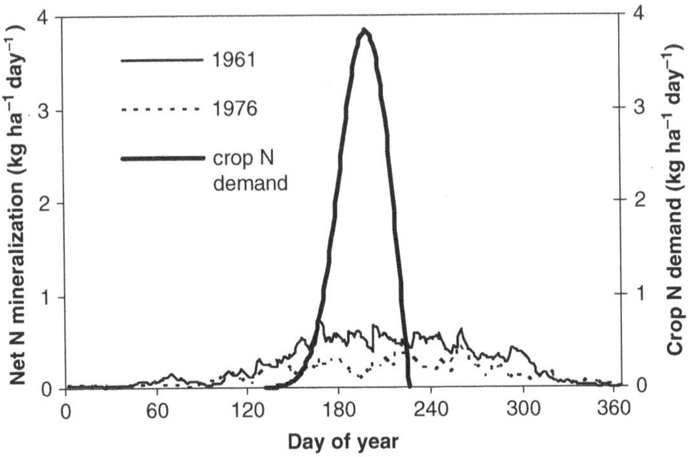

Fig. 5.1 Amounts of nitrogen mineralised from soil organic matter during two extreme years (1961 and 1976) and typical daily N demand of a barley crop for optimum growth. Based on data from a clay soil (0–30 cm depth) in Uppsala, Sweden

period 1956–1999 and transfer functions (Kätterer et al., 2006; Kätterer and Andrén, 2008), and assuming that most of the organic matter in soil is stabilised (Kätterer and Andrén, 2001).

In the example above, the barley crop will be able to take up 10–20 kg of N ha^{-1} accumulated before sowing plus a similar amount of N derived from mineralisation during crop growth. In total, crop uptake of 40–50 kg N ha^{-1} results in poor growth without any further addition of plant-available nitrogen. At least 30% of the total mineralisation per year takes place after crop harvest. Moreover, meeting crop demand through fertilisation with organic manures is no guarantee of sufficient crop supply. Again, climate controls the decomposition of organic components and although manures stimulate biological activity in soil, the course of nutrient release is even less predictable. Organic manures may mineralise or immobilise N depending on the chemical properties of the material, particle size, spatial distribution in soil and time of application. The possibility of matching N release from organic manures with crop demand is therefore quite difficult. Again, N is mineralised when no crop is present and can easily be lost from the soil as gases or through leaching during the cold and wet autumn (see Chapter 7 of this book; Bergström et al., 2008).

It can be argued that cold-temperate climatic conditions have no representativeness and therefore limited relevance as proof of evidence. In fact, analysis of biological activity in soils in different climatic zones in Africa (Andrén et al., 2007) has shown that there is better synchrony between release and uptake of nutrients under warm climatic regimes. Crops are sown at the beginning of a rainy period and nutrients are only released during crop growth. However, synchrony between nutrient release and crop demand under certain climatic regimes does not mean that sufficient nutrients are released. Crops may still suffer from undernourishment or imbalances when only supplied with nutrients from soil biological actions.

Our conclusion is therefore that the argument by Rusch (1978) against inorganic fertilisers is not corroborated by scientific evidence, as only water-soluble nutrient sources can supply crops sufficiently and in synchrony with their demand.

5.2.4 Can Enhancing the Soil Biological Community Improve Plant Nutrient Availability in Organic Systems?

It is often assumed that the soil biological community is enhanced by organic management, developing a greater capacity to supply plants with nutrients from poorly soluble inorganic and organic sources. However, as indicated in Fig. 5.1, the main factors driving decomposer activity in soil are moisture and temperature. Only a larger input of organic material stimulates the biological community and leads to increased nutrient availability. Consequently, if the input of organic material is not larger in organic systems than in their conventional counterparts, or if organic yields are lower than in conventional production, the soil biological community and its activity will not be enhanced relative to conventional systems (Ryan, 1999).

Soil organic matter content has variously been reported to be higher in organic systems (Reganold, 1988; Wander et al., 1994; Liebig and Doran, 1999), lower in organic systems (Lützow and Ottow, 1994; Petersen et al., 1997), or unchanged (Derrick and Dumaresq, 1999; Burkitt et al., 2007) compared with conventional systems. This variety of results reflects different crop sequences and/or addition of different amounts and types of organic inputs and tillage operations (Robertson and Morgan, 1996). Higher addition of organic matter in either system is naturally followed by a larger microbial biomass (Gunapala et al., 1998; Fliessbach and Mäder, 2000). There is no evidence that a larger microbial biomass changes basic relationships in soils (Kirchmann et al., 2004), such as those between concentrations of soil available nutrients and plant nutrient uptake and growth (Ryan, 1999; Ryan and Ash, 1999; Ryan et al., 2000). However, addition of organic matter can in some instances suppress pathogenic organisms through enhancing the presence of groups of antagonistic soil organisms (Sivapalan et al., 1993; Workneh and van Bruggen, 1994).

Overall, our conclusion is that the general claim that organic practices automatically stimulate an enlarged soil biological community and that this can partly substitute for inorganic fertilisers is inaccurate.

5.2.5 Different Roles of Organic and Inorganic Fertilisers

We want to stress that organic and inorganic fertilisers have different functions in soil and complement each other. While input of organic manures contributes to the build-up of soil organic matter, increases the cation exchange capacity, supports soil structure, helps to chelate micronutrients, increases soil moisture retention, etc., inorganic fertilisers supply crops with nutrients at times when their demand is large.

All functions are of importance but only in conventional agriculture is both management of soil fertility and crop supply with plant-available fertilisers an explicit strategy. Focusing solely on soil fertility management and prohibiting the use of artificial fertilisers means setting aside the demands of crops. As a result, crops in organic agriculture are often grown in nutrient-deficient conditions far below their production potential (see Chapter 3 of this book; Kirchmann et al., 2008a, b). From the view of crops, use of artificial fertilisers does not mean by-passing natural processes but complementing nutrient release to fulfil crop demand.

Reliance on untreated minerals and organic fertilisers on organic farms often results in poor use efficiency of the nutrient source and lower yields than can be achieved with artificial fertilisers, particularly in regions where native soil fertility is low. For instance, field experiments with approved organic fertilisers in Europe showed that meat-bone meal and chicken manure increased grain yields only moderately (by 600–1500 kg ha^{-1}) compared with an unfertilised control at application rates of 40–120 kg N ha^{-1} (Lundström and Lindén, 2001). Crop utilisation of N was only 30% compared with 60–80% for inorganic fertilisers (Mattsson and Kjellquist, 1992). In a study of spring wheat fertilised with meat-bone meal, N-utilisation was only 13% (Wivstad et al., 1996).

5.3 Nutrient Supply in Organic Systems

5.3.1 Are Organic Farms Self-sustaining?

There is a widespread belief, as proposed by Steiner (1924), that self-sustaining farms are the real core of sound agricultural production. The need to import nutrients to a farm is considered a sign of failure of the system (Steiner, 1924). Traditional agricultural systems in Europe and elsewhere are often held up as the ideal in this context. The view that optimal measures to maintain nutrient levels in agricultural soils involve a high degree of on-farm recycling of nutrients, with any small losses balanced by soil weathering, is common (IFOAM, 2006).

However, despite internal cycling of nutrients on farms, any agricultural production brings about unavoidable nutrient losses through leaching, runoff, gaseous emissions, etc., but the largest removal of plant nutrients from the farm is through export and sale of harvested crops and animal products. The farm is an open system and even efficient internal recycling is not sufficient to balance nutrient budgets (see Table 5.2). Over the long-term, any agricultural system will become depleted in nutrients if lost or sold nutrients are not replaced. The view of a farm as a self-sustaining unit (Steiner, 1924) is in contradiction to the 'law of nutrient replacement', where nutrient removal must be restored to maintain soil fertility and avoid nutrient mining of soils. The idea of organic agriculture having a closed nutrient cycle on farms is not based on reality.

Examples of nearly self-sustaining farms or agricultural systems are rare, unless there is regular addition of nutrients from an external source. Situations where this

Table 5.2 Farm-gate and soil balances for P and K in organically managed systems (n = 37)

System, farm and country	Farm-gate balance[a] ($kg\ ha^{-1}\ yr^{-1}$)		Type of purchase of nutrients	Reference
	P	K		
With animals				
Talhof, Germany	−2	+1	Straw, fishmeal, feed grains, seaweed	Kaffka and Koepf (1989)
Brynllys, U.K.	+2	+6	Straw, silage, hay, concentrated feedstuff	Fowler et al. (1993)
Lea Hall, U.K.	+21	+53	Straw, poultry manure, bulk feed, concentrated feedstuff	Fowler et al. (1993)
Boschheidehof, Germany	−3	−65	Feedstuff, kieserite	Nolte and Werner (1994)
Kowai site, New Zealand	−7	n.a.	None	Nguyen et al. (1995)
Temuka site, New Zealand	+4	n.a.	Phosphate rock + elemental S	Nguyen et al. (1995)
Templeton site, New Zealand	−4	n.a.	Fish manure, phosphate rock + elemental S	Nguyen et al. (1995)
Kirchweger farm, Austria	+2	+12	Straw, concentrated feedstuff, mineral feed	Wieser et al. (1996)
Farm A, Austria	+1	+1	Straw, concentrated feedstuff, mineral feed	Wieser et al. (1996)
Farm B, Austria	−1.5	0	Straw, concentrated feedstuff, mineral feed	Wieser et al. (1996)
Farm C, Austria	−1	+19	Straw, concentrated feedstuff, mineral feed	Wieser et al. (1996)
Farm D, Austria	+3	0	Straw, concentrated feedstuff, mineral feed	Wieser et al. (1996)

Table 5.2 (continued)

System, farm and country	Farm-gate balance[a] (kg ha^{-1} yr^{-1})		Type of purchase of nutrients	Reference
	P	K		
Farm E, Austria	−1	+1	Straw, concentrated feedstuff, mineral feed	Wieser et al. (1996)
Farm F, Austria	+2	+3	Straw, concentrated feedstuff, mineral feed	Wieser et al. (1996)
Farm G, Austria	−2	−1	Straw, concentrated feedstuff, mineral feed	Wieser et al. (1996)
Farm H, Austria	+6	+2	Straw, concentrated feedstuff, mineral feed	Wieser et al. (1996)
Öjebyn, Sweden	+3	0	Concentrated feedstuff including minerals	Fagerberg et al. (1996)
Skilleby, Sweden	−3	−6	Feedstuff	Granstedt (2000)
Öjebyn, Sweden	+1	+59	Straw, feedstuff	Gustafson et al. (2003)
10 Biodynamic farms, Australia	−17	n.a.	Phosphate rock	Burkitt et al. (2007)
Without animals	**Soil balance**[b]			
Three Skåne trials, Sweden	−10	−28	None	Ivarson and Gunnarsson (2001)
Apelsvoll site, Norway	−14	−20	None	Eltun et al. (2002)
Farm no. 5, U.K.	−8	−21	None	Berry et al. (2003)
Farm no. 8, UK	−1	−52	Phosphate rock	Berry et al. (2003)
Mellby site, Sweden	−6	−2	Potash	Torstensson et al. (2006)
Lanna site, Sweden	−7	−15	None	Aronsson et al. (2007)

[a]The farm-gate balance does not necessarily take into account losses from the soil, which further decreases the figures given.
[b]Soil balances sometimes include losses through leaching and runoff but not necessarily.

occurs without human intervention, for example silts deposited in annual floods, appear rare (Newman, 1997) and are certainly not common enough to produce adequate food for an increasing world population. Agricultural history in Europe also tells an instructive story of how former agricultural practices in northern Europe aimed to counteract depletion of nutrient contents in cropped soils through labour-intensive removal and transport of organic matter from adjacent ecosystems to these soils. Surface layers and litter from natural ecosystems (heathlands, wet grasslands, peat, meadows) containing plant nutrients and enriched in organic material were transferred to arable soils. The mechanism behind the build-up and maintenance of Plaggen (organic matter enriched) soils was the transfer of plant nutrients through soil and litter from natural ecosystems to arable soils for several hundred years (Pape, 1970; Pott, 1990). Despite this enormous transfer, organic matter and nutrient status in arable soils remained low (Springob and Kirchmann, 2002), whereas natural ecosystems were depleted in nutrients. Thus, nutrient depletion of natural ecosystems was necessary to compensate for losses and removal from agricultural soils in order to maintain yields. In other words, already in former times, removal and losses of nutrients were compensated for by equivalent input to avoid depletion of arable soils.

5.3.2 Nutrient Balances of Organic Farms

We examined P and K budgets of organically managed farms to evaluate whether the law of nutrient replacement is being followed. Nitrogen was excluded from the evaluation as nitrogen budgets are dependent on a number of factors, e.g. whether multi-year or single legume crops are grown in the rotation, etc.

This review of nutrient balances (see Table 5.2) revealed that purchased feedstuffs and straw played a key role as nutrient sources for mixed organic systems. Out of 19 mixed farms, 17 imported feedstuffs and 13 imported straw. On average, organically managed farms with animals had a slight surplus of $+1$ kg P and $+5$ kg K ha^{-1} yr^{-1}, whereas farms without animals had negative budgets amounting to -7 kg P and -22 kg K ha^{-1} yr^{-1}, on average, which means that the import was not sufficient to balance removal.

Organically managed farms without animals and without any nutrient import resulted in even larger deficits of P amounting to -10 kg P ha^{-1} yr^{-1}. These deficits are in the same order of magnitude as amounts removed from the farm by e.g. a well-grown organic barley crop (3.5 Mg ha^{-1}) of kg 10 P and 15 kg K ha^{-1} yr^{-1} in the form of grain.

These data clearly show that the strategy of being independent of nutrient purchase according to the ideal of a self-sustaining system has no scientific support. Organic systems are not sustainable without purchase of nutrients, and use of off-farm products is necessary to counteract depletion and keep deficits of P and K to a minimum. However, despite removed nutrients being replaced through purchase, there is concern that the level of available plant nutrients in soil may decrease. Nutrients become less plant-available in organically managed soils over time, a subject which is addressed below.

5.3.3 Conventional Agriculture as a Nutrient Supplier of Organic Systems

Organic farmers can purchase a number of approved products as nutrient sources instead of artificial fertilisers (see Table 5.3). These products can consist of minerals, food and industrial wastes, and products derived from conventional agriculture. A primary question here is the type of nutrient source generally purchased by organic farmers. Another point of interest is whether organic agriculture mainly purchases products from conventional production and is thus dependent on this type

Table 5.3 Use frequency of approved materials in organic agriculture

Approved materials	Use frequency	
	Crop-animal farms (n = 35)	Crop farms (n = 11)
Agricultural products		
Concentrate and feedstuffs[a]	17 (49%)	
Manure	3 (9%)	3 (27%)
Mineral fodder	9 (26%)	
Nutrient-fortified soils[b]		
Seed	35 (100%)	11 (100%)
Straw	14 (40%)	
Wastes		
Ash of wood, straw, peat, cereals		
Digested residues		
Fish meal	2 (6%)	
Meat and bone-meal		1 (9%)
Basic slag	1 (3%)	
Vinasse		
Minerals		
Calcium carbonate		
Dolomite		
Elemental sulphur	2 (6%)	
Gypsum		
Halite		
Kainit		
Kieserite	1 (3%)	
Phosphate rock	17 (49%)	1 (9%)
Potassium magnesia		
Rock powder		

[a]Purchase of fodder by organic farmers from conventional agriculture has been banned by an EU-regulation since 2005 (European Communities, 1999).
[b]The residual effect of inorganic fertiliser application on soil fertility before organic cropping principles has not been considered.

of agriculture. To answer these questions, we evaluated the type and frequency of nutrient sources used in the studies cited in Table 5.2 and case studies of Swedish farms (Nyberg and Lindén, 2000), see Table 5.3.

The compilation in Table 5.3 shows that in total, 75% of all organic crop-animal farms purchase fodder products from conventional agriculture. In Australia and New Zealand, however, rock phosphate applied to pastures was purchased by about 50% of organic farms. Import of straw was more frequent (40%) to organic farms than import of manures (9%). Only on organic farms without animals did imported manures play a major role as a nutrient source (27%). The EU Directive (91/676/EC) permits purchase and application of manures and composts to organically managed systems from conventional farms equivalent to 170 kg N ha^{-1} yr^{-1} unless there are changes in the regulations. Such an amount means that crops in organic agriculture can be completely supplied with manures from conventional production. Seeds seem to originate from conventional production throughout, although information is incomplete. Our data corroborate what Watson et al. (2002b) referred to as a 'cause for concern in relation to the sustainability of dairy farm systems because of their dependence on imported feedstuffs and bedding for P and K, and for N on the very variable fixation of legumes or imports of manures or compost'.

Data in Table 5.3 refer to conditions when it was permissible for organic farmers to purchase feedstuffs from conventional production. These conditions have changed within the EU countries since then. A recent regulation aims to prohibit the use of conventionally grown fodder within organic animal production from 2005 onwards (European Communities, 1999). Little information is available about the current situation but regarding Sweden there is little organically produced fodder available on the market. Indeed, exemptions have been granted to continue use of conventional fodder for organic animal production.

Our conclusion here is that despite official regulations, organic agriculture is reliant on nutrients derived from conventional farming and is not sustainable with respect to nutrient supply. This dependence, and thus non-sustainability, is seldom recognised or pointed out. Moreover, the reliance of organic systems on production systems fertilised with inorganic fertilisers cannot be maintained if a large proportion of conventional farms convert to organic agriculture. If organic farming were to become the dominant form of agriculture, there would be no surplus of fodder, straw or manure. The transfer of nutrients from conventional to organic production would stop and only untreated minerals would remain purchasable for organic farms.

In Sweden, it is also permissible to apply meat meal, bone meal, and wastes derived from food industries (Swedish Control Organisation for Alternative Crop Production, KRAV, 2008), which is a further reliance on nutrients from conventional production. An assessment of the nutrient import through these sources to organic production (information provided by the main Swedish retailer Svenska Lantmännen in Enköping) indicates that noteworthy quantities of nutrients are transferred to organic fields amounting to about 3 kg N, 1.4 kg P and 5 kg K ha^{-1} yr^{-1}.

A further form of nutrient supply from conventional to organic agriculture is through residual nutrient reserves in soil built up prior to conversion to organic cropping principles. The fertility of most conventionally managed soils has been increased through long-term applications of inorganic P and K fertiliser. Previous use of inorganic fertiliser for one or several decades improved the nutrient status of conventionally managed soils. However, the history of nutrient application prior to organic crop production is rarely mentioned and most often not considered in the results. In fact, few organic farms have never received any inorganic fertilisers.

Applying organic cropping principles on previously nutrient-fortified soils has at least two consequences. The plant availability of nutrients in soil again declines to low levels similar to those before application of artificial inorganic fertilisers, since organic agriculture can only apply less soluble nutrient sources. Thus, the pool of plant-available nutrients is mined and not replaced. Furthermore, yields may fail to increase and may even decline over time (see Chapter 3 of this book; Kirchmann et al., 2008b).

5.3.4 Changes in P and K Status of Organically Managed Soils

Although the law of nutrient replacement may be followed in organic agriculture, addition of nutrients is only one of two essential conditions to maintain soil fertility. The second most important condition is that the nutrients added are more plant-available than the existing nutrients in soil. If not, fertility will decline despite nutrient replacement and maintenance of nutrient stocks in soil. Our review of studies on this subject indicates that lower plant availability of nutrients in organically managed soils over time is common.

Reduced concentrations of plant-available P and K were measured in nutrient-rich soils in Norway within five years of conversion to organic practices by Løes and Øgaard (1997). The same authors reported that extractable P on five Norwegian dairy farms decreased over time (Løes and Øgaard, 2001). Similarly, in Denmark Askegaard and Eriksen (2000) reported that K was limiting for growth of barley and clover ley crops on sandy soils after only a few years of organic farming. Gosling and Shepherd (2005) investigated extractable K and P in English soils managed organically and found significantly lower contents on the oldest organic farm compared with conventional fields. Torstensson et al. (2006) found that at a Swedish site, organic management meant a reduction in plant-available nutrients after only a few years. In the organic rotation at that site, potassium carbonate had to be applied in order to avoid complete crop failure of potatoes.

Evans et al. (2006) found very low concentrations of plant-available P on organic farms in New Zealand. Similar results were reported from organic farms in Australia by Burkitt et al. (2007). Rock phosphate, commonly used to maintain a positive P balance on organic farms in Australia and New Zealand, resulted in a low plant availability of P relative to neighbouring conventional farms (Nguyen et al., 1995; Derrick and Dumaresq, 1999; Ryan et al., 2000). In southern Australia, low P inputs to biodynamic farms (2 kg ha^{-1} yr^{-1}) contributed to a negative P balance (−17 kg ha^{-1} yr^{-1}) compared with a small positive balance on conventional farms (2.3 kg ha^{-1} yr^{-1}), which received 19 kg P ha^{-1} yr^{-1} (Burkitt et al., 2007). For instance, Derrick and Ryan (1998) found that compared with a conventional neighbouring farm, cereal grain from an organic farm was P-deficient and had poor seedling vigour. Thus, P deficiency can even indirectly feed back through the systems and limit yield (Ryan et al., 2004).

Overall, there are no indications that plant availability of P and K in soil can be maintained through organic management. Moreover, there is a risk that a lower plant availability over time may also be true for nutrients other than P and K.

5.4 Is the 'Naturalness' of Plant Nutrient Sources a Relevant Quality Criterion?

The literature on organic agriculture describes 'naturalness' to be a prerequisite for sound food production (see Chapter 2 of this book; Kirchmann et al., 2008a). We argue, however, that 'naturalness' of compounds cannot be used as quality criterion. We discuss limitations and disadvantages related to the use of two natural nutrient sources, legumes and phosphate rock, as compared to the use of artificial fertilisers.

5.4.1 Use of Legumes as a Natural Nitrogen Source

One plant macronutrient that enters agricultural systems without direct import is nitrogen through legume cropping. In fact, legumes can provide larger amounts of N than recommended N fertiliser rates. Using legumes in a crop rotation to obtain free N seems at the first glance straight-forward. However, only in systems with legumes grown year after year, such as permanent pastures or grassland, is there a continuous input. In all other cropping systems, legumes must provide N even for the following non-leguminous crop. On organic farms without animals, years with green manure legumes are part of the rotation to substitute for the exclusion of N fertilisers, which requires land to be set aside for green manure crops.

Thus, the actual cost of replacing N fertiliser with green manure years is a reduction in crop production due to years with non-harvested crops. A comparison of N supply in an organic and conventional cropping system without animals (Fig. 5.2) illustrates the difficulties associated with the exclusive use of legumes as a major N source.

According to data from Torstensson et al. (2006), the N input in an organic rotation can amount to 40–50 kg N ha^{-1} yr^{-1} for a legume sown together with a main crop. Only full green manure years provide large amounts (140–160 kg N ha^{-1} yr^{-1}) which are higher than those applied with inorganic N fertiliser. Over a rotation, Torstensson et al. (2006) found mean amounts of N fixed through legumes to be about 70 kg N ha^{-1} yr^{-1} in the organically managed system, whereas 97 kg N ha^{-1} yr^{-1} was applied in the conventional. In the conventional system, N

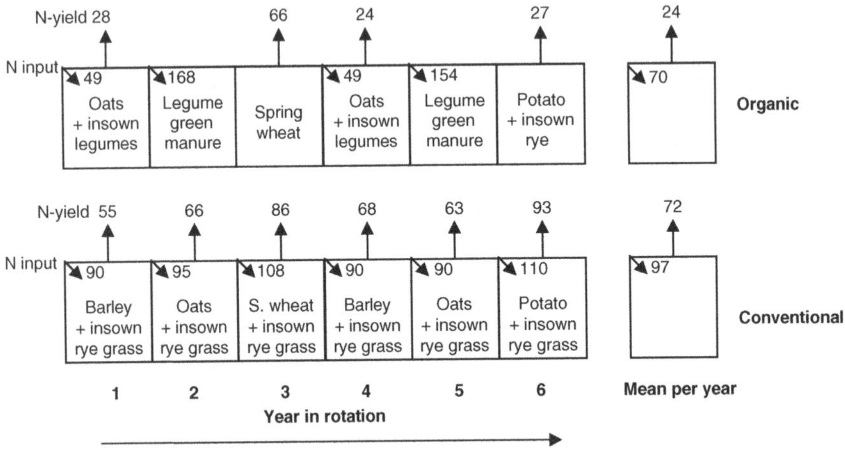

Fig. 5.2 An example of input and yield of nitrogen using legumes in an organic crop rotation and fertiliser application in a comparable conventional rotation. Data from Torstensson et al. (2006) and Torstensson (pers. comm. 2008)

input matched the demands of the crop. In the organic system, N input was not adapted to the need of the following crop being either too high or too low.

Furthermore, N release from green manure legumes is not necessarily synchronised with the demands of the following crop. As shown in Chapter 3 of this book (Kirchmann et al., 2008b), crop utilisation of N from organic manures is lower than from artificial N fertilisers. The possibility of a following crop utilising legume N is relatively low, often less than 30% (e.g. Marstorp and Kirchmann, 1991; Wivstad et al., 1996; Dahlin et al., 2005) due to decomposition of legume material at times when there is no crop demand. Even over the long term, the agronomic efficiency of legume N is much lower than of artificial N fertiliser (Kirchmann et al., 2007). Consequently, leaching of legume N is often higher than from artificial fertiliser (see Chapter 7 of this book; Bergström et al., 2008).

5.4.2 Low Plant Availability of P from Phosphate Rock

Of all major plant nutrients, P is the one for which global reserves are most limited and which must be used efficiently. The estimated life-time of economically viable P reserves varies between 50–100 years (Driver et al., 1999; Isherwood, 2003), while for known total world reserves it is around 340 years (Steward et al., 2005).

In organic agriculture, untreated minerals such as phosphate rock are used instead of water-soluble P fertiliser. However, one of the major drawbacks related to the direct use of phosphate rock is its poor solubility. Most sedimentary and igneous phosphate rocks have a low reactivity in near-neutral and calcareous soils and only a small group of so-called ‘reactive’ phosphate rock types have a similar solubility to artificial phosphate fertilisers (Buresh et al., 1997). The high reactivity of these specific phosphate rock types is explained by differences in their molecular structure, whereby a large proportion of the phosphate ions is substituted by carbonates. The majority of phosphate rock types, however, are highly insoluble and supply small quantities of P to crops, as was discovered a long time ago (e.g. Robertson, 1922). As mentioned above, under certain soil conditions – low soil pH, low exchangeable Ca and low P concentrations in the soil solution – phosphate rock may release more P than the natural phosphate in soil (e.g. Robinson and Syers, 1990; Rajan et al., 1991) and its use can be effective (e.g. Mengel, 1997). However, acid conditions in agricultural soils are undesirable for several reasons and are normally corrected by liming. Thus, the application of phosphate rock to soil is only useful if the material has a significantly higher solubility than the natural phosphate in soil (Hedley and McLaughlin, 2005).

In an attempt to increase the solubility of phosphate rock, materials have been added to composts in order to increase the plant availability of P through biological reactions taking place during composting. During the initial, mesophilic phase of composting lasting a couple of days, organic acids are formed, resulting in a slight decline in the pH of the compost, which favours the dissolution of phosphate rock. Accordingly, phosphate rock should become more plant-available during this stage of composting, which in fact has been corroborated by Singh and Amberger (1998). However, during further composting accompanied by an additional rise in temperature, the organic acids are rapidly broken down and a drastic rise in compost pH is recorded (Epstein, 1997; Beck-Fries et al., 2003). In mature composts, pH values range from 7 to 9, which is too high to dissolve phosphate rocks (e.g. Mahimairaja et al., 1995). Even nitrification in mature composts, producing protons, has no major effect on the dissolution of P as the pH remains above neutral (Mahimairaja et al., 1993). Consequently, field trials with compost amended with phosphate rock compared with compost without phosphate rock revealed no significant difference in yield (e.g. Nyirongo et al., 1999). These findings are in agreement with our understanding that chemical reactions in compost rather than biological reactions determine the P solubility of phosphate rock and that composting does not provide favourable conditions for P dissolution due to high pH values.

An area where organic farmers rely heavily on rock phosphate to supply P to organic crops is southern Australia, where growing season rainfall may be as low as 250 mm and soil extractable bicarbonate P less than 20 mg kg^{-1}. Crop nitrogen requirements are met through N_2-fixation in a preceding legume-based pasture ley. However, in this environment, rock phosphate provides no immediate benefits to crop P-nutrition or growth (Dann et al., 1996; Ryan et al., 2004). The resulting P-deficient status, often coupled with high weed levels, is the primary cause of substantially lower yields in organic systems compared with conventional (Kitchen et al., 2003; Ryan et al., 2004). In many cases, larger amounts of phosphate rock P than water-soluble P fertilisers are applied (e.g. Rajan et al., 1991; Johnston, 2005; Evans et al., 2006) to compensate for low water solubility. Use of reactive phosphate rock slightly improves yields under highly acid soil conditions (Evans et al., 2006). The results also indicate that maintaining the supply of plant-available P with phosphate rock would require more resources. For instance, the long-term use efficiency of phosphate rock applied at 4 Mg ha^{-1} (646 kg P) amounted to only 7% P, compared with 36% from soluble P fertilisers (519 kg P) over 18 years (Kirchmann et al., 2007). The low residual effectiveness of phosphate rock has been discussed by Rajan et al. (1996).

The inefficiency and thus the demand for larger amounts of phosphate rock P can also be criticised from the perspective of conserving a non-renewable and limited global resource. Using only untreated phosphate rock can lead to a faster exploitation rate than with processed P fertilisers, shortening the life-time of phosphate reserves.

5.4.3 Contaminating Soil-Plant Systems with Cadmium Through P Sources

A further disadvantage of using phosphate rock is the potential contamination of soils with cadmium (Cd), as Cd cannot be removed from phosphate rock without chemical treatment. Most phosphate rock contains Cd as impurities substituting Ca in the lattice of hydroxyl apatite. As the oxidation state of both Cd and Ca is 2+ and the Pauling radius of Cd (97 pm) is similar to that of Ca (109 pm), these similarities along with a preference for six-fold coordination facilitate the substitution of Cd into specific Ca sites in phosphate minerals. Cadmium impurities in phosphate rock and phosphorus fertilisers can lead to accumulation in the food chain, since Cd is highly mobile in the soil-plant-system, and there is a risk that excessive Cd absorption can affect human health (McLaughlin and Singh, 1999).

A review of the Cd content of phosphate rock of different origins shows that concentrations vary from 1 up to 600 mg Cd kg^{-1} P, with most reserves having concentrations higher than 50 mg Cd kg^{-1} P (McLaughlin et al., 1996). The largest reserves in the world are minerals of sedimentary origin located in Africa and the USA and have moderate to high Cd concentrations, with only minor reserves (e.g. in Northern Russia and Finland) having low Cd contents.

Applying untreated phosphate rock at a rate of e.g. 250 kg P assuming a mean concentration of 100 mg Cd kg^{-1} P would mean an addition of 25 g Cd ha^{-1}. Relating this figure to the mean Cd content of Swedish arable land, which is 0.23 mg Cd kg^{-1} soil (Eriksson et al., 1997), the Cd content of the upper 10 cm soil layer would be increased by 10%. Andersson and Siman (1991) showed that Cd concentrations in crops and seeds consistently increased with increasing P application when the P source contained 70–150 mg Cd kg^{-1} P. At the same time, medical research indicates that the safety margins in food may be smaller than believed so far (Buchet et al., 1990).

The technical know-how to extract Cd and other impurities efficiently and at a low cost from phosphate rock has been developed recently (Cohen, 2007a, b). Thus, in contrast to phosphate rock, artificial P fertilisers low in Cd can be produced even from highly cadmium-rich P sources in the future.

The discussion makes clear that there are major shortcomings using natural nutrient sources. 'Naturalness' is not a guide-lining authority guaranteeing higher quality.

5.5 Is Nutrient Recycling Enhanced by Organic Farming?

Two nutrient flows can be distinguished as regards recycling. One flow is of nutrients within or between farms, the so-called farm cycle, while the other is of nutrients exported from farm to society and recycled back to the farm, the so-called food cycle.

In Europe, there is a long tradition of recycling nutrients on farms and of reapplying them to arable land. Careful collection of animal wastes, conservation during storage and careful application to soil to minimise losses were well-founded practices long before artificial fertilisers became available. However, practices concerning recycling of wastes other than agricultural waste have been less successful in Europe (Kirchmann et al., 2005). Advanced recycling practices for wastes from society have only been developed in the Far East. For example, King (1911) described nutrient recycling through extensive transport of urine, pulverised human excreta, ash, compost and mud by human- or animal-drawn carts and boats from large cities in China, Korea and Japan back to agricultural land. It is obvious from King's description that careful collection and storage are labour-intensive practices but also that equitable redistribution of nutrients from society back to agricultural land enabled more sustainable production than in Europe or the USA.

There is no doubt that the sustainability of most agricultural systems could be improved through an increased emphasis on recycling and greater return of nutrients in municipal wastes and off-farm products. The central question here is whether organic practices improve the recycling of nutrients in general.

Once again, the other condition required for sustainability of agricultural systems must be borne in mind, namely that the plant availability of recycled nutrients must be higher than that of nutrients in soil in order to maintain production levels. Thus, although a high degree of nutrient recycling is positive and minimises losses from agricultural systems, this is no guarantee of efficient utilisation by crops and maintenance of yields.

5.5.1 Recycling On-Farm Nutrients

Organic farmers recognise on-farm animal manures as a valuable source of nutrients and place much emphasis on proper use of manures in a crop rotation. The common opinion is that organic farmers are therefore better at taking care of on-farm wastes as they are forced to do so due to shortage of nutrients. However, this belief is in contrast to how manure must be treated according to the rules in biodynamic and biological organic agriculture.

Biodynamic farmers must compost solid animal manures (Steiner, 1924), resulting in high losses of N in the form of ammonia (Kirchmann, 1985). This practice, which is central for biodynamic agriculture, leads to the least efficient recycling of N and organic matter compared with other forms of manure treatment. The concept of only surface-applying manure or green manure crops as proposed by Rusch (1978) in biological organic farming also favours losses of ammonia N.

In fact, efficient methods to decrease ammonia volatilisation from animal wastes have been developed within conventional agriculture (Bussink and Oenema, 1998; Gustavsson, 1998). These methods involve anaerobic storage of solid manures, covering slurry and urine tanks, incorporation of animal wastes into soil within four hours of spreading, direct soil injection, and regulation of manure storage capacity and spreading periods. Furthermore, best management practices for manures are easier to apply in conventional agriculture where science-proven rather than nature-based ideas guide decision-making. Good intentions in organic agriculture are insufficient if fundamental practices cannot be changed due to a dogmatic foundation. Still, several on-farm measures to minimise nutrient losses from manures and to utilise nutrients in an optimal way in relation to crop needs are independent of whether the production is organic or conventional.

Advocates of organic agriculture often point out that mixed crop and animal farming and a balance between animal and crop production is a principle guaranteeing efficient on-farm recycling of nutrients (e.g. Granstedt, 2000). This principle is similar to the idea that farms should be self-sustaining, not requiring an input of fodder or nutrients. Mixed farming is a straight-forward solution to avoid nutrient imbalances caused by a large import of feedstuffs, which contributes to an excessive supply at the local or regional level. However, specialisation of single farms does not necessarily mean a surplus of nutrients and inefficient recycling. An excess of manure can be re-distributed within an acceptable area to adjacent farms. Another measure is to prevent a further increase in animal density by limiting the number of farm animals in relation to arable land available regionally (Sims et al., 2005). However, specialisation such that farms in a whole region carry out the same production can have a negative environmental impact. On the other hand, re-introducing mixed farming on a large scale may be difficult and may put the farm's economy at risk.

5.5.2 Recycling Off-Farm Nutrients

Production and sale of agricultural products and thereby outflow of nutrients from farms is an essential part of the food supply. In contrast to on-farm recycling, nutrients exported to the food cycle are far more difficult to return to arable land. The bottlenecks for recycling of nutrients in food and municipal wastes include the high cost of re-transportation of wastes from towns and cities back to rural areas and often a low value and attractiveness due to the presence of organic pollutants and unwanted or high metal concentrations.

Current regulations for organic agriculture restrict recycling of toilet wastes. As toilet wastes contain the largest proportion of nutrients exported from farms to the

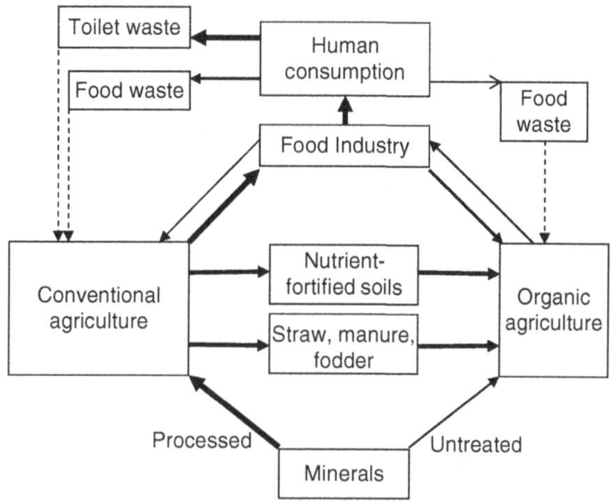

Fig. 5.3 Nutrient flows when organic agriculture is integrated into society

food cycle, significant amounts of nutrients simply cannot be returned to agricultural soils under organic production (Fig. 5.3). Under Swedish conditions, Kirchmann et al. (2005) determined that about 17 kg N and 2 kg P per hectare and year are present in toilet wastes that cannot be recycled to organically managed fields.

Recycling of nutrients from the food cycle back to agriculture could be greatly increased through widespread adoption of new recovery technologies enabling the extraction of nutrients from wastewater, biogas residues and other municipal wastes in pure form without organic matter and unwanted metals. For example, P can be extracted through precipitation as pure calcium phosphate (van Dijk and Braakensiek, 1984; Eggers et al., 1991; Seckler et al., 1996a, b, c; Angel, 1999) or pure magnesium ammonium phosphate (struvite) (Battistoni et al., 1997; Liberti et al., 2001; Ueno and Fujii, 2001). Methods for extraction and recovery of P as ammonium phosphate from urban wastes and ashes have also been developed (Cohen, 2007a, b). Overall, recovery of nutrients in the form of concentrated, inorganic products and redistribution without organic matter may help to overcome the main bottlenecks for recycling of nutrients from municipal wastes. This may be of particular importance in countries such as Australia, where agricultural production is located long distances from population centres and processing industries. However, as the new nutrient recovery technologies will enable production of water-soluble, inorganic products, only conventional farmers may be able to use these products and thereby improve nutrient cycling.

5.6 Conclusions

The organic agriculture practice of excluding artificial fertilisers and being self-sustaining through cycling of nutrients was examined. Based on this literature review, the following conclusions can be drawn:

- The organic principle of excluding water-soluble fertilisers and fertilising the soil rather than directly feeding the crop with nutrients has no basis in science. Untreated minerals often have a low efficiency due to their insolubility. This limitation is not overcome through a change in the soil biological community in organically managed soils. It is also important to note that input of fully water-soluble nutrients to soil also takes place on organic farms through organic materials.
- While the soil biological community on organic farms may differ from that on conventional farms, it cannot compensate for the lack of readily available nutrients in fertilisers. The release of organically bound nutrients through soil biological activity is not necessarily synchronised with crop demands. Changes in the soil biological community do not overcome this limitation.
- Two major conditions determine the sustainability of an agricultural system. Firstly, plant nutrients removed or lost from soil must be replaced to avoid depletion. Secondly, the availability of nutrients to plants must be maintained. Both these conditions are far more difficult to fulfil through organic agriculture than through conventional fertilisation practices.
- Organic agriculture imports feedstuffs, straw, manures and food wastes, mainly originating from conventional production. Today, organic agriculture is largely dependent on nutrient transfer from conventional agriculture and is thereby subsidised by the nutrients in artificial fertilisers. In terms of closing nutrient cycles, organic farming is limited by restrictions on using municipal wastes either directly or as water-soluble, inorganic extracts.
- Overall, there is no scientific support for the idea of 'naturalness' as a guiding quality principle. The 'naturalness' of compounds is not a guarantee of their superiority. We advocate a flexible approach where farming systems are designed to meet specific environmental, economic and social goals, unencumbered by philosophical views on nature. We invite the organic movement to reconsider their opinion towards the use of modern fertilisers.

References

Andersson, A., and Siman, G., 1991, Levels of Cd and some other trace elements in soils and crops as influenced by lime and fertilizer level, *Acta Agric. Scand.* **41**: 3–11.

Andrén, O., Kihara, J., Bationo, A., Vanlauwe, B., and Kätterer, T., 2007, Soil climate and decomposer activity in sub-Saharan Africa estimated from standard weather station data – a simple climate index for soil carbon balance calculations, *Ambio* **36**: 379–386.

Angel, R., 1999, Removal of phosphate from sewage as amorphous calcium phosphate, *Environ. Technol.* **20**: 709–720.

Aronsson, H., Torstensson, G., and Bergström, L., 2007, Leaching and crop uptake of N, P and K from organic and conventional cropping systems, *Soil Use Manage.* **23**: 71–81.

Askegaard, M., and Eriksen, J., 2000, Potassium retention in an organic crop rotation on loamy sand as affected by contrasting potassium budgets, *Soil Use Manage.* **16**: 200–205.

Battistoni, P., Fava, G., Paven, P., Musacco, A., and Cecchi, F., 1997, Phosphate removal in anaerobic liquors by struvite crystallization without addition of chemicals: preliminary results, *Water Res.* **31**: 2925–2929.

Beck-Fries, B., Smårs, S., Jönsson, H., Eklind, Y., and Kirchmann, H., 2003, Composting organic household wastes at different oxygen levels: gaining an understanding of the emission dynamics, *Compost Sci. Util.* **11**: 41–50.

Bergström, L., Kirchmann, H., Aronsson, H., Torstensson, G., and Mattsson, L., 2008, Use efficiency and leaching of nutrients in organic and conventional systems in Sweden, in: *Organic Crop Production – Ambitions and Limitations*, H. Kirchmann and L. Bergström, eds., Springer, Dordrecht, The Netherlands.

Bernal, M.P., Roig, A., and Garcia, D., 1993, Nutrient balances in calcareous soils after application of different rates of pig slurry, *Soil Use Manage.* **9**: 9–14.

Berry, P., Stockdale, E., Sylvester-Bradley, R., Philipps, L., Smith, K., Lord, E., Watson, C., and Fortune, S., 2003, N, P and K budgets for crop rotations on nine organic farms in the UK, *Soil Use Manage.* **19**: 112–118.

Brady, N.C., and Weil, R.R., 2008, *The Nature and Properties of Soils*, Pearson Education International, Upper Saddle River, NJ, USA, 965p.

Buchet, J.P., Lauwerys, R., Roels, H., Bernard, A., Bruaux, P., Claeys, F., Ducoffre, G., DePlaen, P., Staessen, J., Amery, A., Lijnen, P., Thijs, L., Rondia, D., Sartor, F., Saint Remy, A., and Nick, L., 1990, Renal effects of cadmium body burden of the general population, *Lancet* **336**: 699–702.

Buresh, R.J., Smithson, P.C., and Hellums, D.T., 1997, Building soil phosphorus capital in Africa, in: *Replenishing Soil Fertility in Africa*, R.J. Buresh, P.A. Sanchez, and F. Calhoun, eds., SSSA Special Publications No. 51, Madison, WI, USA, pp. 111–149.

Burkitt, L.L., Smal, D.R., McDonald, J.W., Wales, W.J., and Jenkin, M.L., 2007, Comparing irrigated biodynamic and conventionally managed dairy farms. 1. Soil and pasture properties, *Aust. J. Exp. Agric.* **47**: 479–488.

Bussink, D.W., and Oenema, O., 1998, Ammonia volatilization from dairy farming systems in temperate areas: a review, *Nutr. Cycl. Agroecosyst.* **51**: 19–33.

Christensen, B.T., 1985, Decomposability of barley straw: effect of cold-water extraction on dry weight and nutrient content, *Soil Biol. Biochem.* **17**: 93–97.

Cohen, Y., 2007a, Phosphorus recovery from urban wastes and ashes, Acta universitatis Agriculturae Suecia Doctoral thesis no. 2007:34. Uppsala, Sweden.

Cohen, Y., 2007b, Ion recovery by ion-exchange, Swedish patent no. 0700698-4.

Dahlin, S., Kirchmann, H., Kätterer, T., Gunnarsson, S., and Bergström, L., 2005, Possibilities for improving nitrogen use from organic materials in agricultural cropping systems, *Ambio* **34**: 288–295.

Dann, P.R., Derrick, J.W., Dumaresq, D.C., and Ryan, M.H., 1996, The response of organic and conventionally grown wheat to superphosphate and reactive rock phosphate, *Aust. J. Exp. Agric.* **36**: 71–78.

Derrick, J.W., and Ryan, M.H., 1998, Influence of seed phosphorus content on seedling growth in wheat: implications for organic and conventional farm management in south east Australia, *Biol. Agric. Hortic.* **16**: 223–237.

Derrick, J.W., and Dumaresq, D.C., 1999, Soil chemical properties under organic and conventional management in southern New South Wales, *Aust. J. Soil Res.* **37**: 1047–1055.

Driver, J., Limbach, D., and Steen, I., 1999, Why recover phosphorus for recycling and how? *Environ. Tech.* **20**: 652–662.

ECETOC, 1994, *Ammonia Emissions to Air in Western Europe*, European Centre for Ecotoxicology and Toxicology of Chemicals, Technical report no. 62, Brussels, Belgium.

Eggers, E., Dirkzwager, A., and van der Honing, H., 1991, Full-scale experiments with phosphate crystallization in a Crystalactor®, *Water Sci. Technol.* **23**: 819–824.

Eltun, R., Korsaeth, A., and Nordheim, O., 2002, A comparison of environmental, soil fertility, yield, and economical effects in six cropping systems based on an 8-year experiment in Norway, *Agric. Ecosyst. Environ.* **90**: 155–168.

Eriksson, J., Andersson, A., and Andersson, R., 1997, *Current Status of Swedish Arable Soils*, Swedish Environmental Protection Agency, Report 4778, Stockholm, Sweden.

Epstein, E., 1997, *The Science of Composting*, CRC Press, Boca Raton, FL., USA, 487p.
European Communities, 1999, Council Regulation (EC) No. 1804/1999 of July 1999 supplementing Regulation (EEC) No. 2092/91 on organic production of agricultural products and indications referring thereto on agricultural products and feedstuffs to include livestock production, *Official J. Eur. Communities* 24.98.1999. Brussels L 222 1–28.
Evans, J., McDonald, L., and Price, A., 2006, Application of reactive phosphate rock and sulfur fertilisers to enhance the availability of soil phosphate in organic farming, *Nutr. Cycl. Agroecosyst.* **75**: 233–246.
Fagerberg, B., Salomon, E., and Jonsson, S., 1996, Comparisons between conventional and ecological farming systems at Öjebyn, *Swedish J. Agric. Res.* **26**: 169–180.
Feller, C.L., Thuries, L.J-M., Manley, R.J., Robin, P., and Frossard, E., 2003, "*The principal of rational agriculture*" by Albrecht Thaer (1752–1828). An approach to the sustainability of cropping systems at the beginning of the 19th century, *J. Plant Nutr. Soil Sci.* **166**: 687–698.
Ferm, M., 1998, Atmospheric ammonia and ammonium transport in Europe and critical loads: a review, *Nutr. Cycl. Agroecosyst.* **51**: 5–17.
Fliessbach, A., and Mäder, P., 2000, Microbial biomass and size-density fractions differ between soils of organic and conventional agricultural systems, *Soil Biol. Biochem.* **32**: 757–768.
Fowler, S.M., Watson, C.A., and Wilman, D., 1993, N, P and K on organic farms: herbage and cereal production, purchases and sales, *J. Agric. Sci.* **120**: 353–360.
Gilkes, R.J., and Mangano, P., 1983, Poorly soluble, iron-aluminium phosphates in ammonium phosphate fertilizers: their nature and availability to plants, *Aust. J. Soil Res.* **21**: 183–194.
Gosling, P., and Shepherd, M., 2005, Long-term changes in soil fertility in organic arable farming systems in England, with particular reference to phosphorus and potassium, *Agric. Ecosyst. Environ.* **105**: 425–432.
Granstedt, A., 2000, Increasing the efficiency of plant nutrient recycling within the agricultural system as a way of reducing the load to the environment – experience from Sweden and Finland, *Agric. Ecosyst. Environ.* **80**: 169–185.
Gunapala, N., Venette, R.C., Ferris, H., and Scow, K.M., 1998, Effects of soil management history on the rate of organic matter decomposition, *Soil Biol. Biochem.* **30**: 1917–1927.
Gustafson, G.M., Salomon, E., Jonsson, S., and Steineck, S., 2003, Fluxes of K, P and Zn in a conventional and an organic dairy farming system through feed, animals, manure and urine – a case study at Öjebyn, Sweden, *Eur. J. Agron.* **20**: 89–99.
Gustavsson, J., 1998, Swedish measures to reduce ammonia emissions, *Nutr. Cycl. Agroecosyst.* **51**: 81–83.
Härdter, R., Rex, M., and Orlovius, K., 2004, Effects of different Mg fertilizer sources on the magnesium availability in soil, *Nutr. Cycl. Agroecosyst.* **70**: 249–259.
Hedley, M., and McLaughlin, M., 2005, Reactions of phosphate fertilizers and by-products in soils, in: *Phosphorus: Agriculture and the Environment*, J.T. Sims and A.N. Sharpley, eds., Agronomy Monograph No. 46, Madison, WI, USA, pp. 181–252.
Henriksen, T.M., and Breland, T.A., 1999, Evaluation of criteria for describing crop residue degradation in a model of carbon and nitrogen turnover in soil, *Soil Biol. Biochem.* **31**: 1135–1149.
Hultberg, H., and Grennfeldt, P., 1992, Sulphur and sea salt deposition as reflected by throughfall and runoff chemistry in forested catchments, *Environ. Pollut.* **75**: 215–222.
IFOAM, 2006, *The Four Principles of Organic Farming*. The International Federation of Organic Agriculture Movements, www.IFOAM.org, Bonn, Germany, assessed 5/6-2006.
Isherwood, K.F., 2003, *Fertilizer Consumption and Production: Long-term World Prospects*. The Int. Fertilizer Soc., Proc. No. 507, York, UK.
Ivarson, J., and Gunnarsson, A., 2001, Försök med konventionella och ekologiska odlingsformer, 1987–1998, Sveriges Lantbruksuniversitet, Medd. Södra Jordbruksdistr. nr. 53, Sverige (In Swedish).
Johnston, A.E., 2005, Phosphorus nutrition of arable crops, in: *Phosphorus: Agriculture and the Environment*, J.T. Sims and A.N. Sharpley, eds., Agronomy Monograph No. 46, Madison, WI, USA, pp. 495–520.

Kaffka, S., and Koepf, H., 1989, A case study on the nutrient regime in sustainable farming, *Biol. Agric. Hortic.* **6**: 89–106.

Kätterer, T., and Andrén, O., 2001, The ICBM family of analytically solved models of soil carbon, nitrogen and microbial biomass dynamics – descriptions and application examples, *Ecol. Modell.* **136**: 191–207.

Kätterer, T., Andrén, O., and Jansson, P-E., 2006, Pedotransfer functions for estimating plant available water and bulk density in Swedish agricultural soils, *Acta Agric Scand. Sec. B* **56**: 263–276.

Kätterer, T., and Andrén, O., 2008, Predicting daily soil temperature profiles in arable soils in cold temperate regions from air temperature and leaf area index, *Acta Agric. Scand. Section B, in press*, DOI: 10.1080/09064710801920321.

King, F.H., 1911, *Farmers of Forty Centuries or Permanent Agriculture in China, Korea and Japan*. Democrat Printing Co, Madison WI., USA.

Kirchmann, H., 1985, Losses, plant uptake and utilisation of manure nitrogen during a production cycle, *Acta Agric. Scand. Suppl.* **24**: 7–77.

Kirchmann, H., and Bergqvist, R., 1989, Carbon and nitrogen mineralization of white clover plants (*Trifolium repens*) of different age during aerobic incubation with soil, *Z. Pflanzenernähr. Bodenk.* **152**: 283–288.

Kirchmann, H., and Witter, E., 1992, Composition of fresh, areobic and anaerobic farm animal dung, *Bioresour. Technol.* **40**: 137–142.

Kirchmann, H., and Pettersson, S., 1995, Human urine – chemical composition and fertilizer use efficiency, *Fertil. Res.* **40**: 149–154.

Kirchmann, H., Haberhauer, G., Kandeler, E., Sessitsch, A., and Gerzabek, M.H., 2004, Effects of level and quality of organic matter input on soil carbon storage and microbial activity in soil – Synthesis of a long-term experiment, *Global Biogeochem. Cycles* **18**: 38–46.

Kirchmann, H., Nyamangara, J., and Cohen, Y., 2005, Recycling municipal wastes in the future: from organic to inorganic forms? *Soil Use Manage.* **21**: 152–159.

Kirchmann, H., Bergström, L., Kätterer, T., Mattsson, L., and Gesslein, S., 2007, Comparison of long-term organic and conventional crop-livestock systems on a previously nutrient-depleted soil in Sweden, *Agron. J.* **99**: 960–972.

Kirchmann, H., Thorvaldsson, G., Bergström, L., Gerzabek, M., Andrén, O., Eriksson, L.-O., and Winninge, M., 2008a, Fundamentals of organic agriculture – Past and present, in: *Organic Crop Production – Ambitions and Limitations*, H. Kirchmann and L. Bergström, eds., Springer, Dordrecht, The Netherlands.

Kirchmann, H., Bergström, L., Kätterer, T., Andrén, O., and Andersson, R., 2008b, Can organic crop production feed the world? in: *Organic Crop Production – Ambitions and Limitations*, H. Kirchmann and L. Bergström, eds., Springer, Dordrecht, The Netherlands.

Kitchen, J.L., McDonald, G.K., Shepherd, K.W., Lorimer, M.F., and Graham, R.D., 2003, Comparing wheat grown in South Australian organic and conventional farming systems, I. Growth and grain yield, *Aust. J. Agric. Res.* **54**: 889–901.

KRAV, 2008, Regulations of the Swedish Control Organization for Alternative Crop production, www.krav.se, Uppsala, Sweden, assessed 31/3-2008.

Liberti, L., Petruzzelli, D., and De Flurio, L., 2001, Rem nut ion exchange plus struvite precipitation process, *Environ. Technol.* **22**: 1313–1325.

Lide, D.R., 1999, *Handbook of Chemistry and Physics*, 80th ed., CRC Press, Boca Raton, FL, USA.

Liebig, M.A., and Doran, J.W., 1999, Impact of organic production practices on soil quality indicators, *J. Environ. Qual.* **28**: 1601–1609.

Løes, A.K., and Øgaard, A.F., 1997, Changes in the nutrient content of agricultural soil on conversion to organic farming in relation to farm-level nutrient balances and soil contents of clay and organic matter, *Acta Agric. Scand. B, Soil Plant Sci.* **47**: 201–214.

Løes, A.K., and Øgaard, A.F., 2001, Long-term changes in extractable soil phosphorus (P) in organic dairy farming systems, *Plant Soil* **237**: 321–332.

Lundström, C., and Lindén, B., 2001, Nitrogen effects of human urine and fertilisers containing meat bone meal (Biofer), or chicken manure (Binidan) as fertilisers applied to winter wheat,

spring wheat and spring barley in organic farming. Swedish University of Agricultural Sciences, Department of Agricultural Research, Skara, *Series B Crops and Soils, Report* no. **8**. Skara, Sweden.
Lützow, M., and Ottow, J.C.G., 1994, Effect of conventional and biological farming on microbial biomass and its nitrogen turnover in agriculturally used Luvisols of the Friedberg plains, *Zeitschrift Pflanzenernähr. Bodenk.* **157**: 359–367.
Mahimairaja, S., Bolan, N.S., and Hedley, M.J., 1993, Absorption of ammonia released from poultry manure to soil and bark and the use of absorbed ammonia in solubilizing phosphate rock, *Compost Sci. Util.* **1**: 101–112.
Mahimairaja, S., Bolan, N.S., and Hedley, M.J., 1995, Dissolution of phosphate rock during composting of poultry manure: an incubation experiment, *Fertil. Res.* **40**: 93–104.
Marstorp, H., and Kirchmann, H., 1991, Carbon and nitrogen mineralization and crop uptake of nitrogen from six green manure legumes decomposing in soil, *Acta Agric. Scand.* **41**: 243–252.
Mattsson, L., and Kjellquist, T., 1992, Nitrogen fertilization to winter wheat on farms with and without animal husbandry. Swedish University of Agricultural Sciences, Department of Soil Sciences, *Div. Plant Nutr. and Soil Fertil., Report* no. **189**. Uppsala, Sweden.
McLaughlin, M.J., Tiller, K.G., Naidu, R., and Stevens, D.P., 1996, Review: the behavior and impact of contaminants in fertilizers, *Aust. J. Soil Res.* **34**: 1–54.
McLaughlin, M.J., and Singh, B.R., 1999, *Cadmium in Soils and Plants*, Developments in Plant and Soil Sciences, vol 85, Kluwer Academic Publishers, Dordrecht, The Netherlands, 271p.
Mengel, K., 1997, Agronomic measures for better utilization of soil and fertilizer phosphates, *Eur. J. Agron.* **7**: 221–233.
Newman, E.I., 1997, Phosphorus balance on contrasting farming systems, past and present. Can food production be sustainable? *J. Appl. Ecol.* **34**: 1334–1347.
Nguyen, M.L., Haynes, R.J., and Goh, K.M., 1995, Nutrient budgets and status in three pairs of conventional and alternative mixed cropping farms in Canterbury, New Zealand, *Agric. Ecosyst. Environ.* **52**: 149–162.
Nielsson, F.T., 1987, *Manual for Fertilizer Processing, Fertilizer Science and Technology Series*, Vol. 5, Marcel Dekker Inc., New York, USA, 525p.
Nolte, C., and Werner, W., 1994, Investigations on the nutrient cycle and its components of a biodynamically-managed farm, *Biol. Agric. Hortic.* **10**: 235–254.
Nyberg, A., and Lindén, B., 2000, Documentation of ecological farms in Western Sweden 1996–1998. Swedish University of Agricultural Sciences, Department Agricultural Research Skara, *Series B Crops and Soils, Report* no. **6**. Skara, Sweden.
Nyirongo, J.C.V.B., Mughogho, S.K., and Kumwenda, J.D.T., 1999, Soil fertility studies with compost and igneous phosphate rock amendments in Malawi, *Afr. Crop Sci. J.* **7**: 415–422.
Pape, J.C., 1970, Plaggen soils in the Netherlands, *Geoderma* **4**: 229–255.
PDA, 2008, *Potash for Organic Growers*, The Potash Development Association, www.pda.org.uk, York, UK, assessed 29 May 2008.
Petersen, S.O., Debosz, K., Schjønning, P., Christensen, B.T., and Elmholt, S., 1997, Phospholipid fatty acid profiles and C availability in wet-stable macro-aggregates from conventionally and organically farmed soils, *Geoderma* **78**: 181–196.
Pott, R., 1990, *Lueneburger Heide*. Ulmer, Stuttgart, Germany, 256p (In German).
Rajan, S.S.S., Fox, R.L., Saunders, W.M.H., and Upsdell, M., 1991, Influence of pH, time and rate of application on phosphate rock dissolution and availability to pastures, *Fertil. Res.* **28**: 85–93.
Rajan, S.S.S., Watkinson, J.H., and Sinclair, A.G., 1996, Phosphate rocks for direct application to soil, *Adv. Agron.* **57**: 77–159.
Reganold, J.P., 1988, Comparison of soil properties as influenced by organic and conventional farming systems, *Am. J. Altern. Agric.* **3**: 144–154.
Reinertsen, S.A., Elliott, L.F., Cochran, V.L., and Campbell, G.S., 1984, Role of available carbon and nitrogen in determining the rate of wheat straw decomposition, *Soil Biol. Biochem.* **16**: 459–464.
Robertson, G.S., 1922, *Basic Slag and Rock Phosphates*, Cambridge Agricultural Monographs, Cambridge University Press, UK, 120p.

Robertson, F.A., and Morgan, W.C., 1996, Effects of management history and legume green manure on soil microorganisms under 'organic' vegetable production, *Aust. J. Soil Res.* **34**: 427–440.

Robinson, J.S., and Syers, J.K., 1990, A critical evaluation of the factors influencing the dissolution of Gafsa phosphate rock, *J. Soil Sci.* **41**: 597–605.

Rusch, H.P., 1978, *Bodenfruchtbarkeit. Eine Studie biologischen Denkens*, 3rd Printing. Haug Verlag, Heidelberg, Germany, 243p (In German).

Ryan, M., 1999, Is an enhanced soil biological community, relative to conventional neighbours, a consistent feature of alternative (organic and biodynamic) agricultural systems? *Biol. Agric. Hortic.* **17**: 131–144.

Ryan, M.H., and Ash, J.E., 1999, Effects of phosphorus and nitrogen on growth of pasture plants and VAM fungi in SE Australian soils with contrasting fertiliser histories (conventional and biodynamic), *Agric. Ecosyst. Environ.* **73**: 51–62.

Ryan, M.H., Small, D.R., and Ash, J.E., 2000, Phosphorus controls the level of colonisation by arbuscular mycorrhizal fungi in conventional and biodynamic irrigated dairy pastures, *Aust. J. Exp. Agric.* **40**: 663–670.

Ryan, M.H., Derrick, J.W., and Dann, P.R., 2004, Grain mineral concentrations and yield of wheat grown under organic and conventional management, *J. Sci. Food Agric.* **84**: 207–216.

Ryan, M.H., and Tibbett, M., 2008, The role of arbuscular mycorrhizas in organic farming, in: *Organic Crop Production – Ambitions and Limitations*, H. Kirchmann and L. Bergström, eds., Springer, Dordrecht, The Netherlands.

Seckler, M., Bruinsma, O., and van Rosmalen, G., 1996a, Phosphate removal in a fluidized bed. I Identification of physical processes, *Water Res.* **30**: 1585–1588.

Seckler, M., van Leeuwen, M., Bruinsma, O., and van Rosmalen, G., 1996b, Phosphate removal in a fluidized bed. II Process optimization, *Water Res.* **30**: 1589–1596.

Seckler, M., Bruinsma, O., and van Rosmalen, G., 1996c, Calcium phosphate precipitation in a fluidized bed in relation to process conditions: black box approach, *Water Res.* **30**: 1677–1685.

Sims, J.T., Bergström, L., Bowman, B.T., and Oenema, O., 2005, Nutrient management for intensive animal agriculture: policies and practices for sustainability, *Soil Use Manage.* **21**: 141–151.

Singh, C.P., and Amberger, A., 1998, Organic acids and phosphorus solubilization in straw composted with rock phosphates, *Bioresour. Technol.* **63**: 13–16.

Sivapalan, A., Morgan, W.C., and Franz, P.R., 1993, Monitoring populations of soil microorganisms during a conversion from a conventional to an organic system of vegetable growing, *Biol. Agric. Hortic.* **10**: 9–27.

Springob, G., and Kirchmann, H., 2002, C-rich sandy Ap horizons of specific historical land-use contain large fractions of refractory organic matter, *Soil Biol. Biochem.* **34**: 1571–1581.

Steiner, R., 1924, *Geisteswissenschaftliche Grundlagen zum Gedeihen der Landwirtschaft*. Steiner Verlag, 5. Auflage 1975. Dornach, Schweiz (In German).

Steward, W.M., Hammond, L.L., and Van Kauwenbergh, S.J., 2005, Phosphorus as a natural resource, in: *Phosphorus: Agriculture and the Environment*, J.T. Sims and A.N. Sharpley, eds., Agronomy Monograph No. 46, Madison, WI, USA, pp. 3–22.

Thaer, A.D., 1837–1839, *Grundsätze der rationellen Landwirtschaft*. Neue unveränderte Ausgabe. 4 Bände in 2. Gedruckt von E.F. Arnold, Stuttgart, Germany (In German).

Torstensson, G., Aronson, H., and Bergström, L., 2006, Nutrient use efficiency and leaching of organic and conventional cropping systems in Sweden, *Agron. J.* **98**: 603–615.

Ueno, Y., and Fujii, M., 2001, Three years experience of operating and selling recovered struvite from a full-scale plant, *Environ. Technol.* **22**: 1373–1381.

Ulén, B., 1984, Nitrogen and phosphorus to surface water from crop residues, Department of Soil Sciences, Division of Water Management, Report 18, Swedish University of Agricultural Sciences, Uppsala, Sweden, pp. 39–44.

Ullmann's Agrochemicals, 2007, Wiley-VCH Verlag GmbH and Co, Weinheim, Germany, 932p.

van Dijk, J., and Braakensiek, H., 1984, Phosphate removal by crystallization in a fluidized bed, *Water Sci. Technol.* **17**: 133–142.

von Liebig, J., 1840, *Die organische Chemie in ihrer Anwendung auf Agrikultur und Physiologie*. Fr. Vieweg and Sohn, Braunschweig, Germany (In German).

Wander, M.W., Traina, S.J., Stinner, B.R., and Peters, S.E., 1994, Organic and conventional management effects on biologically active soil organic matter pools, *Soil Sci. Soc. Am. J.* **58**: 1130–1139.

Watson, C.A., Atkinson, D., Gosling, P., Jackson, L.R., and Rays, F.W., 2002a, Managing soil fertility in organic farming systems, *Soil Use Manage.* **18**: 239–247.

Watson, C.A., Bengtsson, H., Ebbesvik, M., Løes, A-K., Myrbeck, A., Salomon, E., Schroder, J., and Stockdale, E.A., 2002b, A review of farm-scale nutrient budgets for organic farms as a tool for management of soil fertility, *Soil Use Manage.* **18**: 264–273.

Wieser, I., Heß, J., and Lindenthal, T., 1996, Nutrient balances on organically managed grassland farms in Upper Austria, *Die Bodenkultur* **47**: 81–88.

Wivstad, M., Salomonsson, L., and Salomonsson, A.C., 1996, Effects of green manure, organic fertilisers and urea on yield and grain quality of spring wheat, *Acta Agric. Scand. B, Soil Plant Sci.* **46**: 169–177.

Workneh, F., and van Bruggen, A.H.C., 1994, Microbial density, composition, and diversity in organically and conventionally managed rhizosphere soil in relation to suppression of corky root of tomatoes, *Appl. Soil Ecol.* **1**: 219–230.

Chapter 6
Synthesis of the Apelsvoll Cropping System Experiment in Norway – Nutrient Balances, Use Efficiencies and Leaching

Audun Korsaeth and Ragnar Eltun

Abstract In this overview, a synthesis of the first 10 years of the Apelsvoll cropping system experiment, located in southeast Norway, is given. All major flows of N, P and K in six different cropping systems, each covering 0.18 ha of separately tile-drained plots, were either measured or estimated. The effects of the cropping system on the soil nutrient pools (total-N, -P, and-K) were evaluated using mass balances, and the usefulness of such balances to predict nutrient losses via drainage (leaching) and surface water (runoff) was investigated. The cropping systems were further evaluated in terms of leaching, runoff and use efficiencies. The experiment included conventional arable cropping (CON-A), integrated arable cropping (INT-A), organic arable cropping (ORG-A), conventional forage cropping (CON-F), integrated forage cropping (INT-F), and organic forage cropping (ORG-F). All the arable systems had a calculated reduction in the soil N content. The soil N content of the forage system CON-F was sustained over 10 years, whereas that of INT-F showed a slight reduction. The greatest reductions were calculated for the organic systems ORG-A and ORG-F, amounting to approximately 3.3% of the initial soil N pool (to a depth of 1 m). The two arable systems CON-A and INT-A had increases in their calculated soil P contents, with up to 70 kg P ha^{-1} (CON-A) over the 10 years. The other extreme was found in the two organic systems, ORG-A and ORG-F, in which there were accumulated net reductions of more than 120 kg soil P ha^{-1} over the same period. CON-F appeared to sustain its soil P content, whereas INT-F had a calculated reduction less than half that of ORG-F. Calculated net changes in soil K followed a similar pattern as for soil P. The calculated net changes in soil N were confirmed, at least partly, by soil measurements. Despite the calculated reductions in total soil P and K, a similar pattern was not found in either plant-available P (P-AL) or in acid-extractable K (K-HNO3). Nitrogen was mainly lost via drainage water runoff, P mainly via surface runoff, and K almost equally by these two pathways. On an annual basis, N losses via drainage water and surface runoff were in the range of 17.6–35.2 kg N ha^{-1} yr^{-1}, and increased in the order

A. Korsaeth (✉)
Bioforsk Arable Crops Division, Norwegian Institute for Agricultural and Environmental Research, Apelsvoll, N-2849 Kapp, Norway
e-mail: audun.korsaeth@bioforsk.no

H. Kirchmann, L. Bergström (eds.), *Organic Crop Production – Ambitions and Limitations*, DOI 10.1007/978-1-4020-9316-6_6,

ORG-A ≈ INT-F ≈ ORG-F < INT-A ≈ CON-F < CON-A. Annual P losses were in the range of 0.36–0.49 kg P ha^{-1} yr^{-1} and the annual K losses were in the range of 6.22–9.69 kg K ha^{-1} yr^{-1}. There were fewer differences between the systems in terms of P and K losses, but INT-A appeared to have the lowest overall P and K losses. The balance between N supply and N removal (N applied in fertilizer, animal waste and symbiotic fixation minus N in harvested crops) averaged over the 10-year period explained up to 87% of the variation in total N leaching and N runoff. In contrast to N, mass balances of P and K were poorly correlated with leaching and runoff losses of these nutrients. The integrated systems had the largest production to loss ratios for N, P and K within each group of production (compared with systems having the same crop rotation).

Keywords Cropping system · Leaching · Mass balance · Nitrogen · Potassium · Phosphorus · Runoff · Soil mining · Symbiotic N fixation · Use efficiency

6.1 Introduction

Research on the interface between agriculture and the environment in Norway was at a time dominated by traditional approaches, where the effects of single components of a system were studied (e.g. Lyngstad, 1990; Uhlen, 1989). Later on came studies including continuous monitoring of drainage and runoff losses from specific agricultural areas (Vagstad et al., 1997) and modelling work based upon the results of these studies (Vatn et al., 1996). The Apelsvoll Cropping System Experiment was established in 1988 using an alternative approach; holistic system analyses, using whole farms or cropping systems as experimental units. Oberle and Keeney (1991) listed the following objectives for such systems research: (1) to gain deeper understanding of how the components and processes of a system interact and fit together; (2) to help in solving complex problems; (3) to identify and rank site-specific information needs; and (4) to aid in evaluating and predicting the effects of changes in agricultural policy and management practices, production enterprises, climate and other factors.

Over the years, the Apelsvoll experiment has provided data for many studies covering a range of different topics, such as: yields and yield quality (Eltun, 1996a, b), nutrient leaching and runoff losses (Eltun, 1995; Eltun and Fugleberg, 1996), pesticide drainage and runoff (Korsæth and Eltun, 1996), economic aspects (Repstad and Eltun, 1997), phosphorus and potassium balances (Løes et al., 1998), soil microbial biomass (Breland and Eltun, 1999), nitrogen balances (Korsaeth and Eltun, 2000), insect predators (Andersen and Eltun, 2000), environmental indices (Eltun et al., 2002), economic risk assessment (Lien et al., 2006), soil structure (Riley et al., 2006) and earthworm populations (Pommeresche et al., 2006).

In this overview, a synthesis of the first 10 years of the experiment (1990–1999) is given for major nutrient flows of N, P and K, with focus on mass balances, leaching/runoff losses and use efficiencies.

6.2 Material and Methods

6.2.1 Experimental Site and Treatments

A 3.3 ha large experiment with tile-drained plots was established in 1988/1989 at Apelsvoll Research Centre in central southeast Norway (60°42′ N, 10°51′ E, altitude 250 m). The climate of the region is humid continental with a mean annual precipitation of 600 mm and a mean annual temperature of 3.6 and 12.0 °C in the growing season from May to September. The experimental area, which slopes 2–8% towards northeast, was used as pasture from 1935 to 1975. During the following years, up to the establishment of the experiment in 1988, the field was cropped with a rotation including 10% root crops, 40% small grains and 50% ley, using an average of 10 tonnes cattle slurry ha^{-1} yr^{-1} plus regular amounts of inorganic fertiliser. The first year after tile-draining the experimental site (1989), the area was cropped with barley (*Hordeum distichum* L.). The major soil group of the experimental area is imperfectly drained brown earth (Oxiaquic Cryoboroll, USDA; Gleyed melanic brunisoils, Canada Soil Survey) with dominantly loam and silty sand textures. More detailed soil characteristics have been presented elsewhere (Riley and Eltun, 1994).

Six cropping systems, each with 2 replicates, were distributed on 0.18 ha plots within a randomised complete block design at the experimental site (Fig. 6.1). Each cropping system has an 8 year crop rotation with eight subplots, so that all crops in the rotation are present every year. They include:

CON-A: conventional arable cropping without slurry
INT-A: integrated arable cropping without slurry
ORG-A: organic arable cropping with slurry

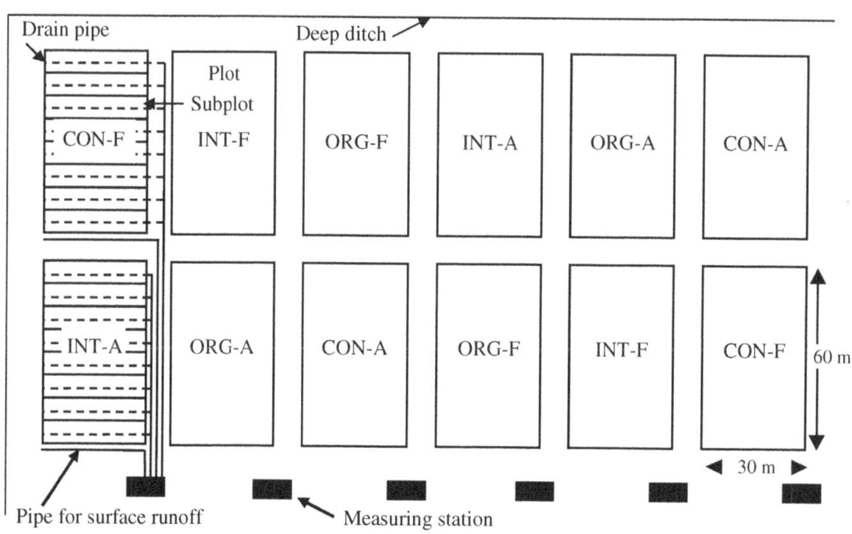

Fig. 6.1 Layout of the experimental site

Table 6.1 Characteristics of the cropping systems at Apelsvoll Research Centre, south-east Norway, 1990–1999

Management	Cropping system					
	Conventional arable (CON-A)	Integrated arable (INT-A)	Organic arable (ORG-A)	Conventional forage (CON-F)	Integrated forage (INT-F)	Organic forage (ORG-F)
Crop rotation	Barley[a]	Barley[a]	Barley[b]	Barley[b]	Barley[b]	Barley[b]
	Winter wheat	W. wheat	Clover grass	1st year ley	1st year ley	1st year ley
	Oats	Oats	S. wheat[c]	2nd year ley	2nd year ley	2nd year ley
	Barley	Barley	Potatoes	3rd year ley	3rd year ley	3rd year ley
	Potatoes	Potatoes	Barley[b]	Swedes[d]	Swedes[d]	Swedes[d]
	Spring wheat	S. wheat	Clover grass	S. wheat	S. wheat	Green fodder
	Oats	Oats	W. wheat[c]	Oats	Oats	S. wheat[c]
	Barley	Barley	Oats[c]	Green fodder	Green fodder	Oats and peas
Fertiliser[e]	+	+	0	+	+	0
Slurry[f]	0	0	0.4	2.0	1.5	1.2
Soil tillage	Spring ploughing[g]	Spring harrowing	Spring ploughing	Spring ploughing[g]	Spring ploughing	Spring ploughing
Plant protection	Chemical	Integrated[h]	Mechanical	Chemical	Integrated[h]	Mechanical

[a] Early potatoes in the period 1990–1994.
[b] With undersown clover grass/ley.
[c] With undersown crop.
[d] Fodder beet in the period 1990–1994.
[e] Inorganic fertilizer.
[f] Cattle slurry as animal units per ha.
[g] Autumn ploughing in the period 1990–1994.
[h] Chemical protection, but with reduced amounts of pesticides and application times compared with the conventional systems.

CON-F: conventional forage cropping with slurry
INT-F: integrated forage cropping with slurry
ORG-F: organic forage cropping with slurry

Characteristics of the cropping systems are shown in Table 6.1.

Each plot is separately drained with PVC pipes at 1 m depth and 7.5 m spacing (one drain for each subplot). Surface runoff is collected at the lower end of each plot and led to a sedimentation tank. Grass covered border zones separate the plot. Drainage water and surface water runoff from the sedimentation tank was transported in sealed plastic pipes to measuring stations equipped for discharge measurements (by tipping buckets) and for volume proportional water sampling. A more detailed description of the experimental design is given in Eltun (1994).

6.2.2 Measurements, Estimates and Calculations

Cattle slurry was sampled 1–2 weeks before application each year and analysed for total nitrogen (N), total phosphorus (P) and total potassium (K). The harvested crops and straw residues (when removed) were weighed (four samples from each subplot) and analysed for N, P and K (up to 1995, whereafter the analyses were not performed). In 1996–1999 the nutrient content of crop and straw was set as the average of the previous 6 years (for each crop and cropping system separately). The proportion of legumes was determined visually before harvest. The water samples (drainage water and surface runoff) were analysed on a monthly basis for N, ammonium-N, nitrate-N, total-P, phosphate-P (all years) and K (1990–1996). Precipitation was sampled on a monthly basis from a rain gauge placed at the experimental area, and analysed for total-N (wet atmospheric deposition). Soil samples (0–30 cm depth) were taken in 1988 and at the same sites in 2003, and analysed for total N (Kjeldahl method), plant available P (P-AL; extracted by a mixture of acetic acid and ammonium lactate, Égner et al., 1960) and acid soluble potassium (K-HNO_3; measured by flame photometry after extraction with 1 N nitric acid). It was decided to use the soil samples from 2003 since these were the first soil analyses taken after 1996, which were comparable with those taken in 1988. In 2000, some adjustments were made in the experimental design (Korsæth et al., 2001). The main changes were that alternate subplot pairs were merged, reducing the number of subplots from eight to four, and that the system INT-F was changed to an organic system with 75% ley. These changes were assumed to have only a minor influence on the changes in soil chemical properties during the 15-year period (1988–2003). Data from the period after 1999 will be published elsewhere.

The input of nutrients with seeds was estimated using measured contents of N, P and K of harvested grain and potatoes from the CON-A system. Literature values were selected for the other seeds. Volatilisation of $NH_3{-}N$ from cattle slurry was estimated individually for each subplot and application time in accordance with Korsaeth and Eltun (2000), by a method based on Horlacher and Marschner (1990).

Table 6.2 Measured changes in selected chemical properties in topsoil (0–30 cm) from 1988 to 2003

Cropping system	Total N (g 100 g^{-1})				P-AL (mg 100 g^{-1})				K-HNO_3 (mg 100 g^{-1})			
	1988	2003	Diff.[a]	p[b]	1988	2003	Diff.	p	1988	2003	Diff.	P
CON-A	0.28	0.20	−0.08	0.009	8.40	6.43	−1.97	0.011	30.5	29.5	−1.00	0.210 n.s.
INT-A	0.25	0.22	−0.03	0.395 n.s.	6.85	7.32	0.47	0.470 n.s.	26.5	34.0	−7.50	0.008
ORG-A	0.26	0.23	−0.03	0.008	6.40	4.10	−2.30	0.007	27.0	27.3	0.03	0.909 n.s.
CON-F	0.26	0.24	−0.02	0.213 n.s.	8.54	5.45	−3.09	0.007	36.0	32.0	−4.00	0.060 n.s.
INT-F	0.28	0.22	−0.06	0.022	8.60	4.15	−4.45	0.029	38.5	24.0	−14.5	0.032
ORG-F	0.27	0.22	−0.05	0.008	7.20	3.80	−3.40	0.030	25.0	31.0	6.00	0.061 n.s.

[a] Change in soil nutrient content from 1988 to 2003 (2003 values minus 1988 values).
[b] Level of significance for the pair-wise t-test of differences.
n.s.: Not significant at 5% level.

Symbiotic N fixation was estimated by a model described by Korsaeth and Eltun (2000). Dry atmospheric deposition was set to 2 kg N ha^{-1} yr^{-1} based on Tørseth and Manø (1996). Atmospheric depositions of P and K were assumed to be negligible. Denitrification was set to 7% of the net amount of mineral N applied, based on experiments executed under fairly similar conditions (Ryden, 1985; Svensson et al., 1991; Maag, 1995). Net losses of NH_3 to the atmosphere by volatilisation from crops were set to 2 kg N ha^{-1} yr^{-1} according to Holtan-Hartwig and Bøckman (1994).

Nutrient runoff and drainage losses occurring during the agrohydrological year, lasting from 1 May to 30 April the following year was attributed to the cropping that occurred in the same period. Nutrient mass balances and precipitation were related to drainage and runoff losses using linear regression, as described by Korsaeth and Eltun (2000).

For each cropping system, all major flows of N, P and K were either measured or estimated and nutrient mass balances calculated. Using a mass balance approach, enabled analysis of possible changes in the respective soil nutrient pools. The calculated mass balances were compared with measured changes in soil chemical properties from 1988 to 2003 (Table 6.2).

Analyses of variance were performed on the annual nutrient loads, using a split-plot model with cropping system as major plot and year as subplot. The Tukey test was used for pairwise comparisons of means (MINITAB version 14.13). To analyse differences in soil nutrient content between years, pairwise *t* tests of differences was performed for each cropping system separately.

6.3 Results and Discussion

6.3.1 Major Nutrient Flows and Changes in Soil Nutrient Pools

6.3.1.1 Nitrogen

The N input was in the range of 63–100% of the N output (Table 6.3). All the arable systems had a calculated reduction in the soil N content. The INT-A and ORG-A had reductions of about 10 and 30 kg N ha^{-1} yr^{-1} more than CON-A, respectively. The soil N content of the forage system CON-F was sustained, whereas that of INT-F showed a slight reduction. The greatest reductions were calculated for the organic systems ORG-A and ORG-F, which accumulated to 423 and 424 kg N ha^{-1} over the 10-year period, respectively. This corresponds to approximately 3.3% of the initial soil N pool (to a depth of 1 m).

The calculated changes corresponded reasonably well with the measured changes in topsoil N content (Table 6.2). CON-F showed no significant reduction in topsoil N, whereas the other systems had a reduced N pool. The INT-A was an exception, as no changes in topsoil N content could be measured, in spite of substantial N deficit in the calculations (Table 6.3). One explanation for this diverging finding may be the use of reduced tillage in INT-A. Reduced tillage is commonly reported to increase

Table 6.3 Measured and estimated nitrogen flows (kg N ha^{-1} yr^{-1}) in the cropping systems at Apelsvoll, mean 1990–1999[a]

N-flow	Cropping system					
	CON-A	INT-A	ORG-A	CON-F	INT-F	ORG-F
Fertiliser	119	75.2	0	108	54.2	0
Cattle slurry	0	0	21.8	90.7	70.1	59.2
Wet atm. deposition	6.20	6.20	6.20	6.20	6.20	6.20
Dry atm. deposition	2.00	2.00	2.00	2.00	2.00	2.00
Symbiotic N-fixation	0	0	35.9	23.4	36.3	53.9
N in seeds	9.77	9.77	7.59	2.50	2.50	3.00
Sum input	137	93.2	73.5	233	171	124
Harvest[b]	103	80.0	89.5	169	143	126
NH_3 —N volat. cattle slurry	0	0	2.84	22.1	17.3	16.8
NH_3 —N volat. crop	2.00	2.00	2.00	2.00	2.00	2.00
Denitrification	8.31	5.27	1.20	11.5	7.08	2.56
Surface N-runoff	2.23	1.59	1.83	1.99	1.61	1.93
Drainage N-leaching	32.9	26.7	18.4	26.4	16.0	17.2
Sum output	149	116	116	233	187	167
ΔN[c] (input minus output)	***−12.0***	***−22.4***	***−42.3***	***0.61***	***−15.9***	***−42.4***

[a] Timestep used is the agrohydrological year (May–April), thus covering the period May 1990–April 2000.
[b] Including N removed with straw (all plots with undersown clover grass/ley, and all cereal plots in 1990–1992).
[c] Changes in soil total-N.

the content of nutrients in the upper part of the topsoil (e.g. Alvarez et al., 1995). A translocation of nutrients from the layer below 30 cm to the topsoil (through plant N uptake followed by litter fall) could have masked a possible reduction in the soil N pool of INT-A.

The calculated changes in soil N confirm the results reported for the first 8 years of the experiment (Korsaeth and Eltun, 2000). The dominating negative values reflect the high initial level of mineralisable N in the soil at this particular site (14.3 tonnes N ha^{-1}). It is difficult to avoid a net reduction of soil organic N in soils with a high content of mineralisable organic N, especially in arable cropping systems (Christensen, 1990; Uhlen, 1991; Heenan et al., 1995; Thomsen and Christensen, 1998). At the other extreme, most cultivation regimes will result in a net accumulation of soil organic N on soils with very low initial soil organic N levels (Fettell and Gill, 1995; Poulton, 1995; Raun et al., 1998). Since the absolute level of ΔN depends on the cultivation history of a field as well as its present regime, the value for one particular case is not a valid criterion for its sustainability. Thus, the systems sustainability cannot be evaluated by the absolute values of ΔN found in our study. On the other hand, the ranking of the cropping systems is useful as an indication of the degree to which each cultivation regime is able to sustain soil organic N levels.

6.3.1.2 Phosphorus

The P input was within the range of 49–131% of the P output (Table 6.4). The two arable systems CON-A and INT-A had an increase in the calculated soil P content,

Table 6.4 Measured and estimated phosphorus flows (kg P ha^{-1} yr^{-1}) in the cropping systems at Apelsvoll, mean 1990–1999[a]

P-flow	Cropping system					
	CON-A	INT-A	ORG-A	CON-F	INT-F	ORG-F
Fertiliser	27.2	19.2	0	11.5	7.52	0
Cattle slurry	0	0	5.19	17.7	13.6	11.4
P in seeds	2.10	2.10	1.56	0.42	0.39	0.41
Sum input	29.3	21.3	6.75	29.6	21.5	11.8
Harvest[b]	21.8	18.7	18.9	28.7	26.1	23.5
Surface P-runoff	0.35	0.28	0.35	0.33	0.31	0.38
Drainage P-leaching	0.11	0.08	0.12	0.12	0.10	0.11
Sum output	22.3	19.1	19.4	29.1	26.5	24.0
ΔP[c] (input minus output)	***7.03***	***2.18***	***−12.6***	***0.51***	***−4.98***	***−12.1***

[a]Timestep used is the agrohydrological year (May–April), thus covering the period May 1990–April 2000.
[b]Including P removed with straw (all plots with undersown clover grass/ley, and all cereal plots in 1990–1992).
[c]Changes in soil total-P.

particularly CON-A, where the increase accumulated to 70 kg P ha^{-1} over the 10 years. The other extreme was found in the two organic systems, ORG-A and ORG-F, which had net reductions of more that 120 kg soil P ha^{-1} over the same period. The CON-F appeared to sustain its soil P content, whereas INT-F was intermediate in the group of forage systems, with a reduction less than half that of ORG-F.

Could these calculated changes be confirmed by soil analyses? Soil total P was not measured at the start of the experiment, but the amounts of plant available P (P-AL) were measured both initially and in 2003 (Table 6.2). All systems showed a decline in P-AL from 1988 to 2003, except INT-A, which had an unaltered P-AL level (Table 6.2). These results agree reasonably well with calculated changes of the integrated and the organic systems. For the conventional systems there was a mismatch, in particular for CON-A, where a reduction in soil P-AL was measured despite a relatively large calculated increase in soil P.

The amount of plant available P in soil is, of course, not a direct measure of total soil P. Both crop rotation and P and N fertilization are reported to affect the distribution of P among labile and stabile inorganic and organic P forms (McKenzie et al., 1992). In a Swedish long-term fertility study, Djodjic et al. (2005) observed a slow decrease in P-AL in a treatment with a balanced P input (P rates to replace P removal by crops), and they assumed that this reduction was due to a progressive shift towards less plant available P forms through the adsorption and fixation of P in soil. This phenomenon most likely explains some of the observed decline in P-AL in this study.

In 1988, the experimental area was fallowed, whereas the samples taken in 2003 were taken after growth of different crops. The lack of plant removal in 1988 may also be a reason for the higher P-AL contents compared to those in 2003. The use of P-AL measurements to validate changes in soil total P may therefore not be justified.

6.3.1.3 Potassium

The calculated K input was within the range of 63–149% of the calculated K output (Table 6.5). Calculated net changes in soil K followed a similar pattern as for soil P, with a net increase in soil K for CON-A and INT-A, roughly sustained K content in CON-F, a net reduction in INT-F and the largest net reduction in the two organic systems, ORG-A and ORG-F. A net reduction in soil K is commonly found in studies of organic farming (Öborn et al., 2005). As an example, Torstensson et al. (2006) calculated an annual deficit of 36 kg K ha^{-1} in an organic cropping system on sandy loam in southern Sweden, with animal manure application and a 6-year crop rotation with three cereal crops, potatoes and two forage crops. This relatively large deficit occurred even when the system was fertilized with 25 kg K ha^{-1} yr^{-1}.

The calculated net reduction of soil K in the present study was poorly reflected by the measured changes in topsoil K – HNO_3 (0–30 cm) from 1988 to 2003. The variation in the initial K – HNO_3 content of the soil was large (Riley and Eltun, 1994). This combined with difficulties of finding exactly the same location for soil sampling in 2003 as in 1988 may explain some of the observed discrepancy. In an earlier report from the experiment, Løes et al. (1998) also found a mismatch between calculated and measured changes in soil K. They related this to measuring errors (i.e. problems of obtaining true dependent samples) due to large natural variation in the soil K content. Moreover, during the 15 years between the samplings considered here, it is most likely that there was considerable lateral transport of soil, mainly due to tillage, which probably increased the problem of obtaining true dependent samples for comparisons. For INT-A, reduced tillage probably affected the topsoil content of K – HNO_3 as already discussed for P-AL.

The accumulated soil K reduction was largest in ORG-F, calculated to 498 kg K ha^{-1} during the 10 years. Öborn et al. (2005) emphasized the uncertainties

Table 6.5 Measured and estimated potassium flows (kg K ha^{-1} yr^{-1}) in the cropping systems at Apelsvoll, mean 1990–1999[a]

K-flow	Cropping system					
	CON-A	INT-A	ORG-A	CON-F	INT-F	ORG-F
Fertiliser	77.2	58.3	0	51.6	41.7	0
Cattle slurry	0.00	0.00	42.0	126.1	97.6	82.4
K in seeds	10.7	10.7	7.76	0.81	0.81	0.82
Sum input	87.9	69.0	49.7	178	140	83.2
Harvest[b]	50.1	42.9	62.7	166	143	123
Surface K-runoff[a]	3.95	3.05	4.02	4.97	4.44	5.25
Drainage K-leaching[a]	5.08	3.17	3.82	4.10	3.00	4.45
Sum output	59.1	49.1	70.6	175	150	133
ΔK[c] (input minus output)	***28.8***	***19.9***	***−20.9***	***3.20***	***−10.3***	***−49.8***

[a] Timestep used is the agrohydrological year (May–April). Runoff and leaching data for K was available for the period May 1991–April 1996.

[b] Including K removed with straw (all plots with undersown clover grass/ley, and all cereal plots in 1990–1992).

[c] Changes in soil total-K.

in K mass balance calculations, and pointed at the mismatch between the frequently reported negative K balances over time and the very few instances of obvious K deficiencies reported. They listed various possible explanations for this, including errors in mass balances through the use of standard values. The K concentration in crops may vary significantly depending on management practice and site condition, including K status of the soil (Öborn et al., 2005). In the present study the nutrient content in crops was measured over the first 6 years of the experiment, but estimated values were used for the last 4 years. A possible overestimation of the K content of crops and removed straw in ORG-F could have explained some of the large calculated deficit. The calculated K balance for ORG-F showed, however, an even larger deficit for the period 1990–1996 ($-64\,\text{kg K ha}^{-1}\,\text{yr}^{-1}$, not including leaching and runoff losses, Løes et al., 1998). The question then is when plant K deficiencies will be observed in the experiment.

6.3.2 Nutrient Losses via Drainage Water and Surface Runoff

6.3.2.1 Nitrogen Losses

During the ten agrohydrological years May 1990–April 2000 (120 months) presented here, drainage discharge and surface runoff occurred in 96 and 44 months, respectively. The monthly mean concentration of N was roughly five times higher in drainage water than in surface runoff. Differences between cropping systems in terms of N concentrations were largest in drainage water (Fig. 6.2, upper plots). The highest N concentrations were measured in drainage water from the two conventional systems (CON-A and CON-F) in spring after fertilization and in late autumn and early winter. The overall lowest N concentrations in drainage water were measured in the two organic systems (ORG-A and ORG-F) and INT-F. The pattern of the monthly N loads followed to a large extent the pattern of the N concentrations (Fig. 6.2, middle plots). However, a diverging pattern was seen in early spring. Low temperatures during January–March led to accumulation of snow, which delayed most of the vertical N transport until snow melting in April. As a result, the averaged monthly N loads were largest in April. There were, however, also substantial losses during late autumn and early winter. The N loads were least in the periods July–September and January–March.

On an annual basis, the N losses via drainage water and surface runoff increased in the order ORG-A $\approx$ INT-F $\approx$ ORG-F $<$ INT-A $\approx$ CON-F $<$ CON-A (Fig. 6.2, lower plots). The latter system lost about twice that of the group of cropping systems with the least drainage and runoff N losses (ORG-A, INT-F and ORG-F).

Total N lost from CON-A via drainage water and surface runoff was about the same as that measured in an arable conventional cropping system (6-year crop rotation: barley, oat, spring wheat, barley, oat, potato) on a sandy loam in southern Sweden (Torstensson et al., 2006). In contrast to the present study, Torstensson et al. (2006) reported that the N losses were as high in an organic system (with animal manure, crop rotation: barley, forage crop, forage crop, oat, pea/barley, potato). There are considerable differences between the experiments, both in management,

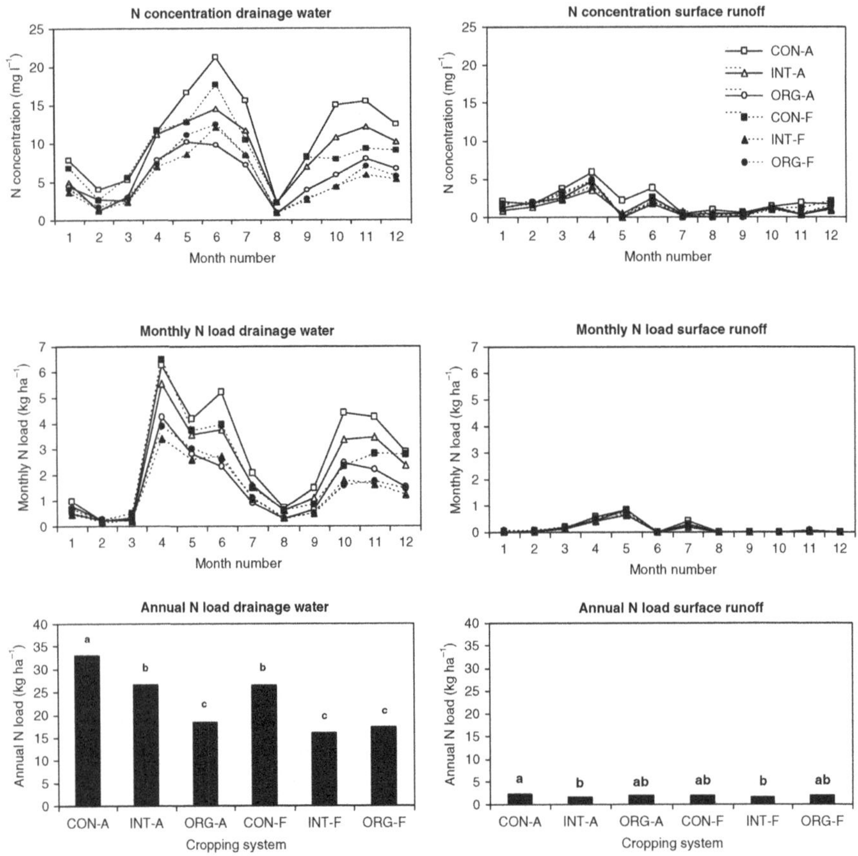

Fig. 6.2 Measured concentrations (*upper plots*) and loads (*middle plots*) of total N as monthly averages (*lines*), and annual loads (*lower plots, bars*) of total N averaged over the agrohydrological years May 1990–April 2000, for drainage water (*left plots*) and surface runoff (*right plots*) from the six cropping systems at Apelsvoll. Bars with the same letter are not significantly different ($P = 0.05$)

soil type and climatic conditions. Whereas the simple N balance (N input minus harvested N) of the conventional system in Sweden was quite similar to that of the organic system (about 10 kg N ha^{-1} larger), the corresponding difference in the present study was 36 kg N ha^{-1} (CON-A vs. ORG-F). The positive relation between N balance and N leaching has been reported earlier (Korsaeth and Eltun, 2000) and will be highlighted more below (Section 6.3.3.1). In addition to more similar N balance of the organic and the conventional system, larger precipitation and higher soil temperatures during the experiment of Torstensson et al. (2006) compared with the present study, may have contributed to the relatively large N leaching from their organic system. The application of N in organic forms involves risks for N release at times without plant N demands. At Apelsvoll the soil is normally frozen during winter, and N mineralization is thus almost absent (Korsaeth et al., 2002). Higher soil temperatures during late autumn and winter, as found in southern Sweden, enhance the risk for N leaching, and this risk is further increased through the larger amounts of precipitation (Korsaeth et al., 2003).

6.3.2.2 Phosphorus Losses

In contrast to N, the P concentrations were largest in surface runoff (Fig. 6.3, upper plots), with the highest values measured during winter and early spring. The peak in June is the result of only one incident; surface runoff in June occurred only in 1995. After a wet May, more than 11 mm of rain came within a few minutes on the afternoon on 1 June 1995. This unusually high intensity led to surface runoff and large erosion losses. The incident caused P losses up to 1.4 kg P ha^{-1} (CON-A),

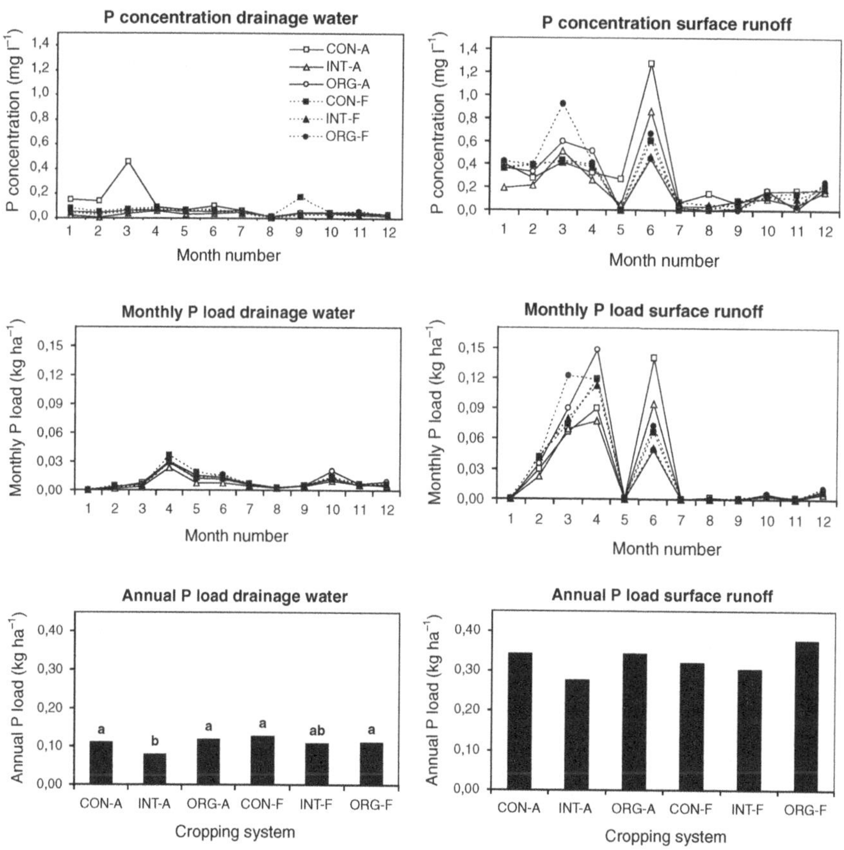

Fig. 6.3 Measured concentrations (*upper plots*) and loads (*middle plots*) of total P as monthly averages (*lines*), and annual loads (*lower plots, bars*) of total P averaged over the agrohydrological years May 1990–April 2000, for drainage water (*left plots*) and surface runoff (*right plots*) from the six cropping systems at Apelsvoll. Bars with the same letter are not significantly different ($P = 0.05$)

which is about three times the average annual P loss via drainage water and surface runoff.

The monthly P loads were very small during the cropping season (June 1995 was the exception), and autumn through early winter, with increasing losses from February to a peak in April (Fig. 6.3, middle plots). There was a strong tendency that the cropping systems with ley in the rotation (ORG-A, CON-F, INT-F and ORG-F) had higher P loads in the surface runoff in March and April compared to systems without ley in the rotation. This may be explained by cell damage of the grasses due to freezing and thawing processes during winter (Miller et al., 1994), with a subsequent loss of soluble components (Timmons et al., 1970; White, 1973), such as $PO_4{-}P$. This assumption was supported by the fact that the loads of $PO_4{-}P$ in surface water was 60–80% larger from cropping systems with ley in the rotation than those without any fresh plant material on the surface during winter (data not shown). Against this background, the use of "vegetation filters" to reduce P losses is questionable.

There were no significant differences among cropping systems in terms of annual P load in surface runoff, but INT-A tended to have the lowest surface losses of P (Fig. 6.3, lower plots). In drainage water, INT-A had significantly lower P load than the other systems, except for INT-F. The tillage of INT-A differed from all the other cropping systems, as spring harrowing was used instead of ploughing. Reduced tillage is commonly found to reduce the risk of erosion and losses of particle bound nutrients (Puustinen et al., 2005). On the other hand, the relatively short slope length and small slope angle of the experimental field implies that the risk of erosion is relatively small on the experimental site, and the only surface runoff of importance (except for in 1995) was usually measured during snow melting in spring, before tillage was performed. The results nevertheless suggest that there is an advantage of excluding ploughing in terms of reducing P losses. This is most likely due to the high structural stability measured in INT-A (Riley et al., 2006).

6.3.2.3 Potassium Losses

In contrast to N and P, K was lost in almost equal amounts via drainage water (56%) and surface runoff (44%). The annual pattern in concentrations and loads of K was quite similar to that of P, but at much higher levels (Fig. 6.4, upper and middle plots). The incident with surface runoff in June 1995 caused very high losses also of K, reaching up to 5.3 kg K ha^{-1} (CON-A). The monthly K loads in surface water were largest during the period February-April, and the systems with ley in the rotation showed a clear tendency to the largest losses during this period. This indicates that fresh plant material on the ground during winter also increased the risk of K losses. As for P, INT-A appeared to have the overall least K losses of the systems (Fig. 6.4, lower plots).

The total K losses (sum of K lost via drainage water and surface runoff) were somewhat smaller than those measured in the cropping systems experiment on sandy loam in Sweden (Torstensson et al., 2006), where all K losses occurred via drainage

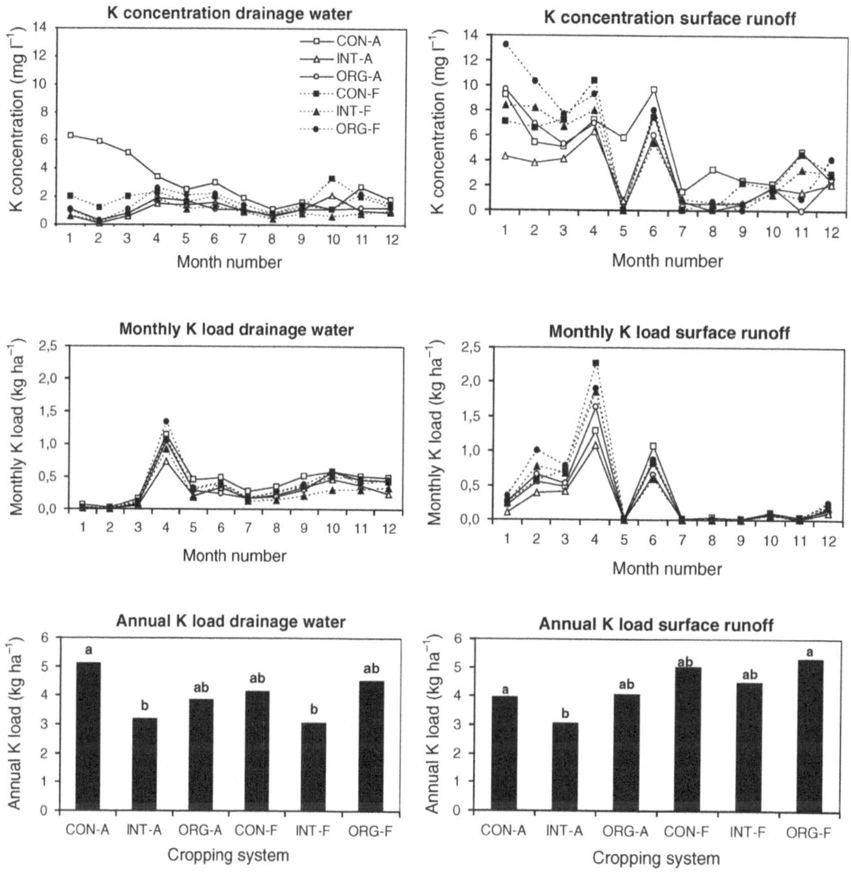

Fig. 6.4 Measured concentrations (*upper plots*) and loads (*middle plots*) of total K as monthly averages (*lines*), and annual loads (*lower plots, bars*) of total K averaged over the agrohydrological years May 1990–April 2000, for drainage water (*left plots*) and surface runoff (*right plots*) from the six cropping systems at Apelsvoll. Bars with the same letter are not significantly different ($P = 0.05$)

water. This is as expected, since the soil type affects K leaching, with greater K leaching losses from sandy soils than from clayey soils (Alfaro et al., 2004).

As the main focus of agricultural research in Europe has changed from productivity to environmental issues, K has become a "forgotten" nutrient element (Öborn et al., 2005). However, the results presented here showed that K leaching is a major K flow out of a system, thus contributing to soil K mining.

6.3.3 Nutrient Balances and Leaching/Runoff Losses

One major path for nutrient losses from agriculture is the transport via drainage water and surface runoff. Reliable measurements of nutrient leaching and runoff losses are difficult to obtain and require rather expensive experimental setup (e.g., Bergström and Brink, 1986; Uhlen, 1994). Finding a simple but robust way to estimate such losses would thus be a great advantage when evaluating cropping systems.

Evaluating the first 8 years of the present experiment, we found that annual N balances could be used to explain most ($r^2 = 0.58$–0.87) of the variation in annual N lost via drainage water and surface runoff (sum of total N), when combined with data on annual precipitation (Korsaeth and Eltun, 2000). The average N balance calculated for these 8 years predicted on its own 86% of the variation in N lost via drainage water and surface runoff from the arable systems. In the present overview, the same method was used for the whole dataset (10 years) and the method was expanded to include total P and K as well.

6.3.3.1 Nitrogen

Nitrogen mass balance predicted up to 28% of the variation in N losses via drainage water and surface runoff, but precipitation was a better single predictor (Table 6.6). All models performed better using data from the arable systems. The simplest N balance, including N supply via fertilizer, slurry and symbiotic fixation, and N removal via harvested produce only (N $\text{balance}_{\text{simple}} = N_{\text{fertiliser}} + N_{\text{slurry}} + N_{\text{fixation}} - N_{\text{harvest}}$), was a better overall predictor than the N balance including all major N flows included in Table 6.3 (N $\text{balance}_{\text{complex}} = N_{\text{fertiliser}} + N_{\text{slurry}} + N_{\text{wet dep}} + N_{\text{dry dep}} + N_{\text{seed}} + N_{\text{fixation}} - N_{\text{harvest}} - N_{\text{NH3 vol. slurry}} - N_{\text{NH3 vol. crop}} - N_{\text{denitrification}}$). An exception was the case of the combined model using N balance and precipitation to predict N lost via drainage water and surface runoff from the forage systems.

Nitrogen leaching and runoff from agricultural fields depend primarily on the mobility of the soil N present, and surplus water to transport soluble N out of the field. Even when mobile soil N is present, no leaching or runoff occurs if surplus

Table 6.6 Coefficients of determination (R^2) for linear regressions using the N balance calculations, precipitation (Prec_t) and precipitation from the previous year (Prec_{t-1}) as predictors, and the sum of surface N runoff and drainage N leaching as the dependent variable

Data	Predictor 1	Predictor 2	Predictor 3	R^2 (p-value)	
				Arable	Forage
y[a]	N $\text{balance}_{\text{simple}}$[b]			0.28 (<0.001)	0.20 (0.001)
y	N $\text{balance}_{\text{complex}}$[c]			0.28 (<0.001)	0.14 (0.010)
y	Prec_t			0.33 (<0.001)	0.31 (<0.001)
y	N $\text{balance}_{\text{simple}}$[b]	Prec_t		0.48 (<0.001)	0.43 (<0.001)
y	N $\text{balance}_{\text{complex}}$[c]	Prec_t		0.47 (<0.001)	0.47 (<0.001)
y	N $\text{balance}_{\text{simple}}$[b]	Prec_t	Prec_{t-1}	0.69 (<0.001)	0.59 (<0.001)
y	N $\text{balance}_{\text{complex}}$[c]	Prec_t	Prec_{t-1}	0.69 (<0.001)	0.59 (<0.001)
$\bar{y}$[d]	N $\text{balance}_{\text{simple}}$[b]			0.87 (0.007)	0.54 (0.098)
$\bar{y}$	N $\text{balance}_{\text{complex}}$[c]			0.86 (0.008)	0.50 (0.118)

[a] Data from each agrohydrological year (May–April).

[b] N $\text{balance}_{\text{simple}} = N_{\text{fertiliser}} + N_{\text{slurry}} + N_{\text{fixation}} - N_{\text{harvest}}$.

[c] N $\text{balance}_{\text{complex}} = N_{\text{fertiliser}} + N_{\text{slurry}} + N_{\text{wet dep.}} + N_{\text{dry dep}} + N_{\text{seed}} + N_{\text{fixation}} - N_{\text{harvest}} - N_{\text{NH3 vol. slurry}} - N_{\text{NH3 vol. crop}} - N_{\text{denitrification}}$.

[d] Data averaged over all years.

water for N transport is lacking. This means that some of the leachable N may be left in the soil in a dry year, thereby increasing the potential for N leaching the following year. Such "delayed N leaching" would cause reduced efficiency in the annual regressions. By combining annual N balances, total precipitation in the same year and that of the previous year in a three-predictor model, some of this effect was obviously reduced (Table 6.6). In this way 69 and 59% of the variation in N leaching + runoff between systems and years could be explained for the arable and the forage systems, respectively (Table 6.6).

By using average values, the effect of the climatic variation was eliminated (Table 6.6). The differences between the arable systems were well described by this model, but not those between the forage systems. Plotting the results, however, revealed a different pattern between the two groups of cropping systems. The forage systems appeared to have a higher N balance threshold, below which the N leaching + runoff was insensitive to the N balance (Fig. 6.5). This may be explained by the effect of the perennial ley. Perennial grassland, with a long growing period, takes up inorganic N at times when it would otherwise be exposed to leaching (e.g. Bergström, 1987; Gustafson, 1987). The presented measurements of the N concentrations in drainage water (month number 10–12, Fig. 6.2) substantiate this hypothesis, as the largest discrepancies between arable and forage systems were found in autumn, after crop harvest. The results are further supported by Rotz et al.

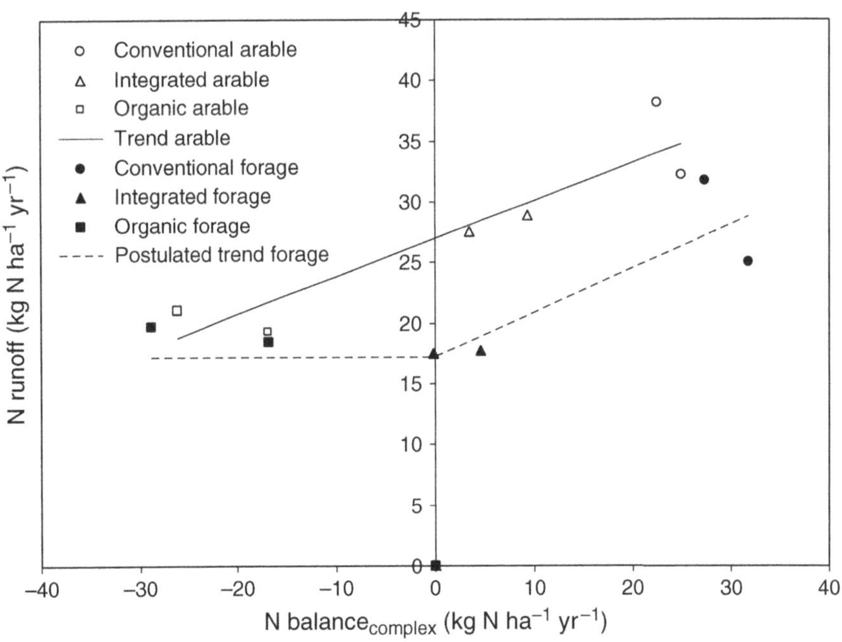

Fig. 6.5 Total N runoff (drainage and surface runoff) plotted against estimated N balance (N balance$_{complex}$ = $N_{fertiliser} + N_{slurry} + N_{wet\ dep} + N_{dry\ dep} + N_{seed} + N_{fixation} - N_{harvest} - N_{volat.\ slurry} - N_{volat.\ crop} - N_{denitrification}$). Data averaged over the agrohydrological years May 1990–April 2000

(2005), who found that N surplus and NO_3-N leaching were positively related in a range of grassland management systems at the Karkendamm experimental farm in Northern Germany. Their data also appear to have a balance threshold (Fig. 6.4b, Rotz et al., 2005), similar to that in the present study, below which N leaching + runoff is insensitive to the N balance.

6.3.3.2 Phosphorus

The relation between P balance (P balance $= P_{fertiliser} + P_{slurry} - P_{harvest}$) and P leaching/runoff was weak but significant (Table 6.7). Annual precipitation affected the P losses from the forage systems, but not from the arable systems. Precipitation the previous year was not related to P leaching/runoff. The best model combining annual P balance and precipitation could explain 19% of the variation in P runoff (forage systems, Table 6.7). Averaging the data over years did not improve the results (data not shown).

The method of using P balances to predict P leaching/runoff gave very poor results, particularly compared with the results for N. One reason for this may be that the proportion of soluble ions was smaller for phosphorus, with about 48% of total P as $PO_4{-}P$, as compared with N, which consisted of about 85% $NO_3{-}N$ (sum of drainage water and surface runoff). Thus, P losses would be less dependant on the annual amounts of surplus water, but depend more on the frequency and magnitude of erosion incidents, caused by heavy rain and/or fast snow melting. The large P losses in June 1995 illustrate this point well.

The relation between P balance and P leaching/runoff is complicated by the strong tendency of soil to bind P (Ekholm et al., 2005). If a P surplus were to enhance the amounts of plant available P in soil, larger P losses would be expected. In a 7-year lysimeter study, Ulén (1999) indeed found a very strong relation between P-AL in the topsoil and the concentration of dissolved P in drainage water. Most of any P surplus will, nevertheless, accumulate in the soil (Hooda et al., 2001).

Table 6.7 Coefficients of determination (R^2) for linear regressions using the P balance calculations, precipitation ($Prec_t$) and precipitation from the previous year ($Prec_{t-1}$) as predictors, and the sum of surface P runoff and drainage P leaching as the dependent variable

Data	Predictor 1	Predictor 2	Predictor 3	R^2 (p-value)	
				Arable	Forage
y[a]	P balance[b]			0.07 (0.047)	0.10 (0.013)
y	$Prec_t$			0.03 (0.131)	0.13 (0.004)
y	P balance	$Prec_t$		0.11 (0.124)	0.19 (<0.003)
y	P balance	$Prec_t$	$Prec_{t-1}$	0.13 (0.609)	0.26 (0.275)
$\overline{y}$[c]	P balance			0.07 (0.608)	0.26 (0.307)

[a] Data from each agrohydrological year (May–April).
[b] P balance $= P_{fertiliser} + P_{slurry} - P_{harvest}$.
[c] Data averaged over all years.

6.3.3.3 Potassium

There was no significant relationship between K balance (K balance $= K_{fertiliser} + K_{slurry} - K_{harvest}$) and K leaching/runoff. The system with the highest calculated net change in soil, ΔK (CON-A), lost almost the same amounts of K via drainage water and surface runoff as the system with the lowest ΔK (ORG-F) (Table 6.5).

Askegaard and Eriksen (2000) tested the fate of very large K rates by applying 988 kg K ha^{-1} to plots sown with a ryegrass (*Lolium perenne* L.) catch crop. They reported that the resulting K leaching accounted for only 0.2% (1.5 kg K ha^{-1}) of the applied K, and concluded that K leaching is a result of the fertilizer history, rather than of the current K balance of a field. Nevertheless, there is an indirect link between K balance and K leaching/runoff. Studying 1300 records from individual fields on arable soils in southern England, Heming (2004) found that 10–50% of the change in the soil exchangeable K could be explained by the K balance.

6.3.4 Nutrient Efficiencies

An ideal agricultural system should both maximise production and minimize any undesirable effects on the environment. The relation between the production of a system and its nutrient loss (production to loss ratio) should thus be as high as possible. In this chapter, production is presented in terms of dry matter (DM) or as harvested N, P or K. Comparisons between systems should only be made for systems with the same crop rotation, since the potential for dry matter production and nutrient yields differs considerably between crops. An alternative method, allowing for comparisons among all the systems, would be to transform all plant produce into edible units, such as energy consumable for humans. This would imply that forage crops would need to be converted into milk and meat equivalents. A study using this approach is under development.

From a resource economy point of view, a further goal is to maximize the efficiencies of input factors such as fertilizer and slurry. Here, nutrient use efficiency (NUE) was calculated as the amounts of harvested nutrients divided by the amounts of applied nutrients (fertilizer and/or slurry). It should be noted however that, with this method of calculating nutrient efficiency, the NUE may be strongly affected by other nutrient sinks and sources.

6.3.4.1 Nitrogen

The DM production to N loss ratio was least for the arable systems and largest for the forage systems (Table 6.8). Grass ley has a large production potential in terms of DM yields, explaining the larger ratios of the forage systems. INT-A had a larger production to N loss ratio than that of CON-A. This superiority of integrated over conventional cropping was even more marked for the forage systems.

Using harvested N instead of harvested DM to calculate the production to N loss ratio gave almost the same results. One small difference was that the ratio for

Table 6.8 Relationships between yields, applied and harvested amounts of nutrients and nutrients lost via drainage and surface runoff

Yield ratios	Unit	Cropping system					
		CON-A	INT-A	ORG-A	CON-F	INT-F	ORG-F
DM yields[a]/N runoff[b]	Mg DM kg N^{-1}	0.15	0.17	0.22	0.29	0.44	0.35
Harvested N[c]/N runoff	kg N kg N^{-1}	2.77	2.70	3.85	5.77	7.85	6.31
Harvested N/appl. N[d]	kg N kg N^{-1}	0.82	1.01	3.61	0.82	1.12	2.04
DM yields/P runoff	Mg DM kg P^{-1}	11.6	13.3	9.6	18.4	18.5	13.7
Harvested P/P runoff	kg P kg P^{-1}	44.5	50.0	35.7	61.3	60.9	45.7
Harvested P/appl. P	kg P kg P^{-1}	0.76	0.94	3.22	0.95	1.19	1.96
DM yields/K runoff	Mg DM kg K^{-1}	0.60	0.77	0.57	0.92	1.03	0.69
Harvested K/K runoff	kg K kg K^{-1}	5.08	6.47	7.24	18.0	18.8	12.3
Harvested K/appl. K	kg K kg K^{-1}	0.59	0.69	1.35	0.92	1.00	1.45

[a] Removed straw not included.
[b] Runoff: Total annual load in surface runoff and drainage water.
[c] Nutrients in harvested produce, removed straw not included.
[d] Nutrients applied with fertiliser and/or slurry.

CON-A appeared to increase relative to INT-A. Greater protein content of the grain produced conventionally compared to that produced with integrated management, as reported by Eltun (1996a), is probably the main reason for this change.

The NUE was remarkably similar for each pair of cropping system with the same intensity: CON-A and CON-F had the least apparent efficiency of the applied N, INT-A and INT-F were at the same intermediate level, and the two organic systems had the two largest N efficiencies of the systems.

The large NUE of the organic systems is a result of low N input via slurry combined with high yields, due to considerable N input from N fixation and soil N mining. A high NUE may thus be achieved at the cost of high rates of soil mining, as shown for the organic systems.

6.3.4.2 Phosphorus

The production to loss ratio for P differed more between forage systems than between arable systems (excluding ORG-A from the comparisons due to different crop rotation) (Table 6.8). Differences between systems in terms of production to loss ratio gave the same ranking of the cropping systems regardless of whether the production was expressed as DM or harvested P. INT-A was the best arable system. Within the forage systems, CON-F and INT-F had about the same DM yield per unit lost P, whereas ORG-F had a lower production to loss ratio.

The P use efficiency was largest for the organic systems and least for CON-A (Table 6.8). The relation between these quotients and the calculated changes in soil P (Table 6.4) is apparent. All systems with a P use efficiency > 1 had a decrease in the soil P content and vice versa.

6.3.4.3 Potassium

As for N and P, the integrated systems had the largest DM production to loss ratio (comparing systems with the same crop rotation), and the ranking was unaffected by whether production was calculated as harvested DM or K (Table 6.8). The production to loss ratio for K was smaller for ORG-F relative to INT-F than that for N and P. One reason for this is that the total K loss via drainage water and surface runoff from ORG-F was larger relative to INT-F than the corresponding values for N and P.

6.4 Conclusions

6.4.1 Major Nutrient Flows and Changes in Soil Nutrient Pools

The organic systems were not sustainable as there was substantial soil mining of N, P and K. The relatively high productivity of these systems will thus sooner or later be reduced, due to nutrient deficits. The conventional forage system, which appeared to be balanced for all nutrients tested, is the system which is managed most similarly to the historic cropping practice of the experimental site. This illustrates that the maintenance level for a sustainable nutrient regime depends on the initial nutrient pools, and thus the long-term cropping history of a soil.

The calculated net changes in soil N were confirmed, at least partly, by soil measurements. Measurements of P-AL and K-HNO_3 did not, however, support the respective calculated changes of soil total P and total K. P-AL and K-HNO_3 do not directly reflect soil total P and K content. Compounds containing P and K are susceptible to transformations in soil, which implies that their pools may be changed without changing the total soil content of these nutrients. If soil measurements are to be used to back up calculated changes in soil total P and K, analyses of total P and total K should therefore be preferred. The measurement of significant changes in soil chemical properties, nevertheless poses a challenge, due to large measurement errors (Uhlen, 1991), partly caused by lateral and vertical soil transport over time. Hence, calculation of nutrient mass balances appears to be a better tool to evaluate changes in soil fertility, at least in the short and medium term.

6.4.2 Nutrient Losses via Drainage Water and Surface Runoff

Nitrogen was mainly lost via drainage water, P was mainly lost via surface runoff, whereas K losses were almost equally divided between the two pathways. These findings reflect the relative mobility and binding behaviour in soil of the respective nutrients.

The distribution of the losses within the year is strongly dependent on the hydrological conditions. Hardly any losses occurred during the winter, as the soil was normally frozen. Largest losses for all nutrients (N, P and K) were measured in spring, due to the large amounts of surplus water attributed to snow melting. Nutrient losses occurred also during autumn and early winter, but these losses were most substantial for N. Since a large proportion of P and K is bound to soil particles, losses of these nutrients are mostly connected with soil erosion under the conditions prevailing in this study. Systems that have large amounts of fresh plant material on the surface during winter also have an elevated risk for nutrient losses via surface runoff.

6.4.3 Nutrient Balances and Leaching and Runoff Losses

The N balance obtained by subtracting the amounts of N in harvested crops from the amounts of N applied via fertilizer, animal waste and symbiotic N fixation, provides a simple predictor for N leaching. For annual predictions, it is, however, necessary to consider the amounts of precipitation. For longer periods averaged balances may explain most of the variation in N leaching. Cropping systems with ley in the rotation appear to have a higher N balance threshold than do arable systems, below which the N leaching is unaffected of the N balance. This is most likely due to the ability of grassland to take up inorganic N at times when it would otherwise be exposed to N leaching. In contrast to N, mass balances of P and K were not usable as predictors for P and K leaching, respectively. This is mainly due to differences in the mobility of the dominating forms the nutrients are present in soil, and the stronger tendency of soil to bind P and K compared to N.

6.4.4 Nutrient Efficiencies

The integrated systems had the largest production to loss ratios for N, P and K within each group of production (compared with systems having the same crop rotation). Using moderate amounts of nutrient inputs, but with at least a substantial part of the nutrients in readily plant available forms (i.e. inorganic fertilizer), thus appears to be the best way to combine high yields with low leaching and runoff losses under these conditions.

Acknowledgments The authors greatly acknowledge O. Nordheim for his skilled technical assistance, H. Riley for critically reading the manuscript, the Norwegian Ministry of Agriculture and Yara International ASA for jointly funding the experiment.

References

Alfaro, M.A., Jarvis, S.C., and Gregory, P.J., 2004, Factors affecting potassium leaching in different soils, *Soil Use Manage.* **20**: 182–189.

Alvarez, R., Díaz, R.A., Barbero, N., Santanatoglia, O.J., and Blotta, L., 1995, Soil organic carbon, microbial biomass and CO_2 —C production from three tillage systems, *Soil Till. Res.* **33**: 17–28.

Andersen, A., and Eltun, R., 2000, Long-term developments in the carabid and staphylinid (Col., Carabidae and Staphylinidae) fauna during conversion from conventional to biological farming, *J. Appl. Entomol.* **124**: 51–56.

Askegaard, M., and Eriksen, J., 2000, Potassium retention and leaching in an organic crop rotation on loamy sand as affected by contrasting potassium budgets, *Soil Use Manage.* **16**: 200–205.

Bergström, L., and Brink, N., 1986, Effects of differentiated applications of fertilizer N on leaching losses and distribution of inorganic N in the soil, *Plant Soil* **93**: 333–345.

Bergström, L., 1987, Nitrate leaching and drainage from annual and perennial crops in tiledrained plots and lysimeters, *J. Environ. Qual.* **16**: 11–18.

Breland, T.A., and Eltun, R., 1999, Soil microbial biomass and mineralization of carbon and nitrogen in ecological, integrated and conventional forage and arable cropping systems, *Biol. Fertil. Soils* **30**: 193–201.

Christensen, B.T., 1990, Long-term changes in soil organic matter content of three experiments: effect of fertilization, straw incorporation and crop rotation, in: *Decomposition and Soil Organic Matter*, NJF-utredning/rapport, pp. 123–128.

Djodjic, F., Bergström, L., and Grant, C., 2005, Phosphorus management in balanced agricultural systems, *Soil Use Manage.* **21**: 94–101.

Égner, H., Riehm, H., and Domingo, W.R., 1960, Untersuchung über die chemische Boden-Analyse als Grundlage für die Beurteilung des Nährstoffzustandes der Boden. (Using chemical soil analysis as a basis for assessing soil nutrient status), *Kungl. Landtbrukshögskolans Annaler* **26**: 199–215.

Ekholm, P., Turtola, E., Grönroos, J., Seuri, P., and Ylivainio, K., 2005, Phosphorus loss from different farming systems estimated from soil surface phosphorus balance, *Agric. Ecosyst. Environ.* **110**: 266–278.

Eltun, R., 1994, The Apelsvoll cropping system experiment. I. Background, objectives and methods, *Norw. J. Agric. Sci.* **8**: 301–315.

Eltun, R., 1995, Comparisons of nitrogen leaching in ecological and conventional cropping systems, *Biol. Agric. Hortic.* **11**: 103–114.

Eltun, R., and Fugleberg, O., 1996, The Apelsvoll cropping system experiment VI. Runoff and nitrogen losses, *Norw. J. Agric. Sci.* **10**: 229–248.

Eltun, R., 1996a, The Apelsvoll cropping system experiment. III. Yield and grain quality of cereals, *Norw. J. Agric. Sci.* **10**: 7–22.

Eltun, R., 1996b, The Apelsvoll cropping system experiment. IV. Yield and quality of potatoes, forage crops and entire cropping systems, *Norw. J. Agric. Sci.* **10**: 23–42.

Eltun, R., Korsaeth, A., and Nordheim, O., 2002, A comparison of environmental, soil fertility, yield, and economical effects in six cropping systems based on an 8-year experiment in Norway, *Agric. Ecosyst. Environ.* **90**: 155–168.

Fettell, N.A., and Gill, H.S., 1995, Long-term effects of tillage, stubble, and nitrogen management on properties of a red-brown earth, *Aust. J. Exp. Agric.* **35**: 923–928.

Gustafson, A., 1987, Nitrate leaching from arable land in Sweden under four cropping systems, *Swed. J. Agric. Res.* **17**: 169–177.

Heenan, D.P., McGhie, W.J., Thomson, F.M., and Chan, K.Y., 1995, Decline in soil organic carbon and total nitrogen in relation to tillage, stubble management, and rotation, *Aust. J. Exp. Agric.* **35**: 877–884.

Heming, S.D., 2004, Potassium balances for arable soils in southern England 1986–1999, *Soil Use Manage.* **20**: 410–417.

Holtan-Hartwig, L., and Bøckman, O.C., 1994, Ammonia exchange between crops and air, *Norw. J. Agric. Sci.* **14**: 5–40.

Hooda, P.S., Truesdale, V.W., Edwards, A.C., Withers, P.J.A., Aitken, M.N., Miller, A., and Rendell, A.R., 2001, Manuring and fertilization effects on phosphorus accumulation in soils and potential environmental implications, *Adv. Environ. Res.* **5**: 13–21.

Horlacher, D., and Marschner, H., 1990, Schätzrahmen zur Beurteilung von Ammoniakverlusten nach Ausbringung von Rinderflüssigmist, *Zeitschr. Pflanzenernähr. Bodenk.* **153**: 107–115 (In German).

Korsaeth, A., and Eltun, R., 2000, Nitrogen mass balances in conventional, integrated and ecological cropping systems and the relationship between balance calculations and nitrogen runoff in an 8-year field experiment, *Agric. Ecosyst. Environ.* **79**: 199–214.
Korsaeth, A., Bakken, L.R., and Riley, H., 2003, Nitrogen dynamics of grass as affected by N input regimes, soil texture and climate; lysimeter measurements and simulations, *Nutr. Cycl. Agroecosys.* **66**: 181–199.
Korsaeth, A., Henriksen, T.M., and Bakken, L.R., 2002, Temporal changes in mineralization and immobilization of N during degradation of plant material: implications for the plant N supply and nitrogen losses, *Soil Biol. Biochem.* **34**: 789–799.
Korsæth, A., and Eltun, R., 1996, Tap av plantevernmidler i perioden 1990–1995 fra dyrkingssystemprosjektet på Apelsvoll, *Jord- og Plantekultur* **1**: 250–252 (In Norwegian).
Korsæth, A., Eltun, R., and Nordheim, O., 2001, Forandring av forsøksplanen for dyrkingssystemene på Apelsvoll, *Grønn Forskning* **1**: 316–323 (In Norwegian).
Lien, G., Flaten, O., Korsaeth, A., Schumann, K.D., Richardson, J.W., Eltun, R., and Hardaker, J.B., 2006, Comparison of risk in organic, integrated and conventional cropping systems in Eastern Norway, *J. Farm Manage.* **12**: 385–401.
Lyngstad, I., 1990, A lysimeter study on the nitrogen balance in soil. I. Fate of 15N-labelled nitrate fertilizer applied to barley, *Norw. J. Agric. Sci.* **4**: 119–128.
Løes, A.-K., Hansen, S., Eltun, R., Korsæth, A., and Nordheim, O., 1998, Fosfor og kalium i jord, og næringsbalanser ved økologisk, integrert og konvensjonell dyrking, *Planteforsk Rapport* **02/98**: 33 (In Norwegian).
McKenzie, R.H., Stewart, J.W.B., Dormaar, J.F., and Schaalje, G.B., 1992, Long-term crop-rotation and fertilizer effects on phosphorus transformations. 1. In a chernozemic soil, *Can. J. Soil Sci.* **72**: 569–579.
Miller, M.H., Beauchamp, E.G., and Lauzon, J.D., 1994, Leaching of nitrogen and phosphorus from the biomass of three cover crop species, *J. Environ. Qual.* **23**: 267–272.
Maag, M., 1995, Denitrification losses from sandy and sandy loam soils under spring barley fertilized with slurry and ammonium nitrate, *Acta Agric. Scand. Sect. B* **45**: 234–241.
Oberle, S.L., and Keeney, D.R., 1991, A case for agricultural systems research, *J. Environ. Qual.* **20**: 4–7.
Öborn, I., Andrist-Rangel, Y., Askegaard, M., Grant, C.A., Watson, C.A., and Edwards, A.C., 2005, Critical aspects of potassium management in agricultural systems, *Soil Use Manage.* **21**: 102—112.
Pommeresche, R., Hansen, S., Eltun, R., and Korsaeth, A., 2006, Long-term effects of cropping systems on the earthworm populations in a loam soil, *Bioforsk Tema* **1**: 1–4.
Poulton, P.R., 1995, The importance of long-term trials in understanding sustainable farming systems: the Rothamsted experience, *Aust. J. Exp. Agric.* **35**: 825–834.
Puustinen, M., Koskiaho, J., and Peltonen, K., 2005, Influence of cultivation methods on suspended solids and phosphorus concentrations in surface runoff on clayey sloped fields in boreal climate, *Agric. Ecosyst. Environ.* **105**: 565–579.
Raun, W.R., Johnson, G.V., Sembiring, H., Lukina, E.V., LaRuffa, J.M., Thomason, W.E., Phillips, S.B., Solie, J.B., Stone, M.L., and Whitney, R.W., 1998, Indirect measures of plant nutrients, *Comm. Soil Sci. Plant Anal.* **29**: 1571–1581.
Repstad, K., and Eltun, R., 1997, Economic return with conventional, integrated and ecological farming, *NILF Report* **3**: 60.
Riley, H., and Eltun, R., 1994, The Apelsvoll cropping system experiment. II. Soil characteristics, *Norw. J. Agric. Sci.* **8**: 317–333.
Riley, H., Eltun, R., and Korsaeth, A., 2006, Changes in soil structure between 1988 and 2003 in a comparison of cropping systems in southeast Norway. *Proc. Int. Soil Till. Res. Org.*, 17th Triennal Conference, 28.08-01.09.06, Kiel, Germany.
Rotz, C.A., Taube, F., Ruselle, M.P., Oenema, J., Sanderson, M.A., and Wachendorf, M., 2005, Whole-farm perspectives of nutrient flows in grassland agriculture, *Crop Sci.* **45**: 2139–2159.
Ryden, J.C., 1985, Denitrification loss from managed grassland, in: *Denitrification in the Nitrogen Cycle*, H.L. Golterman, ed., Plenum Press, New York, pp. 121–134.

Svensson, B.H., Klemedtsson, L., Simkins, S., Paustian, K., and Rosswall, T., 1991, Soil denitrification in three cropping systems characterized by differences in nitrogen and carbon supply. I. Rate distribution frequencies, comparison between systems and seasonal N-losses, *Plant Soil* **138**: 257–271.

Thomsen, I.K., and Christensen, B.T., 1998, Cropping system and residue management effects on nitrate leaching and crop yields, *Agric. Ecosyst. Environ.* **68**: 73–84.

Timmons, D.R., Holt, R.F., and Latterell, J.J., 1970, Leaching of crop residues as a source of nutrients in surface runoff water, *Water Resour. Res.* **6**: 1367–1375.

Torstensson, G., Aronsson, H., and Bergström, L., 2006, Nutrient use efficiencies and leaching of organic and conventional cropping systems in Sweden, *Agron. J.* **98**: 603–615.

Tørseth, K., and Manø, S., 1996, Overvåking av langtransportert forurenset luft og nedbør. *Atmoskfærisk tilførsel* 1996. Norsk institutt for luftforskning, Kjeller, Norway (In Norwegian).

Uhlen, G., 1989, Nutrient leaching and surface runoff in field lysimeters on a cultivated soil. Nutrient balances 1974–1981, *Norw. J. Agric. Sci.* **3**: 33–46.

Uhlen, G., 1991, Long-term effects of fertilizers, manure, straw and crop rotation on total-N and total-C in soil, *Acta Agric. Scand. Sect. B* **41**: 119–127.

Uhlen, G., 1994, The leaching behaviour and balances of nitrogen and other elements under spring wheat in lysimeter experiment 1985–1992, *Acta Agric. Scand. Sect. B* **44**: 201–207.

Ulén, B., 1999, Leaching and balances of phosphorus and other nutrients in lysimeters after application of organic manures or fertilizers, *Soil Use Manage.* **15**: 56–61.

Vagstad, N., Eggestad, H.O., and Høyås, T., 1997, Mineral nitrogen in agricultural soils and nitrogen losses: Relation to soil properties, weather conditions, and farm practices, *Ambio* **26**: 266–272.

Vatn, A., Bakken, L.R., Bleken, M.A., Botterweg, P., Lundeby, H., Romstad, E., Rørstad, P.K., and Vold, A., 1996, Policies for reduced nutrient losses and erosion from Norwegian agriculture, *Norw. J. Agric. Sci.* **23**: 1–319.

White, E.M., 1973, Water leachable nutrients from frozen or draied prairie vegetation, *J. Environ. Qual.* **2**: 104–107.

Chapter 7
Use Efficiency and Leaching of Nutrients in Organic and Conventional Cropping Systems in Sweden

Lars Bergström, Holger Kirchmann, Helena Aronsson, Gunnar Torstensson and Lennart Mattsson

Abstract In the past few years, organic farming has been proposed as a possible way of reducing N leaching from agricultural soils and improving the use efficiency of plant nutrients. This is, to a large extent, considered to be attributed to the fact that synthetic fertilisers are not allowed in such systems and the N inputs mainly originate in various types of organic manures. In this overview, results from a number of Swedish field studies are presented in which crop yields, nutrient-use efficiencies and leaching in organic and conventional systems are evaluated. Some studies were conducted in lysimeters and others in large tile-drained field plots. In two lysimeter experiments, leaching of N derived from either poultry manure or red clover (*Trifolium pratense* L.) green manure were compared with fertiliser N, all labeled with 15-N. In the lysimeters on which poultry manure was applied, 32% of N applied leached during three years, whereas only about 3% leached in ammonium nitrate fertilised lysimeters. In plots on a sandy soil, annual N leaching loads averaged over the whole 6-yr crop rotation reached 39 kg N ha^{-1} in the organic rotations and 25 kg N ha^{-1} in the conventional rotation. Phosphorus-leaching loads were overall small in all systems, whereas K leaching was highest in the conventional rotation (i.e., on average, 27 kg ha^{-1} yr^{-1}). In terms of crop yields, they were reduced by 20–80% in the organic rotations compared to the same crops in the conventional rotations. This was explained in terms of N deficiency, weed competition, and infestation of crop diseases in the organic systems. These results suggest that organic crop production uses agricultural soils less efficiently, with no benefit for water quality.

Keywords Crop yields · Fertilisers · Manures · Water quality · Weed pressure

L. Bergström (✉)
Department of Soil and Environment, Swedish University of Agricultural Sciences, P.O. Box 7014, SE-75007 Uppsala, Sweden
e-mail: lars.bergstrom@mark.slu.se

H. Kirchmann, L. Bergström (eds.), *Organic Crop Production – Ambitions and Limitations*, DOI 10.1007/978-1-4020-9316-6_7,

7.1 Introduction

All forms of agriculture have the ultimate goal of producing abundant and nutritious food at a reasonable cost, with a minimum of disturbance to the environment and in a sustainable manner. In conventional agriculture, this has been achieved by a number of different technologies that have continuously been refined over time. Despite such technological improvements, agricultural activities can cause environmental disturbances, of which contamination of various water bodies by nutrients is among the most serious. During the past couple of decades, eutrophication has been the driving force for development of efficient countermeasures to reduce N leaching losses from agricultural fields. These include cover crops (Aronsson, 2000), use of controlled-release fertilisers (Giller et al., 2004) and conservation tillage practices to reduce P losses (Withers and Jarvis, 1998). In parallel, whole farming concepts (organic, integrated) have been proposed as a solution for more environmentally sensitive production. Organic farming has received a lot of attention among the scientific community, farmers and the general public since it has been suggested that this type of agriculture is the way forward. Today organic farming is practised in many countries around the world and intensive lobbying has also triggered some political actions. For example, a few years ago, the Swedish government set up the goal that by the year 2005, 20% of agricultural soils should be under organic farming. However, this goal was not reached and only approximately 13% of Swedish arable land is currently (2008) cultivated according to organic practices. It is important to note that there is a lack of scientific evidence to support such political actions (Trewavas, 2004). Nevertheless, organic farming is the only farming system to be legally defined in various regulations, due to the fact that it is believed to be superior in terms of sustainability (Watson et al., 2002).

The other farming concept that strives to achieve long-term sustainability is integrated farm management. The fundamental concept behind integrated farming is to combine the best of traditional farming with sensible use of modern technology (Trewavas, 2001). What really distinguishes organic from integrated and conventional cropping systems is the *a priori* exclusion of soluble inorganic fertilisers and synthetic pesticides in the former. Only nutrient inputs in various organic forms (e.g. animal and green manures) or untreated, naturally occurring minerals, often with very low solubility (e.g. apatite for P), can be used in organic farming.

Key components of soil fertility and nutrient-use efficiency in organic cropping systems have been examined in several studies (see Supplement to Soil Use and Management Vol. 18, 2002). However, there is a tendency to consider the entire systems rather than studying single processes in the different systems. This is due to a resistance among researchers who have a positive attitude towards organic farming to make direct comparisons between certain aspects or isolated processes of two systems (e.g. organic and conventional systems; MacKerron et al., 1999) but to be biased towards a holistic approach (Fjelsted Alrøe and Kristensen, 2002). This is certainly true for comparisons of both nutrient leaching (see review by Kirchmann and Bergström, 2001) and crop yield (e.g. Ivarson and Gunnarsson, 2001) in organic and conventional systems. Any estimates made are often based on nutrient budgets (e.g. Halberg et al., 1995; Dalgaard et al., 1998; Hansen et al., 2001), mineral N content in soil (Watson et al., 1993; Kristensen et al., 1994) or modelling (Johnsson et al., 2006) rather than direct measurements. In terms of leaching, the nutrient surpluses calculated in this way are simply indicators of the potential losses from the systems (Dalgaard et al., 1998) with no partition between different losses (e.g. leaching, denitrification or volatilisation in the case of N). It is also quite common to base leaching estimates on nutrient concentrations obtained from soil water samples in porous suction probes and water flux calculations with mathematical simulation models (e.g. Stopes et al., 2002), which also generates uncertain load estimates. Another limitation of a number of leaching studies is that leaching loads are only related to the cropping area, disregarding yield differences (Kirchmann and Bergström, 2001; Corré et al., 2003), which can lead to erroneous conclusions about the environmental benefits of organic systems (e.g. Stockdale et al., 2001). The leaching measurements presented in the present overview were carried out in tile-drained field plots or field lysimeters.

In this overview the following questions are discussed: (i) How are yield levels and plant nutrient uptake affected in organic and conventional systems? (ii) Do leaching losses of nutrients decrease following a change to organic cropping practices in which organic manures are used rather than soluble mineral fertilisers? (iii) Are there significant differences in nutrient-use efficiency between organic and conventional cropping systems? (iv) How does weed pressure in an organic system affect the N balance? The presentation is based on published Swedish field studies carried out in plot experiments (Torstensson et al., 2006; Kirchmann et al., 2007) and lysimeters (Bergström and Kirchmann, 1999, 2004) during periods ranging from 2 to 18 years.

7.2 Brief Description of Sites

7.2.1 Lysimeter Studies in Uppsala

The lysimeters included in this overview contained an undisturbed sandy soil profile (0.3-m diam. and 1.0-m long) collected at the Mellby site described below. The soil columns were collected with a coring technique which ensures that the integrity of the soil is maintained (Persson and Bergström, 1991), and side-wall flow effects are minimized (Bergström et al., 1994). After collection, the soil profiles were placed in a lysimeter station in Uppsala, Sweden (Bergström, 1992), where they were kept for the duration of the experiments.

Two experiments are reported and discussed here; one in which plant uptake and leaching of mineral N fertiliser (NH_4NO_3) were compared with N derived from anaerobically-stored poultry manure, and the other in which comparisons were made with red clover (*Trifolium pratense* L.) green manure. All N sources were ^{15}N-labelled to allow quantification of N deriving from the respective source, and applied at rates of 80–160 kg N ha^{-1} during the initial year. Plant uptake and leaching of added N were then monitored during 2–3 years.

The experiments are described in detail by Bergström and Kirchmann (1999, 2004).

7.2.2 Field Study at Mellby

The Mellby site, where the tile-drained field plots included in this overview are installed, is located in southern Sweden (56°29′N, 13°0′E). The soil at the site is a sandy loam with sand content between 77 and 91% down to 1.0-m depth. The organic matter content is relatively high in the topsoil (4%), but very low further down in the profile. Drainage flows from the plots were recorded continuously and water samples for chemical analysis were collected biweekly. Crop samples were also collected at harvest each year.

Two 6-yr organic crop rotations and one conventional rotation are included in this overview. In one of the organic rotations, animal manure was used (OAM), whereas N was provided from green manures (a legume/grass mixture) in the other (OGM). In the conventional rotation (CON), perennial ryegrass (*Lolium perenne* L.) or winter rye (*Secale cereale* L.) were used as cover crops in an attempt to reduce N leaching. All crop rotations had predominantly spring cereals, except during the final year when potatoes were grown (Table 7.1). The OGM rotation had green manures during two of the six years and the OAM rotation had

Table 7.1 Crop rotations and mean nutrient input at the Mellby and Bjärröd sites (data from Torstensson et al., 2006; Kirchmann et al., 2007)

	Cropping systems at Mellby			Cropping systems at Bjärröd	
	Organic with animal manure	Organic with green manure	Conventional	Organic	Conventional
Crop rotations					
Year 1	Barley	Oats	Barley	Barley	Barley
Year 2	Clover/ grass ley	Green manure	Oats	Red clover	Clover/grass ley
Year 3	Clover/ grass ley	Spring wheat	Spring wheat	Winter wheat	Oilseed rape
Year 4	Oats	Oats	Barley	Beans	Winter wheat
Year 5	Pea/barley	Green manure	Oats	Potato	Oats
Year 6	Potato	Potato	Potato	Peas	Sugarbeet
Mean nutrient input (kg ha^{-1} yr^{-1})					
N[a]	121	71	97	108	147
P	6	0	24	50	29
K	60	25	85	43	82

[a]Nitrogen input data include N_2-fixation. Figures were estimated using above-ground crop data as an input to the STANK model (Version 4:1, Swedish Board of Agriculture), which includes a grassland submodel according to Fagerberg et al. (1990).

forage crops during two years (Table 7.1). The crops in the conventional rotation received on average 97 kg N ha^{-1}, 24 kg P ha^{-1}, and 85 kg K ha^{-1} as inorganic fertilisers. The corresponding inputs in the organic rotations were: 71 (total-N), 0 (total-P) and 25 (K) kg ha^{-1} in the OGM rotation, and 121 (total-N), 6 (total-P) and 60 (K) kg ha^{-1} in the OAM rotation.

A more detailed description of the site and experimental procedures can be found in Torstensson et al. (2006).

7.2.3 Field Study at Bjärröd

The long-term experiment at Bjärröd in southern Sweden (55°42′N, 13°43′E) was carried out on a sandy loam soil (13–14% clay, 23–24% silt and 62–64% sand, throughout the profile to 1.0-m depth). Before the start of the experiment in 1980, the soil was highly P and K depleted, due to the fact that no inorganic fertilisers (or pesticides) had been applied since the mid-1940s.

In this presentation, data from two cropping systems are discussed, one organic and one conventional, both designed to support dairy production but with different crop rotations. Unfertilised plots with a similar crop rotation to the conventional system were used as a reference system. The cropping systems were laid out in tile-drained plots with a net size of 35 × 90 m each. The major management differences between the organic and conventional systems were: (i) growth of legumes every second year in the organic rotation and use of legumes as cover crops; (ii) application of P in the organic system at higher rates than inorganic P fertiliser in the conventional system; (iii) exclusion of oil-seed rape (*Brassica napus* L.) and sugarbeet (*Beta vulgaris* L.) in the organic system, but inclusion of potato (*Solanum tuberosum* L.); (iv) frequent mechanical weeding in the organic system, whereas pesticides were used in the conventional system; and (v) use of solid animal manure in the organic system and slurry in the conventional system. The mean annual inputs of N, P and K were: 147, 29 and 82 kg ha^{-1}, respectively, in the conventional system, and 108, 50 and 43 kg ha^{-1} in the organic system. The 6-yr crop rotations (Table 7.1) were repeated 3 times, giving a total experimental period of 18 years. Crop samples were collected at harvest each year.

More information about the site, crop rotations and other experimental procedures is provided by Kirchmann et al. (2007).

7.3 Yield Levels and Nutrient Uptake by Crops

An important consideration when comparing crop yields and nutrient removal by crops in organic and conventional systems is how the systems are managed. Organic and conventional systems differ in a number of aspects apart from use of fertilisers and pesticides. For example, green manure crops are grown in organic systems on a regular basis to provide succeeding crops with N, whereas this is uncommon in conventional systems. Furthermore, the crop rotations included in the studies reviewed here were quite different, which made comparison of crop yields between the conventional and organic systems difficult.

At the Mellby site, barley (*Hordeum distichum* L.) yield in the OAM system was only 50% of that in the conventional plots, while yield of spring wheat (*Triticum aestivum* L.) in the OGM system was about 80% of that in the conventional system. The average annual yield of all grain crops was significantly ($P < 0.1$) higher in the conventional system than in the organic system with green manures (OGM), as was the amount of N harvested with the crop ($P < 0.05$). This indicates that the N release from the green manure, which was incorporated into soil in late autumn, was not adequate for crop requirements. The most remarkable difference in crop yields and N harvested with crops occurred when potatoes were grown. In the conventional system, the potato yield was 9118 kg DM ha^{-1}, whereas it was only about 1500 kg DM ha^{-1} in the organic systems. The main reason for this large difference was damage by potato blight fungus (*Phytophthora infestans*) in the organic systems, whereas this problem was eliminated by use of fungicides in the conventional system. An additional contributing factor for the low potato yields in the organic systems was most likely K deficiency. The amount of K in harvested potatoes was only about 30 kg K ha^{-1} in the organic systems, whereas it was 175 kg K ha^{-1} in the conventional system.

At Bjärröd, the mean yield level of the organic system, including all crops, amounted to only 50% of the conventional yield (i.e. 3170 vs. 6380 kg DM ha^{-1} yr^{-1}, Table 7.2). For the individual crops, yield of sugarbeet in the conventional rotation was the main reason for the large difference between the systems, illustrating the basic problem when comparing cropping systems with different cropping sequences, as mentioned above. When sugarbeet yield was excluded from the comparison, the conventional system produced on average 5140 kg DM ha^{-1} yr^{-1} and the organic system 62% of that amount. This figure is similar to official Swedish statistics (SCB, 2004), which report mean organic yields amounting to 50–70% of those in conventional cropping systems. As regards the other crops, mean yields of grass ley were not significantly different between the organic and conventional cropping systems, but yields of cereal crops were significantly lower ($P<0.01$) in the organic rotations (Table 7.2). There are several possible reasons for the lower yields in the organic system, e.g. lower inputs of N, greater weed competition, lower nutrient-use efficiencies and poorer control of pests and diseases.

The concentrations of N, P and K in grains of barley and winter wheat at Bjärröd were significantly different ($P < 0.05$) between the conventional and organic systems. Nitrogen concentrations were lower in the organically grown cereals (1.8–2.2%) than in the conventional (2.2–2.7%), whereas P and K concentrations were slightly higher in the organic system (0.41% P and 0.51% K vs. 0.38% P and 0.44% K). Higher N concentrations in the conventional cereals and higher concentrations of P in the organic cereals were expected, since the application rates of N were higher in the conventional system and those of P in the organic. However, the organic system received considerably less K (43 vs. 82 kg K ha^{-1} yr^{-1}), but the

Table 7.2 Average dry matter yields at the Mellby (1997–2002) and Bjärröd (1981–1998) sites. Standard errors (±SE) for the Bjärröd site and relative yield figures are given in brackets (data from Torstensson et al., 2006; Kirchmann et al., 2007)

Site and cropping system	Dry matter yield (kg ha^{-1})			
	All crops	Barley	Winter wheat	Cover/grass[a]
Mellby				
Conventional	6096 (100)	4480 (100)	Not grown	Not grown
Organic with green manure	1951 (32)	Not grown	Not grown	Not grown
Organic with animal manure	5682 (93)	2133 (48)	Not grown	9442
Bjärröd				
Conventional	6380(±755)a[b] (100)	3745(±650)a (100)	6075(±524)a (100)	7480(±755)a (100)
Organic	3170(±436)b (50)	2105(±176)b (56)	4200(±544)b (69)	6140(±146)a (82)
Unfertilised	2080(±336)c	1119(±75)c (30)	3680(±644)c (60)	Not grown

[a]Clover/grass includes weeds.
[b]Within columns, mean values followed by different letters are significantly different at $P < 0.05$.

concentrations of K were still higher in organic cereals. A contributing factor to this could be the much lower cereal yields in the organic system (Table 7.2).

A reduction in crop yields in organic systems, which has been reported in a number of other long-term experiments besides those referred to here (e.g. Leake, 1999, 2000; Aronsson et al., 2007), has to be taken seriously, considering that future population growth will require more food to be produced. In this context, it is critical to increase or at least maintain yields per area without jeopardising environmental quality. Therefore, one can question whether organic crop production is a viable alternative, since large yield reductions for organically grown crops have to be compensated for by an increase in agricultural land. Results obtained in the experiments included in this overview show that crop yields in organic rotations were reduced by 20–80% compared with the same crops in conventional rotations. These examples indicate that organic crop production uses agricultural soils less efficiently, which is further discussed below and also in Chapter 3 of this book (Kirchmann et al., 2008).

7.4 Are There Any Water Quality Benefits Associated with Organic Agriculture?

One of the main arguments for changing over to organic crop production is that it is beneficial for the environment, especially water quality. A large proportion of the general public believes that nutrient loads to surface waters and groundwater decrease in response to management and use of nutrients according to organic principles (Granstedt, 1995). Conventional farms tend to operate at greater input levels of most nutrients than organic systems, as revealed by a compilation of farm-gate balances (Kirchmann and Bergström, 2001). Many expect this to result in larger leaching losses, primarily of N. However, is this really the case?

The main difference between organic and conventional agriculture regarding the use of plant nutrients is the exclusion of soluble inorganic fertilisers in the former. Therefore, in a first step we need to investigate the difference in leaching behaviour between organic manures and inorganic fertilisers. This was done with respect to N in two lysimeter studies in which NH_4NO_3 was compared with different types of poultry manure (Bergström and Kirchmann, 1999), although, only anaerobically stored poultry manure will be considered here, and green manures (Bergström and Kirchmann, 2004). Over the 3-yr period following application of poultry manure and fertiliser in the first year, leaching of N derived from the respective N sources was 3.5 kg ha^{-1} of added NH_4NO_3 and 31.8 kg ha^{-1} of N in poultry manure (Fig. 7.1). This represented about 3.5 and 32% respectively of added N, since 100 kg N ha^{-1} of both N sources were used. In the comparison between NH_4NO_3 and red clover manure over 2 years, the corresponding figures were 4 and 6%, respectively (Fig. 7.1). Thus, leaching of fertiliser N was lower throughout, although the difference in N leaching between the two organic N sources investigated was quite large. This is a clear indication that organic N sources are more vulnerable to leaching than inorganic N fertilisers. Incorporation of animal and green manures

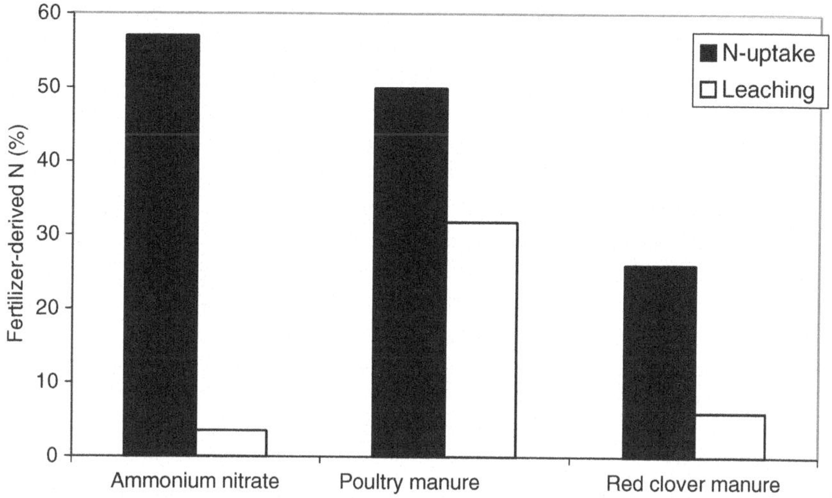

Fig. 7.1 Leaching and crop uptake of N originating from mineral fertiliser, poultry manure and red clover manure, measured in field lysimeters (modified from Bergström and Kirchmann, 1999, 2004)

carries with it a high risk of N loss by leaching, due to the fact that the inorganic N is often released from these sources during periods when there is no crop uptake of N. In cold and humid regions, such as Sweden, this often coincides with climatic conditions in autumn, when both soil temperatures and moisture content are high enough to trigger mineralisation of organic N fractions and annual crops are harvested. The released inorganic N is highly exposed to leaching, due to the frequently large surplus amount of precipitation. A leaching experiment with pig slurry applied in increasing amounts to lysimeters of the same type as those described above clearly corroborated this (Bergström and Kirchmann, 2006). The slurry was applied during 2 of 3 years at annual rates ranging from 50 to 200 kg N ha^{-1}. For comparison, NH_4NO_3 was applied on other lysimeters at a rate of 100 kg N ha^{-1} yr^{-1}. During the 3-yr period, N leaching loads increased with increasing slurry application rate to an average of 139 kg N ha^{-1} at the highest application rate. When slurry-derived N was applied at or above the application rate of inorganic fertiliser-N, the loads were significantly larger ($P<0.05$), but crop yields were not increased. This study further confirms that organic N sources are less efficiently used by crops than inorganic N fertilisers under cold humid conditions and increase the risk of leaching. However, it is important to note that comparative fertiliser experiments, such as those discussed above, only provide indications of how leaching would proceed in an agricultural system.

In the Mellby study, in which realistic organic and conventional systems were compared, annual N leaching loads, averaged over the whole 6-yr crop rotation, were 39 kg ha^{-1} (organic with animal manure; OAM), 34 kg ha^{-1} (organic with green manure; OGM), and 25 kg ha^{-1} (conventional with cover crops; CON)

(Table 7.3). However, as it is difficult to compare leaching data due to different crops being included in the rotations, we limited the comparison to years with simultaneous grain crops (3 out of 6 years in the CON and OGM rotations). In this more strict comparison, the annual average N leaching load was significantly ($P<0.05$) smaller in the CON system than in OGM (13 vs. 22 kg ha^{-1}). As these data were derived from real cropping systems, the results strongly support the hypothesis that organic systems leach more N, which, over the long-term, must have a considerable impact on surface water and groundwater quality. In all systems, the largest leaching loads by far occurred when potatoes were grown, with as much as 98 kg N ha^{-1} leached in the organic system with animal manure (OAM), which represented about 40% of the total load during the 6-yr period. Large N leaching loads with potato crops are quite common (Madramootoo et al., 1992), due to a number of conditions such as: shallow root system, large N fertiliser applications, and intensive tillage operations before planting and after harvest, to mention a few. However, the large N leaching loads from the organic systems at Mellby were primarily due to the extremely low potato yields caused by the blight fungus infestation, as mentioned above. In the Bjärröd experiment, the highest N concentrations in water draining from large lysimeters (34 mg N L^{-1}) also occurred in conjunction with potato crops in the organic system (Kirchmann et al., 2007).

Overall, the annual average leaching loads of P at Mellby were small in all systems, not exceeding 0.2 kg ha^{-1} (Table 7.3). This is mainly attributable to the strong adsorption ability of layers in the soil profile with a high iron content (Ghorayshi and Bergström, 1991). The annual average leaching loads of K were highest in the conventional rotation (Table 7.3), and these were significantly higher ($P < 0.05$) than in the OGM rotation as regards grain crops. This was presumably due to the much larger inputs of K in the conventional system, which annually received on average 85 kg K ha^{-1} compared to 25 kg K ha^{-1} in the OGM rotation.

It is obvious from the results presented above that proportionally less N is removed by crops and more N is thereby potentially available for leaching in organic cropping systems than in conventional, despite an often lower input of N in organic systems. This was also shown in a study carried out on a clay soil in southwest Sweden, in which conventional and organic systems were compared (Aronsson et al., 2007). It is also obvious from these studies that N leaching per hectare is not reduced when organic manures are used instead of inorganic N fertilisers.

When comparing N leached in relation to N in harvested crops, we need to consider certain practical factors in the systems studied. In Swedish organic systems, green manure crops are grown during a whole summer season, precluding harvest of other crops. In the Mellby study, green manure crops were grown during 2 out of 6 years in the OGM rotation, which means that the allocated production area was 33%. When N leaching loads per hectare in each system were divided by the corresponding dry matter yields of harvested crops, the lowest leaching (approx. 4.1 kg N Mg^{-1} dry matter yield) occurred in the conventional system, compared to 6.9 and 17.4 kg N Mg^{-1} dry matter yield for the OAM and OGM systems, respectively (Table 7.3). Comparing N leaching loads in relation to the crop grown, it was obvious that potatoes caused the highest N leaching loads in all systems, especially

Table 7.3 Average annual leaching loads of N, P and K per unit area and per unit harvested dry matter crop yield at the Mellby site (1997–2002) in southern Sweden (data from Torstensson et al., 2006)

Cropping System	N	P	K	N			P			K		
				All crops	Cereals	Potato	All crops	Cereals	Potato	All crops	Cereals	Potato
	kg ha^{-1}			kg Mg^{-1} dry matter yield								
Conventional	25	0.2	27	4.1	3.4	5.9	0.03	0.04	0.01	4.4	5.6	2.3
Organic with green manure	34	0.1	16	17.4	7.8	53.3	0.05	0.02	0.08	8.2	5.3	12.4
Organic with animal manure	39	0.2	12	6.9	6.3	62.6	0.03	0.05	0.08	2.1	3.0	6.5

in the organic systems (62.6 and 53.3 kg N Mg^{-1} dry matter yield for the OAM and OGM systems, respectively; Table 7.3). Furthermore, leaching loads of P per unit dry matter yield were not reduced in the organic systems.

These results clearly suggest that inorganic N fertilisers are used more efficiently than green and animal manures. Furthermore, even viewed from a cropping systems perspective, nutrient losses are less in conventional farming systems, both per unit area and per unit crop yield. Still, we also have to consider the changes in agricultural land required to produce the same amount of food if arable land is cropped according to organic rather than conventional principles. In other words, the results need to be integrated into a land-use perspective to determine the impact of organic and conventional systems on water quality (Fig. 7.2).

The Mellby and Bjärröd studies showed that yield levels in organic production are on average about 60% of those in conventional cropping systems. This difference is confirmed by data reported in the official agricultural statistics of Sweden, which is discussed in Chapter 3 of this book (Kirchmann et al., 2008). Therefore, to produce the same amount of food through organic farming, the organically managed area must be increased by 67% to compensate for the 40% yield loss. This would mean that non-arable land such as extensively managed rangeland, woodland, or other semi-natural land would have to be converted to arable land. In general, such land uses cause lower leaching losses than arable land. Accordingly, leaching from a larger area of organically managed land needs to be compared with leaching from a smaller conventionally managed area plus a non-arable area required to make the organic and conventional areas of equal size. Such a comparison, which takes into account food supply and land demand based on yield reduction, shows that N leaching will be approximately 2.5 times higher from organic than conventional production (Fig. 7.2). In summary, the necessary expansion of agricultural land to maintain food production through organic methods will most likely have negative effects on water quality with regard to N.

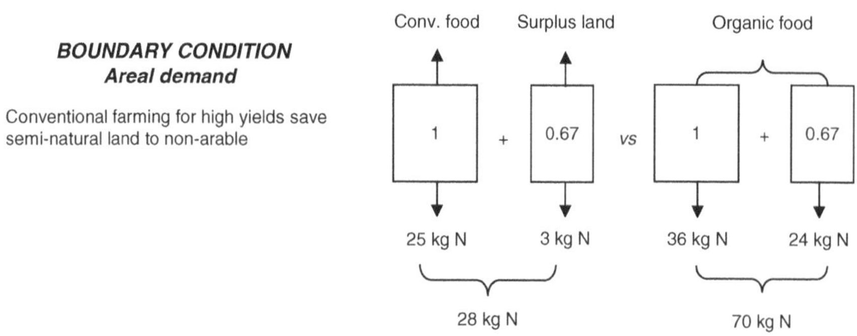

Fig. 7.2 A comparison of farming systems producing similar products but with an organic yield only 60% of that in conventional production. The same area of land required for low-yielding (organic) systems must also be considered for high-yielding (conventional) systems. Yield and leaching figures for arable land are taken from Tables 7.2 and 7.3. Leaching data for semi-natural land are taken from Löfgren and Olsson (1990)

7.5 Nutrient-Use Efficiency in Organic and Conventional Cropping Systems

Short-term measurements of nutrient-use efficiency over a few years can be misleading if the residual effect of previously used nutrient sources is not accounted for. This is critical if organic nutrient sources and untreated minerals, which are frequently used in organic agriculture, are included. In such cases, long-term calculations such as those permitted by the results from the Bjärröd site seem more appropriate. As reported above, crop uptake of nutrients was measured over 18 years in the organic and conventional systems at Bjärröd, to which the examples presented below refer.

The use efficiency of inorganic N and P fertilisers was higher than that of organic sources or untreated minerals (Table 7.4). In particular, crop utilisation of P was as low as 7% in the organic system compared to 36% in the conventional system. It is reasonable to assume that the large application of sparingly soluble apatite-P (once at 646 kg P ha^{-1}) greatly reduced the P use efficiency. When apatite-P was excluded from the calculation, the P use efficiency increased to 20% in the organic system, which is still considerably less than in the conventional system. The use efficiency of K was 63% in both the organic and conventional systems, most likely due to the fact that K was mainly added in easily soluble form in both systems and not at excessive rates. The long-term agronomic efficiency of N (the increase in grain yield compared to the unfertilised control divided by the N input) was significantly lower in the organic system than in the conventional (10 and 17 kg cereal yield increase per kg N added, respectively) (Table 7.4). The lower N use efficiency in the organic system was primarily due to poor synchronicity between mineralisation of N from the organic sources and crop N demand, as discussed above, but was also due to competition for N by weeds (see below).

For the poultry manure and the red clover green manure used in the lysimeter studies, the N use efficiency was 50 and 24%, respectively, over the 3- and 2-yr periods. When NH_4NO_3 was used, the corresponding figure was 57%. Again, lack

Table 7.4 Long-term use efficiency of N, P and K additions at the Bjärröd site (Kirchmann et al., 2007)

Measure of efficiency	Organic system	Conventional system
Agronomic efficiency of N[a]		
Barley	9 kg yield increase kg^{-1} N	18 kg yield increase kg^{-1} N
Winter wheat	10 kg yield increase kg^{-1} N	16 kg yield increase kg^{-1} N
Use efficiency[b]		
P	7%	36%
K	63%	63%

[a]The long-term agronomic efficiency of N by grain crops (1981–1998) was calculated as the increase in grain yield compared with the control divided by the N input.
[b]The long-term use efficiency of P and K by all crops (1981–1998) was derived from nutrient removal (nutrient yield of treatment – nutrient yield of control) divided by the nutrient input.

of synchronicity between mineralisation and crop uptake of N was the main reason for this difference between the N sources.

A key question is whether the shortage of available N when using organic manures is limited to the studies referred to here, or whether it is characteristic of other organic cropping systems as well (see Clark et al., 1999). It is quite obvious that organic cropping systems rely for their N supply on biological N fixation or on purchased organic N sources, since the occurrence of untreated N minerals (Guano, Chilean nitrate) is scarce in the world. Purchased organic N sources (e.g. meat and bone meal, food industry wastes, animal manure) often originate from conventional production, which shows a reliance on conventional production. Irrespective of the organic N source used in organic crop production, several studies have shown that they provide the crop with less N than similar amounts of N applied as mineral fertiliser (e.g. Thomsen et al., 1997; Aronsson et al., 2007; Bergström and Kirchmann, 2006). Thus, a lower N use efficiency seems to be a common problem in organic cropping systems, despite the fact that N inputs tend to be lower in organic systems than in conventional.

7.6 Weed Pressure in the Long-Term

Another possible reason why yields are typically lower in organic cropping systems is weed competition as discussed in Chapter 3 of this book (Kirchmann et al., 2008). There is reason to believe that competition between weeds and crops is higher in organic systems and also that the weed pressure gradually increases in response to lack of herbicide use. This aspect is not well documented, or at least not the long-term effects.

At the Bjärröd site, weed biomass and removal of N by weeds were measured each year in all systems (unfertilised control, organic and conventional). These measurements showed that weed biomass was a significant component of the organic system, amounting to on average 1021 kg DM ha^{-1} yr^{-1}, with peak values of more than 3000 kg DM ha^{-1} yr^{-1}. These peak values occurred during years with legumes (peas and beans) whereas during years with winter wheat the weed biomass was similar to that in the conventional system. It is also notable that the presence of weeds increased over time in the organic system, by about 17 kg DM ha^{-1} yr^{-1}. However, the most dominant weed species varied over time. During the first 9 years, charlock (*Sinapis arvensis* L.), spurry (*Spergola arvensis* L.) and red shank (*Polygonum percicaria* L.) were very frequent, whereas field thistle (*Cirsium arvense* Scop.) and couch grass (*Agropyron repens* L.) tended to dominate during the final 9 years. During years with grassland, dandelions (*Taraxacum vulgare* L.) were common. Assuming an N concentration of 1.5% in dry matter, the presence of on average 1000 kg DM ha^{-1} weed biomass in the organic system contributed to a withdrawal of 15 kg N ha^{-1} yr^{-1} from the main crop.

In the unfertilised and conventional systems, the weed biomass was significantly lower ($P<0.05$) than in the organic system, amounting to on average only

27 kg DM ha^{-1} and yr^{-1}. In contrast to the organic system, weed biomass decreased by about 5 kg DM ha^{-1} yr^{-1} in the unfertilised and conventional systems during the 18-yr period. The uptake and withdrawal of N by weeds in the conventional system was negligible (< 1 kg N ha^{-1} yr^{-1}).

These results suggest that weed biomass can be a yield-decreasing factor in organic cropping systems.

7.7 Conclusions

The results presented in this overview clearly show that the use of green and animal manures in organic cropping systems results in lower crop yields, with no benefit for water quality. Therefore, claims about sufficient food supply and improved water quality associated with the use of manures in organic cropping systems should be viewed with great caution. In terms of N, we were able to identify two main reasons for this: (i) a lack of synchronicity between release of N from legume residues, animal manures and other organic N sources, and demand for N by the main crop; and (ii) a high release of N from such organic sources during periods without a crop. Build-up of a large weed biomass can also be a contributing factor to yield depressions in organic systems, as exemplified by results from long-term comparisons of conventional and organic cropping systems.

From the points listed, we can conclude that the most critical consideration in efforts to reduce N leaching from agricultural soils in cold and humid climatic regions is to supply the crop with N when it is needed and to avoid surplus amounts of N in soil during autumn/winter when no crop is growing. One way of reducing surplus N in soil during autumn/winter is to use cover crops. In one of the studies presented here, the smallest N leaching load by far was recorded in a conventional system with a ryegrass cover crop. This system also had the highest yields of comparable crops in the conventional and organic systems. In fact, the use efficiencies of N and P estimated for an 18-yr period were higher in the conventional system than in the organic.

It can be concluded that the use of countermeasures such as cover crops efficiently reduces N leaching losses within conventional agriculture and maintains high crop yields, but not a change-over to low-yielding organic cropping practices, which require more land.

References

Aronsson, H., 2000, Nitrogen turnover and leaching in cropping systems with ryegrass catch crops. PhD Dissertation. Acta Universitatis Agriculturae Sueciae **214**. Swedish University of Agricultural Sciences.

Aronsson, H., Torstensson, G., and Bergström, L., 2007, Leaching and crop uptake of N, P, and K from a clay soil with organic and conventional cropping systems, *Soil Use Manage.* **23**: 71–81.

Bergström, L., 1992, A lysimeter test system suitable for studying the fate and behaviour of pesticides in soil, in: *Lysimeter Studies of the Fate of Pesticides in the Soil*, F. Führ and R.J. Hance, eds., BCPC Monogr. 53, British Crop Protection Council, Farnham, UK, pp. 73–81.

Bergström, L., and Kirchmann, H., 1999, Leaching of total N from 15N-labeled poultry manure and inorganic N fertilizer, *J. Environ. Qual.* **28**: 1283–1290.

Bergström, L., Jarvis, N., and Stenström, J., 1994, Pesticide leaching data to validate simulation models for registration purposes, *J. Environ. Sci. Health* **A29**: 1073–1104.

Bergström, L.F., and Kirchmann, H., 2004, Leaching of total nitrogen from 15-N-labeled green manures and $^{15}NH_4{}^{15}NO_3$, *J. Environ. Qual.* **33**: 1786–1792.

Bergström, L., and Kirchmann, H., 2006, Leaching and crop uptake of nitrogen and phosphorus from pig slurry as affected by different application rates, *J. Environ. Qual.* **35**: 1803–1811.

Clark, M.S., Horwarth, W.R., Shennan, C., Scow, K.M., Lantni, W.T., and Ferris, H., 1999, Nitrogen, weeds, and water as yield-limiting factors in conventional, low-input and organic tomato systems, *Agric. Ecosyst. Environ.* **73**: 257–270.

Corré, W.J., Schröder, J.J., and Verhagen, A., 2003, Energy use in conventional and organic farming systems. Proceedings 511. International Fertilizer Society, York, UK, 24p.

Dalgaard, T., Halberg, N., and Sillebak Kristensen, I., 1998, Can organic farming help to reduce N-losses? *Nutr. Cycl. Agroecosys.* **52**: 277–287.

Fagerberg, B., Torsell, B., and Nyman, P., 1990, Handledning för dataprogrammet PCVALL, Del III. Modellbeskrivning med biologisk/fysikalisk bakgrund. Dept. of Crop Production Science, Report 21. Swedish University of Agricultural Sciences, Uppsala (In Swedish).

Fjelsted Alrøe, H., and Kristensen, E.S., 2002, Towards a systemic research methodology in agriculture: rethinking the role of values in science, *Agric. Human Values* **19**: 3–23.

Ghorayshi, M., and Bergström, L., 1991, Equilibrium studies of the adsorption of dichlorprop on three Swedish soil profiles, *Swedish J. agric. Res.* **21**: 157–163.

Giller, K., Chalk, P., Dobermann, A., Hammond, L., Heffer, P., Nyamudeza, P., Maene, L., Ssali, L., and Freney, J., 2004, Emerging technologies that will increase the efficiency of use of fertilizer nitrogen, in: *Agriculture and the Nitrogen Cycle: Assessing the Impact of Fertilizer Use on Food Production and the Environment*, AR Mosier et al., eds., Island Press, Washington DC, USA, pp. 35–51.

Granstedt, A., 1995, Studies of the flow, supply and losses of nitrogen and other plant nutrients in conventional and ecological agricultural systems in Sweden, *Biol. Agric. Hortic.* **11**: 51–67.

Halberg, N., Kristensen, E.S., and Kristensen, I.S., 1995, Nitrogen turnover on organic and conventional mixed farms, *J. Agric. Environ. Ethics* **8**: 30–51.

Hansen, B., Fjelsted, H., and Kristensen, E.S., 2001, Approaches to assess the environmental impact of organic farming with particular regard to Denmark, *Agric. Ecosyst. Environ.* **83**: 11–26.

Ivarson, J., and Gunnarsson, A., 2001, Försök med konventionella och ekologiska odlingsformer 1987–1998. Sveriges Lantbruksuniversitet. Meddelande från södra jordbruksförsöksdistriktet Nr 53, Sverige (In Swedish).

Johnsson, H., Mårtensson, K., Torstensson, G., and Persson, K., 2006, Beräkning av normalutlakningen av kväve 2003 för den ekologiskt odlade arealen. Teknisk rapport 105, Div. Water Quality Management, Swedish University of Agriculural Sciences, Uppsala, Sweden (In Swedish).

Kirchmann, H., and Bergström, L., 2001, Do organic farming practices reduce nitrate leaching? *Commun. Soil Sci. Plant Anal.* **32**: 997–1028.

Kirchmann, H., Bergström, L., Kätterer, T., Mattsson, L., and Gesslein, S., 2007, Comparison of Swedish long-term organic and conventional crop-livestock systems on a previously nutrient depleted soil, *Agron. J.* **99**: 960–972.

Kirchmann, H., Bergström, L., Kätterer, T., and Andrén, O., 2008, Can organic crop production feed the world? in: *Organic Crop Production – Ambitions and Limitations,* H. Kirchmann and L. Bergström, eds., Springer, Doordrecht, The Netherlands.

Kristensen, S.P., Mathiasen, J., Lassen, J., Madsen, H.B., and Reenberg, A., 1994, A comparison of the leachable inorganic nitrogen content in organic and conventional farming systems, *Acta Agric. Scand., Sect. B, Plant and Soil Sci.* **44**: 19–27.

Leake, A.R., 1999, A report of the results of CWS agriculture's organic farming experiments, *J. R. Agric. Soc. Engl.* **160**: 73–81.

Leake, A.R., 2000, Climate change, farming systems and soils, *Aspects Appl. Biol.* **62**: 253–259.

Löfgren, S., and Olsson, H., 1990, Tillförsel av kväve och fosfor till vattendrag i Sveriges inland. Naturvårdsverket Report 3692. Swedish Environmental Protection Agency, Stockholm, Sweden (In Swedish).

MacKerron, D.K.L., Duncan, J.M., Hillman, J.R., Mackay, G.R., Robinson, D.J., Trudgill, D.L., and Wheatley, R.J., 1999, Organic farming: science and belief. 1999 Annual Report, Scottish Crop Research Institute (January 2007); http://www.milkismilk.com/news_organic_farming_science_belief.html. Assessed Nov. 2007.

Madramootoo, C.A., Wayo, K.A., and Endright, P., 1992, Nutrient losses through tile drains from potato fields, *Appl. Eng. Agric.* **8**: 639–646.

Persson, L., and Bergström, L., 1991, Drilling method for collection of undisturbed soil monoliths, *Soil Sci. Soc. Am. J.* **55**: 285–287.

SCB, 2004, Production of organic and non-organic farming 2003. Cereals, peas, oilseeds, table potatoes and temporary grasses. Sveriges Officiella Statistik – Statistiska Meddelanden JO 16 SM 0402. Stat. Sweden, Örebro, Sweden (In Swedish).

Stockdale, E.A., Lampkin, N.H., Hovi, M., Keatinge, R., Lennartsson, E.K.M., Macdonald, D.W., Padel, S., Tattersall, F.H., Wolfe, M.S., and Watson, C.A., 2001, Agronomic and environmental implications of organic farming systems, *Adv. Agron.* **70**: 261–327.

Stopes, C., Lord, E.I., Philipps, L., and Woodward, L., 2002, Nitrate leaching from organic farms and conventional farms following best practice, *Soil Use Manage.* **18**: 256–263.

Thomsen, I.K., Kjellerup, V., and Jensen, B., 1997, Crop uptake and leaching of 15N applied in ruminant slurry with selectively labelled faeces and urine fractions, *Plant Soil* **197**: 233–239.

Torstensson, G., Aronsson, H., and Bergström, L., 2006, Nutrient use efficiencies and leaching of organic and conventional cropping systems in Sweden, *Agron. J.* **98**: 603–615.

Trewavas, A., 2001, Urban myths of organic farming, *Nature* **410**: 409–410.

Trewavas, A., 2004, A critical assessment of organic farming-and-food assertions with particular respect to the UK and the potential environmental benefits of no-tillage agriculture, *Crop Prot.* **23**: 757–781.

Watson, C.A., Fowler, S.M., and Wilman, D., 1993, Soil inorganic-N and nitrate leaching on organic farms, *J. Agric. Sci.* **120**: 361–369.

Watson, C.A., Atkinson, D., Gosling, P., Jackson, L.R., and Rayns, F.W., 2002, Managing soil fertility in organic farming systems, *Soil Use Manage.* **18**: 239–247.

Withers, P.J.A., and Jarvis, S.C., 1998, Mitigation options for diffuse phosphorus loss to water, *Soil Use Manage.* **14**: 186–192.

Chapter 8
How Will Conversion to Organic Cereal Production Affect Carbon Stocks in Swedish Agricultural Soils?

Olof Andrén, Thomas Kätterer and Holger Kirchmann

Abstract Soil carbon changes were modelled over 30 years with the focus on cereal crops, since leys are often managed similarly in organic and conventional agriculture. Other crops were not considered due to difficulties in large-scale cropping of oilseed rape and potatoes organically because of pest problems. Four scenarios were used: 0, 8 (current), 20 and 100% organic cereal production. Conversion to organic cereal crop production was found to reduce the amount of carbon stored as organic matter in agricultural soils. Three factors contributed to decrease soil carbon levels in a given field: (i) a yield decrease, resulting in less C input through roots and above-ground crop residues; (ii) lower leaf area causing less water uptake, which resulted in higher water content in soil and an increased decomposition rate of soil organic matter; and (iii) more frequent and intensive mechanical cultivation for weed control, which resulted in increased mixing and exposure of soil organic matter to oxidative processes, speeding up decomposition. Due to lower yields in organic agriculture, more land must be used to produce the same amount. With 20% organic cereal production, land currently in fallow would have to be taken into production, while with 100% of cereals produced organically, all fallow land plus conversion of forest land to agriculture would be required. An 8% level of organic cereal production would lead to losses of 0.3 Tg C over a 30-year period, 20% would cause losses of 1.1 Tg C and 100% would cause losses of 12.8 Tg C. The annual CO_2 losses from 100% organic cereal production would be equivalent to the amount emitted by 675,000 average cars in Sweden annually. Losses of soil carbon under organic cultivation would continue for a much longer period than 30 years until a new equilibrium is reached.

Keywords Carbon stocks · CO_2 emission · Cereals · Land use · Modelling · Soil carbon decline

O. Andrén (✉)
Department of Soil and Environment, Swedish University of Agricultural Sciences, P.O. Box 7014, SE-75007 Uppsala, Sweden
e-mail: olof.andren@mark.slu.se

H. Kirchmann, L. Bergström (eds.), *Organic Crop Production – Ambitions and Limitations*, DOI 10.1007/978-1-4020-9316-6_8,

8.1 Introduction

Comparisons of soil C content on organically managed and conventional farms often end up with contradictory results. Soil organic matter content has been reported to be higher (Reganold, 1988; Wander et al., 1994; Droogers and Bouma, 1996; Liebig and Doran, 1999; Marriott and Wander, 2006), lower (Lützow and Ottow, 1994; Petersen et al., 1997), or similar (Derrick and Dumaresq, 1999; Burkitt et al., 2007) in organic and conventional production systems. Evidence of higher soil C content on organically managed farms is easily interpreted as a proven advantage of organic agriculture in sequestering more carbon in soil (Smith, 2004), but comparing organic and conventional systems requires great care, since a number of production factors may differ. Purchase of straw, manures or organic wastes is common on organic farms (Goulding et al., 2008; Kirchmann et al., 2008a) and affects soil C content. Incorporation of crop residues in one system and removal and sale in another system also affects the amount of C added. In addition, the large quantities of weeds in organic systems can contribute a significant C input. However, the total input of C from plant residues is usually lower in organic management, even when weeds are taken into account (Kirchmann et al., 2007).

A stringent comparison between systems requires the factors mentioned above to be considered. For example, organic farming systems that use organic matter of off-farm origin in the form of approved organic fertilisers such as composts, manures or organic wastes derived from food industries, etc. generally show higher soil C contents than conventional farming systems not importing the same quantities of organic fertilisers (e.g. Clark et al., 1998; Gunapala and Scow, 1998; Bulluck et al., 2002; Marriott and Wander, 2006). However, direct comparisons are not valid in such cases. Similarly, comparisons of organic systems that rear livestock and return animal wastes to the soil (e.g. Wander et al., 1994; Friedel, 2000; Pulleman et al., 2003) with conventional systems without livestock have no relevance. In addition, comparisons of organic systems using catch crops (Foereid and Høgh-Jensen, 2004) with conventional systems without catch crops lack the stringency required for useful comparisons. The main pitfall when comparing organic and conventional livestock systems is that different amounts of animal manure can be applied. Higher applications of animal manure (through purchased inputs) in the organic system than in the conventional can create non-system-specific differences (Faerge and Magid, 2003; Kirchmann et al., 2007). An appropriate basis for scientific comparisons of cropping systems with different management strategies is that the C input is related to the production level and that purchase of off-farm C sources does not differ significantly. It is therefore not possible to compare soil C levels between systems and attribute the differences to organic or conventional practices (i) if straw is returned in one system but removed in another; (ii) if catch crops are used in the organic system but are not used as an equally integrated countermeasure to reduce N leaching in the conventional system; (iii) if application rates of organic manures are not coupled to production levels; and (iv) if off-farm C sources are applied only in the organic system.

In fact, the number of studies required for a valid comparison is limited. Data from long-term field trials fulfilling the conditions outlined above, the Swiss DOK trial (Alföldi et al., 1993; Fließbach and Mäder, 2000) and the Norwegian model farm study (Korsaeth and Eltun, 2008), show that organic farming could not maintain soil N and thereby soil C content to the same level as conventional farming. In the Norwegian Apelsvoll system only trials with forage production could be compared, as the organic arable system received some animal manure but not the conventional arable system (Breland and Eltun, 1999). A survey of soils by Gosling and Shephard (2005) on four farms managed organically for at least 15 years indicated no significant differences concerning soil organic matter levels between organic and conventional management. Similarly, organic matter C content in soils of Norwegian farms that converted from conventional to organic practices was not different after 5 years (Løes and Øgaard, 1997).

In terms of basic production, organic yields are considerably lower than those of conventional systems (Kirchmann et al., 2008b), which means less C input through roots, above-ground crop residues or animal manure. In addition, lower yields involve less transpiration and thereby higher soil water content, which speeds up decomposition of soil organic matter. The breakdown of soil organic matter in organic farming can be further speeded up by frequent mechanical weeding (more soil cultivation since herbicides are forbidden).

In this chapter, we project soil C dynamics under cereals grown on mineral agricultural topsoils 30 years into the future, excluding soils with more than 12% soil C.

8.2 Calculation Procedures

8.2.1 Models, Data, Scenarios and Boundary Conditions

For the analysis we used the dynamic soil carbon model ICBM (Andrén and Kätterer, 1997), which has been calibrated for a number of long-term experiments (Kätterer and Andrén, 1999) and parameterised to cover different soil and crop types and climatic conditions for geographical regions (Andrén et al., 2004, 2007). The model has also been used for reporting changes in Swedish land use according to the Kyoto protocol (Andrén et al., 2008). A spreadsheet version of the ICBM model for use in projections of future scenarios can be downloaded or run directly (Andrén, 2007). The model has two compartments, called young and old soil organic matter, and the decomposition rate of the young and old pool is set to $k_Y = 0.8$ and $k_O = 0.006\text{year}^{-1}$, respectively. The decomposer activity (r_e) is multiplied by k_Y and k_O to determine the decomposition rates of the young and old pool, respectively. A humification coefficient (h) determines the fraction of the input that goes through the young pool and into the old (humus, or refractory), and is assumed to be about 0.12 for most crop residues and about 0.35 for manure (Andrén and Kätterer, 1997).

In this study, pedotransfer functions developed from a Swedish soil database containing water retention properties and textural composition (Kätterer et al., 2005) were used to estimate soil water content at wilting point and at field capacity. The soil water balance was based on an FAO concept (Allen et al., 1998). Soil texture data from the nationwide soil monitoring programme (Eriksson et al., 1997, 1999) were used to estimate hydraulic properties. Carbon contents in Swedish agricultural soils were derived from nationwide soil samplings including regional distribution (Eriksson et al., 1997, 1999) and soil properties and data were used to calculate topsoil carbon mass for different soil types and area. Annual crop yield data for eight production regions in Sweden (Table 8.1) were derived from national Swedish statistics from 1990 to 2004 (e.g. SCB, 2005a). Yield data were used as input to the model to calculate C inputs through crop residues into soil by applying an allometric function with different parameters for each crop type (Andrén et al., 2004). The original calculations included 32 crops, bulked into 9 major crop types for each region. Additions of C to soil through manure applications were estimated from annual regional data obtained from national statistics (Andrén et al., 2004). Annual carbon inputs to soils through crop residues and manure are summarised in the parameter i.

Daily climate data from 22 weather stations managed by SMHI, the Swedish Meteorological and Hydrological Institute (2–4 stations representing each region), were used as driving variables. The input variables were: air temperature (°C), precipitation (mm), and reference evapotranspiration (mm). Air temperature was used to calculate topsoil temperature through an empirical approach (Kätterer and Andrén, 2008).

Decomposer activity in soil expressed as the factor r_e was calculated as the product of three factors, soil water content (r_w), soil temperature (r_t) and soil cultivation (r_c). Soil temperature was assumed to affect decomposer activity according to a quadratic relationship (Kätterer et al., 1998). Water response was assumed to be about 10-fold higher just below field capacity than at wilting point (see Andrén et al., 2004, 2007, 2008).

Separate calculations were made for eight regions, nine crop types and 14 soil types for each year, resulting in 1008 combinations. Averages and sums were calculated, both on a per hectare basis and for the entire area. In this application, we selected only the spring and winter cereal crop types, which constituted two of the nine crop types used in the national C budget calculations (Andrén et al., 2008).

We projected four scenarios 30 years into the future: (i) assuming no organic cereal production (0% scenario); (ii) the current situation of 8% of the cereal area being organically cropped (8% scenario); (iii) assuming organic cereal production to comprise 20% of the cereal area (20% scenario); and (iv) assuming all Swedish cereals to be organically cropped (100% scenario). The 20% scenario was chosen because of the Swedish Government target that 20% of total Swedish agricultural land, including perennial leys, should be under organic agriculture by 2010 (Anonymous, 2005).

Average yields in Swedish organic agriculture are about 50% of those in conventional agriculture for cereals and 75% for leys (SCB, 2004, 2005b, 2006). Significantly lower yields of organically grown crops are also shown by data from a long-term experiment with organic agriculture on low-production soils (Kirchmann et al., 2007). However, since according to SCB (2006) conversion to organic agri-

Table 8.1 Characteristics of Swedish agricultural production regions, 1990–2004. Total area of Swedish agricultural land was 2.76 Mha. Area and yield data from Swedish Official Statistics (SCB, 2005a), climatic data from the Swedish Meteorological and Hydrological Institute, and soil data from Ericsson et al. (1997, 1999)

Region and mean area (ha)	Latitude (°N)	Mean temp. (°C)	Precip. (mm)	Typical crops	Typical soils	Spring barley yield (Mg ha^{-1})	Winter wheat yield (Mg ha^{-1})
South-western coastal, 344,297	55–57	7.8	757	Spring and winter cereals, sugar beet	Sandy loam, loam	5.8	7.9
South-eastern coastal, 324,312	55–58	7.4	612	Grass ley, spring and winter cereals	Sandy loam, loam	4.5	6.5
South central plains, 453,364	58	6.5	570	Spring and winter cereals, grass ley	Sandy loam, loam	4.7	5.9
Central plains, 630,917	59–60	6.6	601	Spring cereals, grass ley	Clay, silty clay	4.1	5.2
Southern forest, 518,783	56–58	6.8	769	Grass ley, spring cereals	Sandy loam, loamy sand	3.5	5
Central forest, 201,789	58–61	5.3	688	Grass ley, spring cereals	Silt loam, silty clay loam	3.3	4.8
North, 166,144	60–65	3.1	600	Grass ley, spring cereals	Silt loam, loam	2.4	NG[a]
North and mountain, 124,642	63–69	2.2	575	Grass ley, spring cereals	Silt loam, sandy loam	2.2	NG[a]

[a]NG = crop not grown in this region.

culture has been more common on low-production areas than on high-production, yield differences are probably exaggerated. In the following, we thus assumed that yields of organically grown cereals amounted to 60% of those of conventionally grown cereals. Since differences in yields are smaller for leys than for cereal crops, C input through crop residues may not be significantly different for organic and conventional leys. Furthermore, an unknown proportion of the organically managed leys had been cropped without inorganic fertilisers before 'conversion' to organic practices and there may not have been any actual change in management other than registration of the area to obtain subsidies for organic agriculture. For this reason, we simplified our projections and considered only the agricultural area used for cereal production in our calculations.

Yield decline through organic production also needs to be viewed from the perspective of food supply. To maintain the same food and feed supply in Sweden, lower yields through organic practices must be compensated for and replaced by additional production. Consequently, more agricultural area is required, independent of whether the production decline is replaced by goods purchased on the world market or balanced by additional production in Sweden. Thus, the most central boundary condition for a system comparison is that the same total amount of cereal crops has to be produced.

In this study, changes in soil carbon stocks of Swedish agricultural soils were assessed in two steps. In a first step, the effect of organic production on soil C dynamics of the current cropped area was demonstrated without considering the need for increased area. In a second step, the yield decline after conversion to organic agriculture was considered and balanced through an equivalent increase in land use for agricultural production.

8.3 Changes in Soil Carbon Mass Due to Organic Cereal Production

8.3.1 Conversion from Conventional to Organic Cereal Production

We calculated changes in the C mass of Swedish agricultural soils cropped with cereals when converted from conventional to organic cereal production. Four conversion scenarios were modelled, as shown in Table 8.2 and Fig. 8.1.

Conversion to organic cereal production was shown to reduce yields and thereby affect the amount of fresh crop residues returned to soil. Conversion to complete organic cereal production (100% scenario) reduced the annual C input through crop residues to soil by roughly 1 Mg ha^{-1}, from 2.7 to 1.7 Mg C ha^{-1} yr^{-1} (Table 8.2). In addition, lower yields meant smaller leaf area and therefore lower water uptake, which affected the soil climate factor (r_e), a multiplier for decomposition rate (Table 8.2). As the water content in soil remained higher under lower production, decomposition rates in the organically cropped soils were 5–10% higher. Concerning the application of animal manure to organic and conventional production sys-

Table 8.2 Amounts of carbon in Swedish arable soils at steady-state conditions when cereals are cropped organically. The parameters used in the ICBM Excel model for winter and spring cereals were 1,197,000 ha for total cereal area and 81.26 Mg C ha^{-1} for initial C mass of 0–25 cm topsoil. Sum of annual C input from crop and manure (i), soil climate factor (r_e), humification factor (h)

Percentage of cereal land cropped organically	C input (Mg C ha^{-1})		Parameters and steady-state values			
	Crop residues	Animal manure	i	r_e	h	Soil C (Mg C ha^{-1})
0% scenario	2.85	0.356	3.21	1.05	0.154	82.16
8% scenario (current situation)	2.76	0.356	3.11	1.09	0.156	77.54
20% scenario	2.62	0.356	2.98	1.11	0.156	73.04
100% scenario	1.71	0.356	2.07	1.15	0.168	52.14

tems, it was assumed that the same amount was used in all scenarios, since changes in the total numbers of cattle through conversion to organic cereal production are difficult to predict.

According to the projections in Table 8.2, a drastic decline in soil C can be expected if all cereals are cropped organically. The 100% scenario shows that in the long-term perspective, i.e. under steady-state conditions, one third of soil C under cereals would be lost, declining from 81.3 to 52.1 Mg soil C ha^{-1} under exclusively organic production. Continuing with the current level of organic cereal production (8% scenario), the soil carbon mass in soil will decrease from 81.3 Mg ha^{-1} to 77.54 Mg ha^{-1} at steady-state. However, as steady-state conditions will take a very long time to reach under Swedish climatic conditions, a more relevant projection of the soil C mass dynamics would be the 30-year period shown in Fig. 8.1.

The decomposition curves in Fig. 8.1 indicate that changes are more rapid during early years, when 'young' soil organic matter is reaching steady-state. Different decomposition rates for young and old soil carbon are used in the model (Andrén and Kätterer, 1997) and for the far more decomposition-resistant old soil C, the period chosen was too short to reach steady-state. Even 30 years after complete

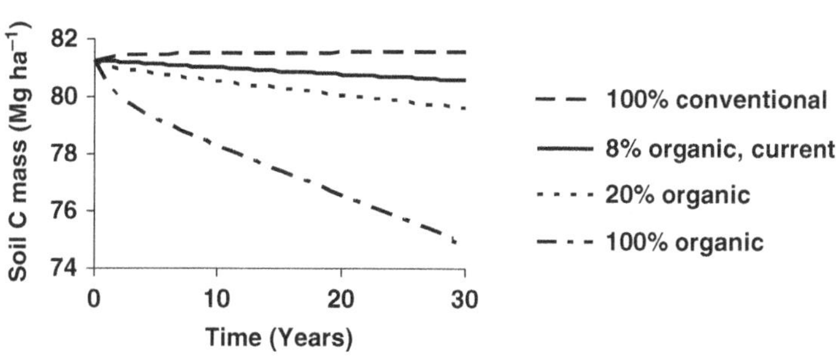

Fig. 8.1 Projecting soil carbon changes in Swedish arable topsoils under cereals starting from current carbon mass (0–25 cm depth, Mg ha^{-1}) for four scenarios over 30 years

Table 8.3 Projected 30-year changes in C content in Swedish soils used for organic or conventional cereal production when the same total amount of cereals is produced. Conditions as outlined in Table 8.2; initial C mass of 0–25 cm forest soil = 81.26 Mg ha^{-1}

Land use scenarios for cereal production	Cereal area (1000 ha)	Relative yield	Average grain yield (Mg ha^{-1})	Total yield (Tg)	Proportion of yield (%)	Changes in soil C after 30 yrs	
						Soil C (Mg ha^{-1})	Total soil C (Tg)
0% scenario							
100% conventional	1088	100	5.1	55.5	100	+0.33	+0.33
8% scenario							
8% organic	90	60	3.0	2.7	4.9	−6.3	−0.6
92% conventional	1036	100	5.1	52.8	95.1	+0.3	+0.3
Sum	1126			55.5	100		−0.3
20% scenario							
20% organic	225	60	3.0	6.8	12.2	−6.3	−1.4
80% conventional	901	100	5.1	45.9	82.7	+0.3	+0.3
Additional 4% conventional using fallow land[b]	56	100	5.1	2.8	5.1	+0.3	0[b]
Sum	1182			55.5	100		−1.1
100% scenario							
100% organic	1126	60	3.0	33.8	60.8	−6.3	−7.1
Additional 17% organic using all fallow land[a]	321	60	3.0	9.6	17.4	−6.3	0[a]
Additional 40% organic using forested land	485	50	2.5	12.1	21.8	−11.7	−5.7
Sum	1932			55.5	100		−2.8

[a]No change in soil C was assumed when shifting from fallow to conventional cereal production.
[b]The fallow area amounted to 321,000 ha in 2005.

conversion to organic production (100% scenario), soil C levels were shown to be still declining and were far from the steady-state level of 52.1 Mg C ha^{-1} (Fig. 8.1).

8.3.2 Conversion from Conventional to Organic Cereal Production Including Additional Land Required to Produce the Same Amount of Cereals

Although differences in soil C stocks between organic or conventional production seem clear enough, the projections in Section 8.3.1 do not consider that the same amount of cereals has to be produced by organic methods as by conventional. Cereal demand remains the same for both systems and therefore lower organic yields per area must be compensated for by cropping of additional land.

A central question is which additional type of land can be used for organic cereal production if the current level of organic production (8% scenario) is increased to 20% or even 100% of total cereal production. Swedish statistics indicate that the total agricultural area classified as fallow increased from 250,000 ha in the year 2000 to about 330,000 ha by 2005 (SCB, 2006). Thus, agricultural land under fallow can be considered for organic production in the first instance. In fact, cropping of parts of the current fallow area with organic cereals would be sufficient to compensate for the production decline in the 20% scenario. However, if all cereals in Sweden were to be produced organically (100% scenario), the current fallow land would not be sufficient and additional land would be required. In that case, the only realistic option would be the use of forested areas formerly used as agricultural land, as permanent pastures are generally difficult to cultivate due to hilly conditions, stone outcrops or other difficult terrain. The conditions for the extended projection and changes in soil C stocks are outlined in Table 8.3.

If the current organic cereal production level (8% scenario) were to increase to 20%, our calculations show a significant decrease of 1.1 Tg C in Swedish arable soils over a 30-year period. In the 100% scenario, there would be a total soil C loss of as much as 12.8 Tg C, with almost 50% originating from the additional area required to compensate for lower yields. If we were to abandon organic agriculture and grow cereals exclusively through conventional methods (0% scenario), we would slightly increase soil C stocks by 0.33 Tg after 30 years (Table 8.3).

8.4 Soil Carbon Dioxide Losses Caused by Organic Cereal Production in Relation to Emissions by Cars

In order to relate the decline in soil C mass and thus CO_2 emissions from Swedish agricultural soils upon conversion to organic cereal production to other emissions in society, we used emissions by cars for comparison. We calculated the number of cars emitting the same amount of CO_2 per year (1.6 CO_2 Tg $year^{-1}$ or 0.42 Tg C $year^{-1}$)

as would be lost from agricultural soil after conversion to 100% organic cereal production.

For this calculation we assumed the following: On average, Swedish cars are run 10,000 km per year and use 0.1 L petrol km^{-1}, resulting in an annual use per car of 1000 L petrol. Petrol has a C content of 84% and a density of 0.74 kg L^{-1}, which means that 622 kg petrol-C are combusted per car and year.

The amount of 1.6 Tg CO_2 emitted per year due to conversion to complete organic cereal production in Sweden is therefore equivalent to the amount emitted by 675,000 cars per year. The comparison illustrates the highly significant contribution of CO_2 emissions from soil. Moreover, as emissions from cars and other anthropogenic sources are likely to decrease through the introduction of new technologies, CO_2 emissions from organically managed soils may make an even more substantial contribution relative to other anthropogenic CO_2 emissions since they will continue over a very long period.

8.5 Conclusions

- Agricultural practices that increase photosynthesis and crop yields will also increase the amount of C stored as organic matter in soil.
- Producing cereals organically will significantly reduce soil C stocks. Lower organic yields result in less crop residues, which in turn results in less soil organic matter formation. Furthermore, reduced crop growth leads to less uptake of water and more moisture in soil, which speeds up decomposition.
- Lower cereal yields in organic production need to be balanced by additional cereal production elsewhere, requiring an additional production area. In a scenario where all cereals are produced organically, all current fallow and some current forest land would have to be converted to agriculture.
- If all cereals are grown organically, soil carbon losses will cause annual CO_2 emissions of 1.6 Tg. This is equivalent to the amount of CO_2 emitted by 675 000 average cars in Sweden during one year.

References

Alföldi, T., Mäder, P., Oberson, A., Spiess, E., Niggli, U., and Besson, J.M., 1993, DOK-Versuch: vergleichende Langzeituntersuchungen in den drei Anbausystemen biologisch-organisch, biologisch-dynamisch und konventionell. III. Boden: Chemische Untersuchungen, 1. und 2. Fruchtfolgeperiode, *Schweizerische Landwirtschaftliche Forschung* **32**: 479–507.

Allen, R.G., Pereira, L.S., Raes, D., and Smith, M., 1998, Crop evapotranspiration – guidelines for computing crop water requirements. FAO Irrigation and Drainage Paper 56, FAO, Rome, 300p.

Andrén, O., and Kätterer, T., 1997, ICBM: The introductory carbon balance model for exploration of soil carbon balances, *Ecol. Appl.* **7**: 1226–1236.

Andrén, O., Kätterer, T., and Karlsson, T., 2004, ICBM regional model for estimations of dynamics of agricultural soil carbon pools, *Nutr. Cycl. Agroecosys.* **70**: 231–239.

Andrén, O., 2007, http://www-mv.slu.se/vaxtnaring/olle/ICBM_Swed_Afr.xls

Andrén, O., Kihara, J., Bationo, A., Vanlauwe, B., and Kätterer, T., 2007, Soil climate and decomposer activity in sub-Saharan Africa estimated from standard weather station data – a simple climate index for soil carbon balance calculations, *Ambio* **36**: 379–386.

Andrén, O., Kätterer, T., Karlsson, T., and Eriksson, J., 2008, Soil C balances in Swedish agricultural soils 1990–2003, with preliminary projections, *Nutr. Cycl. Agroecosys.* **81**: 129–144.

Anonymous, 2005, Ekologisk produktion och konsumtion – Mål och inriktning till 2010, Regeringskansliet, Stockholm, No. 29 (In Swedish).

Breland, T.A., and Eltun, R., 1999, Soil microbial biomass and mineralization of carbon and nitrogen in ecological, integrated and conventional forage and arable cropping systems, *Biol. Fert. Soils* **30**: 193–201.

Bulluck, III, L.R., Brosius, M., Evanylo, G.E., and Ristaino, J.B., 2002, Organic and synthetic fertility amendments influence soil microbial, physical and chemical properties on organic and conventional farms, *Appl. Soil Ecol.* **19**: 147–160.

Burkitt, L.L., Small, D.R., McDonald, J.W., Wales, W.J., and Jenkin, M.L., 2007, Comparing irrigated biodynamic and conventionally managed dairy farms. 1. Soil and pasture properties, *Aust. J. Exp. Agric.* **47**: 479–488.

Clark, M.S., Horwarth, W.R., Shennan, C., and Scow, K.M., 1998, Chnages in soil chemical properties resulting from organic and low-input farming practices, *Agron. J.* **90**: 662–671.

Derrick, J.W., and Dumaresq, D.C., 1999, Soil chemical properties under organic and conventional management in southern New South Wales, *Aust. J. Soil Res.* **37**: 1047–1055.

Droogers, P., and Bouma, J., 1996, Biodynamic vs. conventional farming effects on soil structure expressed by simulated potential productivity, *Soil Sci. Soc. Am. J.* **60**: 1552–1558.

Eriksson, J., Andersson, A., and Andersson, R., 1997, Tillståndet i svensk åkermark *(Current status of Swedish arable soils)*, Swedish Environmental Protection Agency, Stockholm, Report 4778 (In Swedish).

Eriksson, J., Andersson, A., and Andersson, R., 1999, Åkermarkens matjordstyper (*Texture of agricultural topsoils in Sweden*), Swedish Environmental Protection Agency, Stockholm, Report 4955 (In Swedish).

Faerge, J., and Magid, J., 2003, Assessment on organic farming benchmark trials in Denmark, *Acta Agric. Scand.* (Section B) **53**: 64–68.

Fließbach, A., and Mäder, P., 2000, Microbial biomass and size-density fractions differ between soils of organic and conventional agricultural systems, *Soil Biol. Biochem.* **32**: 757–768.

Foereid, B., and Høgh-Jensen, H., 2004, Carbon sequestration potential of organic agriculture in northern Europe – a modelling approach, *Nutr. Cycl. Agroecosys.* **68**: 13–24.

Friedel, J.K., 2000, The effect of farming on labile fractions of organic matter in Calcari-Epileptic Regosols, *J. Plant Nutr. Soil Sci.* **163**: 41–45.

Gosling, P., and Shephard, M., 2005, Long-term changes in soil fertility in organic arable farming systems in England, with particular reference to phosphorus and potassium, *Agric. Ecosyst. Environ.* **105**: 425–432.

Goulding, K., Stockdale, E., and Watson, C., 2008, Plant nutrients in organic farming, in: *Organic Crop Production – Ambitions and Limitations*, H. Kirchmann and L. Bergström, eds., Springer, Dordrecht, The Netherlands.

Gunapala, N., and Scow, K.M., 1998, Dynamics of soil microbial biomass and activity in conventional and organic farming systems, *Soil Biol. Biochem.* **30**: 805–816.

Kätterer, T., Reichstein, M, Andrén, O., and Lomander, A., 1998, Temperature dependence of organic matter decomposition: A critical review using literature data analysed with different models, *Biol. Fert. Soils* **27**: 258–262.

Kätterer, T., and Andrén, O., 1999, Long-term agricultural field experiments in Northern Europe: analysis of the influence of management on soil carbon stocks using the ICBM model, *Agric. Ecosyst. Environ.* **72**: 165–179.

Kätterer, T., Andrén, O., and Jansson, P.-E., 2005, Pedotransfer functions for estimating plant available water and bulk density in Swedish agricultural soils, *Acta Agric. Scand.* (Section B) **56**: 263–276.

Kätterer, T., and Andrén, O., 2008, Predicting daily soil temperature profiles in arable soils from air temperature and leaf area index, *Acta Agric. Scand.* (Section B) 00:00–00. DOI: 10.1080/09064710801920321 (In press).

Kirchmann, H., Bergström, L., Kätterer, T., Mattsson, L., and Gesslein, S., 2007, Comparison of long-term organic and conventional crop-livestock systems on a previously nutrient depleted soil in Sweden, *Agron. J.* **99**: 960–972.

Kirchmann, H., Kätterer, T., and Bergström, L., 2008a, Nutrient supply in organic agriculture – plant availability, sources and recycling, in: *Organic Crop Production – Ambitions and Limitations*, H. Kirchmann and L. Bergström, eds., Springer, Dordrecht, The Netherlands.

Kirchmann, H., Bergström, L., Kätterer, T., Andrén, O., and Andersson, R., 2008b, Can organic crop production feed the world? in: *Organic Crop Production – Ambitions and Limitations*, H. Kirchmann and L. Bergström, eds., Springer, Dordrecht, The Netherlands.

Korsaeth, A., and Eltun, R., 2008, Synthesis of the Apelsvoll cropping system experiment in Norway – nutrient balances, use efficiences and leaching, in: *Organic Crop Production – Ambitions and Limitations*, H. Kirchmann and L. Bergström, eds., Springer, Dordrecht, The Netherlands.

Liebig, M.A., and Doran, J.W., 1999, Impact of organic production practices on soil quality indicators, *J. Environ. Qual.* **28**: 1601–1609.

Løes, A.K., and Øgaard, A.F., 1997, Changes in the nutrient content of agricultural soil on conversion to organic farming in relation to farm-level nutrient balances and soil contents of clay and organic matter, *Acta Agric. Scand.* (Section B) **47**: 201–214.

Lützow, M., and Ottow, J.C.G., 1994, Effect of conventional and biological farming on microbial biomass and its nitrogen turnover in agriculturally used Luvisols of the Friedberg plains, *Zeitschrift Pflanzenernähr. Bodenk.* **157**: 359–367.

Marriott, E.E., and Wander, M.M., 2006, Total and labile soil organic matter in organic and conventional farming systems, *Soil Sci. Soc. Am. J.* **70**: 950–959.

Petersen, S.O., Debosz, K., Schjønning, P., Christensen, B.T., and Elmholt, S., 1997, Phospholipid fatty acid profiles and C availability in wet-stable macro-aggregates from conventionally and organically farmed soils, *Geoderma* **78**: 181–196.

Pulleman, M., Jongmans, A., Marinissen, J., and Bouma, J., 2003, Effects of organic versus conventional arable farming on soil structure and organic matter dynamics in a marine loam in the Netherlands, *Soil Use Manage.* **19**: 157–165.

Reganold, J.P., 1988, Comparison of soil properties as influenced by organic and conventional farming systems, *Am. J. Altern. Agric.* **3**: 144–154.

Smith, P., 2004, Carbon sequestration in croplands: the potential in Europe and the global context, *Eur. J. Agron.* **20**: 229–236.

SCB, 2004, *Production of Organic and Non-Organic Farming 2003*, Statistics Sweden JO 16 SM 0402 (In Swedish, with English summary).

SCB, 2005a, *Yearbook of Agricultural Statistics*, Official Statistics of Sweden, SCB, Örebro, Sweden.

SCB, 2005b, *Production of Organic and Non-organic Farming 2004*, Statistics Sweden JO 16 SM 0502 (In Swedish, with English summary).

SCB, 2006, *Production of Organic and Non-Organic Farming 2005*, Statistics Sweden JO 16 SM 0602 (In Swedish, with English summary).

Wander, M.M., Traina, S.J., Stinner, B.R., and Peters, S.E., 1994, Organic and conventional management effects on biologically active soil organic matter pools, *Soil Sci. Soc. Am. J.* **58**: 1130–1139.

Chapter 9
Energy Analysis of Organic and Conventional Agricultural Systems

Göte Bertilsson, Holger Kirchmann and Lars Bergström

Abstract Energy parameters of a Swedish long-term field experiment comparing organic and conventional agricultural systems were evaluated. There is great potential for misinterpretation of system comparisons as a result of choice of data and how energy data are expressed. For example, reported yields based on single crops and not the whole rotation can result in significantly different interpretations. Energy use per unit yield was lower in organic crop and animal production than in the corresponding conventional system, as previously found in other studies. This is due to the exclusion of N fertiliser, the largest energy input in conventional cropping systems. Energy use per unit yield expresses system efficiency, but the term is insufficient to evaluate the energy characteristics of agricultural systems. Calculation of the most important energy component, net energy production per unit area, showed that conventional systems produced far more energy per hectare than organic systems. The energy productivity (output/input ratio), i.e. the energy return on inputs, was at least six in both types of agriculture, revealing the highly positive energy balance of crop production in general. Lower yields in the organic systems, and consequently lower energy production per unit area, mean that more land is required to produce the same amount of energy. This greater land requirement in organic production must be considered in energy balances. When the same area of land is available for organic and conventional crop production, the latter allows for complementary bio-energy production and can produce all the energy required for farming, such as fuels, N fertilisers, etc., in the form of ethanol. In a complete energy balance, options such as combustion, gasification or use as fodder of protein residues from ethanol production must also be taken into account. There is a common belief that the high fossil fuel requirement in N fertiliser production is non-sustainable. This is a misconception, since the use of N fertilisers provides a net energy gain. If N fertilisers were to be completely replaced by biological N_2 fixation, net energy production would be significantly lower. In addition, N fertiliser production can be based on renewable energy sources such as bio-fuels produced by gasification. Conventional crop production is thus energetically fully sustainable. Energy

G. Bertilsson (✉)
Greengard AB, P.O. Box 552, SE-24495 Dösjebro, Sweden
e-mail: bertilsson@greengard.se

H. Kirchmann, L. Bergström (eds.), *Organic Crop Production – Ambitions and Limitations*, DOI 10.1007/978-1-4020-9316-6_9,

analyses of agricultural systems presented in this chapter illustrate that published data may require recalculation in relation to the background, prevailing trends and boundary conditions, and subsequent re-interpretation. New perspectives on energy use must also be considered.

Keywords Bio-fuel production · Cropping systems · Energy budgets · Energy parameters · Energy use

9.1 Introduction

Rising prices for fossil energy and the need to reduce carbon dioxide emissions are creating a demand for improved energy use in general and certainly also within the agricultural sector. A recurring opinion is that agriculture should become organic and thereby not rely so heavily on fossil fuels (e.g. Jørgensen et al., 2005). Organic agriculture is described as being more sustainable due to the exclusion of N fertilisers, the production of which requires large amounts of fossil fuels (Pimentel, 2006). Energy consumption by different types of agriculture is one parameter used to rate the sustainability of agricultural systems (Eckert et al., 1999).

A common way to express energy use in agriculture is to calculate the energy requirement per unit yield. According to this definition, organic production is more efficient than conventional and thus preferable (e.g. Refsgaard et al., 1998; Dalgaard et al., 2001; Mäder et al., 2002). The exclusion of N fertilisers and the associated energy saving in organic agriculture results in a lower energy use per unit yield. In fact, conventional agriculture has been criticised for its lower energy efficiency and thus lower sustainability than organic agriculture. This criticism deserves attention and evaluation.

A further possibility to characterise the energy use in agriculture is to calculate energy productivity – a ratio describing energy output over input. Calculations of this ratio show a decreasing trend in the energy productivity of conventional agriculture during the 1970s (e.g. Pimentel et al., 1973; Hirst, 1974), but more recent studies indicate a shift during the 1980s from decreasing to increasing energy productivity in agriculture as a whole (Balwinder and Fluck, 1993; Bonny, 1993; Cleveland, 1995; Uhlin, 1998, 1999). However, the calculation of output/input ratio is a poor measure for system comparisons as it only expresses the efficiency and not the total or net energy production. As all common calculations such as balances, flows, prices, etc. are based on amounts, total and net energy production needs to be a central measure. It is therefore astonishing that the most central calculations are disregarded when organic systems are evaluated with regard to energy (Refsgaard et al., 1998; Dalgaard et al., 2001; Mäder et al., 2002; Jørgensen et al., 2005).

A tool preferred by some to evaluate energy use in organic agriculture is 'emergy' calculations. Emergy is defined as the solar energy required for production or services transforming all forms of available energy into a common basis, solar emjoules (Odum, 1996). Emergy calculations have for example been applied to compare the energy efficiency of different ecosystems (Lefroy and Rydberg, 2003; Martin et al., 2006) and to quantify renewable and non-renewable energy use when using a horse or tractor (Rydberg and Jansén, 2002). Although the calculations allow integration of different kinds of energy and facilitate the definition of boundary conditions of a system, the evaluation of data remains critical. Land use requirement, production per unit land and efficiency ratios etc. must be evaluated in a stringent way in order to draw meaningful conclusions, but are unfortunately often disregarded in emergy calculations.

We are accustomed to think that minimising inputs of non-renewable energy is beneficial for saving natural resources. However, reducing the energy input to crop production through exclusion of N fertilisers is not automatically beneficial because cropping systems have a positive energy balance. The potential to bind more solar energy through higher energy input results in a large positive energy balance (Uhlin, 1999). In other words, more energy is produced than is consumed. In practice, the higher energy yield can be used to substitute for energy sources in other processes and thereby save energy in total. Therefore, evaluation of the sustainability of cropping systems cannot be based on characterising efficiency calculations only, but must be based on balancing of total energy amounts.

The overall objective of this chapter is to discuss and critically review the issues outlined above, based on energy calculations using data from a Swedish long-term field study in which organic and conventional forms of production were compared. Specific objectives were to: (i) compare energy parameters of organic and conventional crop and animal production systems; (ii) address pitfalls associated with energy analysis of agricultural systems and the link to land requirement; and (iii) correct misconceptions about energy use for N fertiliser production.

9.2 Framework for Energy Calculations

Energy issues are concrete and easy to understand (Baumann and Tillman, 2004; Brentrup et al., 2004a) but no standard methods for energy calculations are currently available. The steps involved in life-cycle assessment (LCA) can be a useful basis for the discussion of energy use (ISO, 1997). In particular, agricultural systems need special attention with regard to boundary conditions affecting the land area requirement, which is incorporated in the latest revision of the LCA standard (Finkbeiner et al., 2006). Furthermore, energy analysis of cropping systems requires consideration of other aspects that are not commonly recognised, such as status of the applied technology and new ways of using energy (Uhlin, 1999; Hülsbergen and Kalk, 2001; Corré et al., 2003).

Production systems have different characteristics, which have to be considered in energy interpretations. Therefore, there should be a clear description of how and to what extent the following issues have been addressed.

- Organic systems often have lower nutrient inputs and rely on nutrients previously added to soils before conversion to organic agriculture. It may take decades until yields decline to levels reflecting true organic practices. Thus, there is a risk of energy outputs from organic systems being overestimated.

- Valid comparisons of systems must consider the total production of the systems. However, only crop-wise yields are usually reported, disregarding the area used to grow non-harvested green manure crops in organic agriculture. Therefore, total production over a whole crop rotation period and not crop-wise yields needs to be used to estimate the energy production of agricultural systems.
- Energy required by machinery used for cultivation of non-harvested crops or fallow must be included in a correct energy analysis of agricultural systems.
- It is important to use the most recent data in energy calculations. For example, in nitrogen fertiliser production, the best available technology often requires less energy (Ramirez, 2006) and emits less nitrous oxide than older technologies (Jenssen, 2004).
- Other aspects that need to be considered are whether the systems were run according to short-term profit maximisation with no or few restrictions from society or with a long-term view and environmental restrictions (Bergström et al., 2005). It must be established whether good agricultural practice such as rotations, maintenance of soil organic matter and structure, etc. were part of both systems or whether these aspects were considered for only one system?

A more detailed description of these production issues is given in Chapter 3 of this book (Kirchmann et al., 2008).

9.3 Interpretation of Energy Data in Crop Production

A field study located at three sites in southern Sweden (Scania) representing different soil fertility levels was monitored over two six-year rotations starting in 1987 (Ivarson and Gunnarsson, 2001). The study consists of five systems managed according to best agricultural practice, of which four were included in this evaluation:

A. Conventional crop production
B. Conventional with leys and manure (simulated animal production)
C. Organic with leys and manure (simulated animal production)
D. Organic crop production

Yield data for the experimental period are given in Table 9.1. Energy calculations of the systems for the period 1993–1997 made by Törner (1999) are also included in Table 9.1.

In the following we distinguish between crop production systems without animals (A and D) and animal production systems with leys and manure (B and C), as circulation of nutrients and the possibilities for symbiotic nitrogen fixation differ greatly between the two system categories. In line with what has been discussed in Chapter 3 (Kirchmann et al., 2008) and also stressed above, it is necessary to use total production over a rotation instead of crop-wise yields in the analysis.

Table 9.1 Yield and energy use in the Swedish long-term trials. Average data from three sites according to Törner (1999) and Ivarson and Gunnarsson (2001)

System and variable	Mean per crop						Mean per rotation
Conventional crop production (A)	W-wheat	Barley	Rape	W-wheat	Sugar-beet[a]	Peas	
Yield (Mg dry matter ha^{-1})	5.3	4.8	2.7	5.3	8.2	3.1	4.9
Energy for tractor (MJ ha^{-1})	2891	2848	3812	2891	3647	2765	3142
Energy for machinery (MJ ha^{-1})	792	842	1109	792	1303	814	942
Other energy need (MJ ha^{-1})	2113	2092	1588	2113	216	1703	1637
Energy for N fertiliser (MJ ha^{-1})[b]							4536
Total energy use (MJ ha^{-1})							10,258
Organic crop production (D)	W-wheat	Beans	Barley	Green manure	Sugar-beet	Peas	
Yield (Mg dry matter ha^{-1})	2.9	2.2	2.9	0	5.3	2.7	2.7
Energy for tractor (MJ ha^{-1})	3139	3474	3265	1544	3395	3265	3014
Energy for machinery (MJ ha^{-1})	893	1058	990	529	1235	990	949
Other energy need (MJ ha^{-1})	1778	1537	1544	1062	238	1566	1288
Total energy use (MJ ha^{-1})							5251
Conventional animal production (B)	Barley	Grass/clover	Grass/clover	W-wheat	Sugar-beet	Peas	
Yield (Mg dry matter ha^{-1})	4.7	9.0	9.6	4.7	8.0	3.2	6.5
Energy for tractor (MJ ha^{-1})	3028	3937	3722	3772	5900	2972	3889
Energy for machinery (MJ ha^{-1})	734	1955	1894	960	1766	714	1337
Other energy need (MJ ha^{-1})	2759	1722	1709	3078	7340	4160	3461
Energy for N fertiliser (MJ ha^{-1})							5544
Total energy use (MJ ha^{-1})							14,231

Table 9.1 (continued)

System and variable	Mean per crop						Mean per rotation
Organic animal production (C)	Barley	Grass/ clover	Grass/ clover	W-wheat	Sugar-beet	Peas	
Yield (Mg dry matter ha^{-1})	3.3	5.4	7.7	3.4	5.9	3.2	4.8
Energy for tractor (MJ ha^{-1})	3649	2594	3289	3329	4536	3174	3429
Energy for machinery (MJ ha^{-1})	1456	1040	1715	887	1463	788	1225
Other energy need (MJ ha^{-1})	1790	1459	1456	1666	5443	3424	2540
Total energy use (MJ ha^{-1})							7193

[a] Yields of sugarbeet are expressed as dry matter assuming a water content of 80% (yield x 0.2) in order to be comparable with grain crops. Yields of grass/clover and other crops are also presented in terms of dry matter.

[b] The energy use for N fertiliser production was assumed to be 42 MJ kg^{-1} N.

9.3.1 Calculations of Energy Use and Their Limitations

As mentioned above, a common way to evaluate energy utilisation in crop production is to calculate the specific energy use, i.e. the energy required to produce a unit of product. According to Table 9.2, the specific energy use was higher in the conventional system (A) than in the organic system (D) using mean yields over a rotation, 2.1 and 2.0 MJ kg^{-1} harvested dry matter yield, respectively. However, when the green manure year was omitted and crop-wise yields were used for the calculation, the energy use for the organic system decreased to 1.6 MJ kg^{-1} dry matter yield. This is significantly lower than for conventional production and similar to the figure reported by Corré et al. (2003), indicating better efficiency of organic systems.

Another significant difference between the organic and conventional cropping system can be caused by the exclusion of energy used for machinery for non-harvested crops (e.g. Horne et al., 2003). When machinery was excluded for non-harvested crops and crop-wise yields were used in the calculation, the relative energy use of the organic system only amounted to 69% of that in conventional production, which in reality is 85% (Table 9.2).

A central input for the interpretation of energy analysis is the consumption of energy for N fertiliser manufacturing. Technical improvement of fertiliser plants to increase the efficiency has reduced energy demand for N fertiliser production in recent years. This has significant implications on energy calculations for agricultural systems (Table 9.2). Jenssen and Kongshaug (2003) reported that an energy use of 38 MJ kg^{-1} N is most appropriate for modern fertiliser plants. Applying this figure instead of the 42 MJ kg^{-1} N that is commonly used reveals that there is no difference in energy use between the conventional and organic systems (Table 9.2).

Table 9.2 Interpretation of energy use for conventional and organic crop production as influenced by choice of input data. The data used are taken from Table 9.1

Choice of input data	Conventional system	Organic system	Organic/ Conventional
	MJ kg^{-1} dry matter yield[a]		(%)
Total yield vs crop-wise yield			
Total yield (organic = 54% of conv.)	2.09	1.97	94
Crop-wise yield (organic = 67% of conv.)	2.09	1.60	77
Energy use for machinery for non-harvested crops			
Included	1.90	1.61	85
Excluded	1.90	1.31	69
High and low energy consumption for N fertiliser production			
50 MJ kg^{-1} N instead of 42	2.27	1.97	87
38 MJ kg^{-1} N instead of 42	2.01	1.97	98

[a] The energy content of common crops varies between 15 and 18 MJ kg^{-1} dry matter (Fluck, 1992) and we used the lower value of 15 for all calculations.

9.3.2 Other Energy Calculations

Although energy use per kg product is a well-defined and a common estimate for agricultural systems, it provides insufficient information as it is an efficiency parameter. The lower the value, the more efficient is the system. This seems clear enough but the examination of agricultural systems on the basis of energy use is misleading. The limitation of the efficiency term energy use per kg product becomes obvious through the following example.

If the aim is to reduce energy consumption per unit product as much as possible, not only N fertilisers, but also energy consumption by farm machinery should be excluded and the human or animal-powered alternative should be chosen. Energy consumption per kg of wheat would then be reduced by 95%, which would result in a situation similar to that on many African smallholdings. On the other hand, crop production would also be reduced, sometimes by as much as 80%, but this would not be reflected in the energy efficiency term. This clearly shows that energy data must be expressed in relation to the amounts of energy produced as crop yields in order to understand the energy characteristics of agricultural systems. Calculations of total or net energy production describe the amounts of energy gained and illustrate the land requirement to produce the same amount of energy.

The data in Table 9.3 (based on Table 9.1) show that net energy production – total energy produced minus energy input – was almost twice as high in the conventional Swedish long-term cropping system as in the organic. The energy productivity (output/input ratio) shows that the energy return was 7–8 times higher in crop products than in inputs and that it was slightly higher in the organic system. In other words, crop production has a highly positive energy balance due to photosynthetic activity, whereby solar radiation is transformed into biomass. This also shows that minimising inputs and thus saving energy is not necessarily a successful strategy since a high energy input leads to higher returns, which can be well justified.

Table 9.3 Further energy parameters derived from Table 9.1. Values are based on dry matter yields from crop production systems and no removal of crop residues

Energy parameter	Pure crop production	
	Conventional	Organic
Energy use (MJ kg^{-1} dry matter yield)[a]	2.09	1.97
Energy use (MJ ha^{-1})	10,258	5251
Gross energy production (MJ ha^{-1})	73,500	40,500
Net energy production (MJ ha^{-1})	63,242	35,249
Energy productivity (Output/input ratio)	7.2	7.7

[a] The energy content of common crops varies between 15 and 18 MJ kg^{-1} dry matter (Fluck, 1992) and we used the lower value of 15 for all calculations.

9.3.3 Findings from Other Organic Field Studies

Pimentel (2006) calculated the efficiency of energy use in agriculture in the USA. Compared with European conventional agriculture, energy use is often high and yields are low under USA conditions. The results from one experimental site in Pennsylvania (Rodale), where yields of organic maize were similar to those of conventional maize in a maize-soybean rotation due to large inputs of animal manures (see Chapter 3 of this book; Kirchmann et al., 2008), had a major impact on the conclusions drawn in the report. Such relations are not common in Europe, where yields of organically grown crops are consistently lower than those of conventional crops due to much lower nutrient inputs. Output/input ratios reported for major US crops were also very low (2.2), while European studies (Refsgaard et al., 1998; Dalgaard et al., 2001; Brentrup et al., 2004b) report, for instance, at least 6.5 for winter-sown cereals, similar to the data in Table 9.3.

Hülsbergen and Kalk (2001) presented a detailed energy assessment of long-term field studies in Germany showing that less fossil energy was used per hectare in the organic systems but that the energy yield was reduced because of lower yields compared with conventional systems. Similarly, Mäder et al. (2002) reported results from a 21-year experiment in Switzerland with organic farming systems and concluded that organic production consumes less energy per unit product. Yields were on average 20% lower (potatoes 40%). The relatively small yield reduction in the organic systems in that study may be explained by the high animal density, 1.2–1.4 animal units per hectare, and the results are thus not generally applicable.

In a comprehensive review by Corré et al. (2003), it was concluded that: (i) organic agricultural systems require less energy but more land than conventional; (ii) the highest energy use efficiency in agriculture would be achieved by intensive conventional farming including the production of energy crops; and (iii) the composition of the human diet has a larger effect on energy use than the type of agricultural system.

9.4 Interpretation of Energy Data in Mixed Animal-Crop Production

Production involving livestock (the Swedish systems B and C) means growth of leys, removal of straw and beet tops, and addition of manure. An important condition at the Swedish site is that the manure addition to each system was adapted to the productivity of each site and corresponded to roughly one animal unit per hectare.

The organic livestock system produced on average 74% of yields over all crops compared with the conventional animal system (system B vs C, Table 9.1). The energy characteristics (Table 9.4) indicate that the organic livestock system was more favourable than the conventional. The energy use per unit product was significantly lower in the organic system and the output/input ratio was larger than in the corresponding conventional system (Table 9.4). This may be explained by the fact that growth of leys in the beef/milk producing system increased the N input (N_2 fixation) in combination with the return of N through manure and the higher N input led to somewhat higher yield levels than in the crop production system without animal manure. Furthermore, leys in the conventional system require more N fertiliser than other crops and the energy input to the conventional livestock system was thereby higher than that to the conventional crop production system (Table 9.1).

Table 9.4 Energy parameters for animal production systems derived from Table 9.1. The values are based on dry matter yields

Energy parameter	Mixed crop-animal production	
	Conventional	Organic
Energy use (MJ kg^{-1} dry matter yield)[a]	2.18	1.49
Energy use (MJ ha^{-1})	14,231	7193
Gross energy production (MJ ha^{-1})	98,000	72,500
Net energy production (MJ ha^{-1})	83,769	65,059
Energy productivity (Output/input ratio)	5.9	9.0

[a] The energy content of common crops varies between 15 and 18 MJ kg^{-1} dry matter (Fluck, 1992) and we used the lower value of 15 for all calculations.

Despite the lower efficiency of the conventional system, net energy production was considerably larger in that system. As the amount of energy produced is of the utmost importance in sustaining animal production, lower amounts of energy produced must be compensated for. Maintaining the energy production for organic animal husbandry would require an expansion of the area used for agriculture (e.g. Cederberg and Mattsson, 2000; Casey and Holden, 2006). In other words, high input agriculture requires less land per unit energy and this fact must be fully evaluated.

9.5 Does Biological Instead of Artificial Nitrogen Fixation Save Energy?

The high energy demand and use of fossil energy for production of N fertilisers is often pointed out in arguments against the use of artificial N fertilisers in organic agriculture. In fact, if N fertilisers are replaced by biological nitrogen fixation, a major proportion of the fossil fuels used for plant production can be saved (see Table 9.1). The argument for not being dependent on fossil fuels for N fertiliser production and thus saving energy has gained wide-spread acceptance. Biological nitrogen fixation through legumes is seen as the given, sustainable alternative.

A realistic way to assess the energetic value of N fixation by legumes is to compare two similar cropping systems where legumes are an integral part of the organic system and N supply in the conventional system is based on fertiliser N. An isolated single crop comparison would not be convincing, since crops other than legumes must also be produced through organic farming.

Such energy analysis of conventional and organic crop production systems does not support the argument that saving fossil fuels for N fertiliser production is beneficial. This may be difficult to accept, as saving means less consumption, but although N fertiliser production is energy-demanding, the much higher energy yield of the conventional system greatly compensates for the energy demand for production of N fertiliser. For example, the energy production in the Swedish conventional crop production system (A) was 73 GJ ha^{-1} compared with 40 GJ ha^{-1} in the organic system (D) (see Table 9.3), but the energy input for N fertiliser was only 4.5 MJ ha^{-1} (see Table 9.1). Thus, even though the use of inorganic N fertiliser is the largest item in the energy budget of conventional crop production, it is also true that correct use of N fertiliser produces far more energy than is required for fertiliser production. Nitrogen fertiliser acts as a boost to solar energy capture by crops. In fact, crop production is associated with a positive energy balance and not many energy investments in society give such high energy returns. The energy yield through crops and residues is actually 6–15 times larger than the energy required for N fertiliser production (Ratke et al., 2002; Brentrup et al., 2004b).

In Fig. 9.1 (data from Tables 9.1 and 9.3), the input and output items of energy for the conventional (A) and organic crop production system (D) are compared. The extra energy yield obtained by use of fertiliser N was 7-fold higher than the energy required for N fertiliser production. This highly positive balance needs to be fully recognised. The use of energy for N fertiliser production has clearly no negative effect on the energy balance. Still, one may argue that the fossil energy needed to produce N fertiliser is non-renewable and the positive energy balance does not help to conserve fossil resources in the long-term. However, this argument is in fact not applicable for N fertiliser production. The raw materials needed for ammonia fertiliser production are air, water, and energy of almost any kind. If oil and gas reserves were to become depleted and/or very expensive, renewable or remote energy sources could be used instead because of the positive energy balance. In fact, less than 20 years ago, large N fertiliser factories were actually run on hydro-electricity in Glomfjord, Norway. As ammonia is easy and cheap to transport in large tankers, remote energy sources may be the basis for ammonia factories

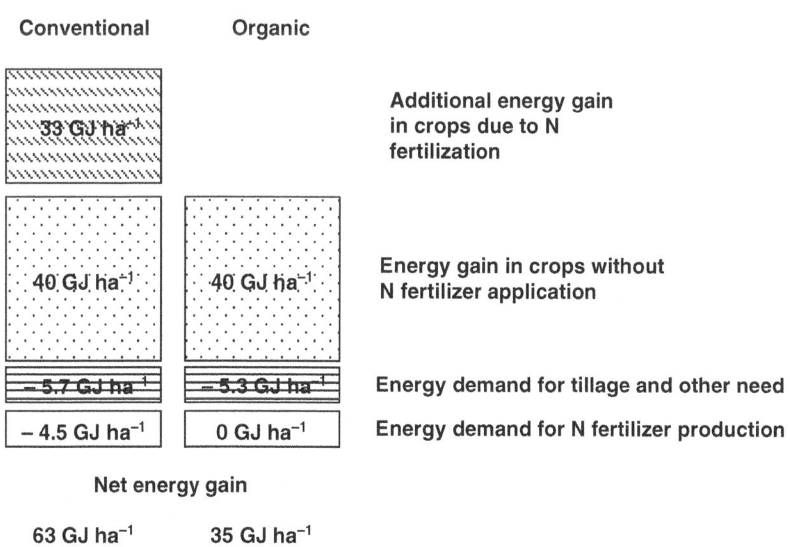

Fig. 9.1 Energy input in relation to energy gain through crops and residues in the conventional and organic crop production systems. Data derived from Tables 9.1 and 9.3

serving the world without competing with other energy demands. In other words, oil and gas depletion may become a global problem but it is not decisive for the production of N fertilisers. Sustainable production of artificial N fertilisers can be achieved using hydrogen produced through electricity or gasification of biomass. In an energy-scarce situation, fertilisers will be needed to create more bio-energy and improve sustainability. Furthermore, the entire global fertiliser industry currently uses less than 2% of global energy consumption (IFA, 2006).

Finally, correcting the argument against producing N fertilisers due to energy saving does not mean that N should be used negligently. Like all powerful tools, fertilisers have to be used correctly and conservatively to enhance crop growth and maintain soil fertility (Carlgren and Mattsson, 2001). Use of fertilisers should not be an excuse for poor agricultural practices. Fertilisers may tempt the user to make shortcuts such as use of unsuitable monocultures and exhaustive tillage, but these are malpractices. The use of fertilisers should never been seen as replacing soil conservation measures, recycling of nutrients or growth of legumes. Fertilisers should be viewed as one important tool for crop production, not eliminating other necessary practices which enable agriculture to be sustainable.

9.6 Additional Bio-Fuel Production Through Conventional Crop Production on Set-Aside Land

Bertilsson (1993) pointed out that high-yielding food production creates the least demand for agricultural land and thereby leaves more options for society for various other uses. Access to land is an important resource and a natural constraint that needs to be considered in energy analyses. In fact, the demand for land for food production may compete with the demand for land for bio-energy production (Van den Broek et al., 2001; Connor and Mingues, 2006).

9.6.1 Combustion

In the comparison in Fig. 9.2 (data from Tables 9.1 and 9.3), the same amount of food is produced through conventional and organic crop production. The additional land area needed when organic production methods are applied can be used for bio-fuel production in conventional agriculture. Combustion of bio-fuels can replace use of fossil fuels for the production of heat in some processes and the need for the corresponding amount of fossil energy would then be reduced through direct substitution.

The energy input required for conventional food and bio-fuel production would amount to 22 GJ yr^{-1} (Fig. 9.2). This figure can be compared with organic production, which produces the same amount of food (4.9 Mg dry matter yield) on the same area and only uses 9.5 GJ yr^{-1} of fossil fuel, but provides no additional bio-energy.

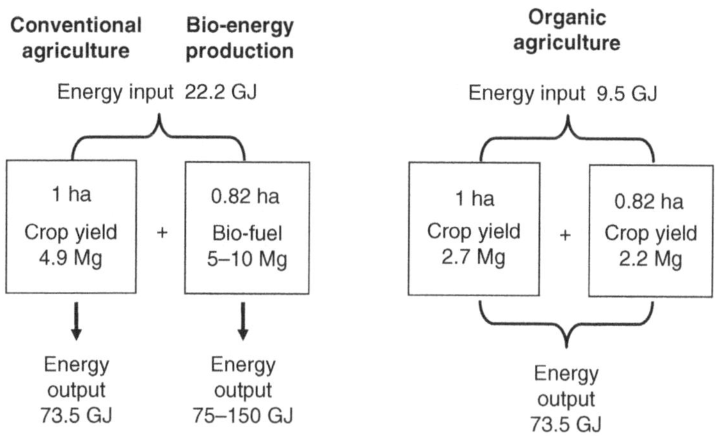

Fig. 9.2 Producing the same amount of food through conventional and organic crop production. The boundary condition was access to the same area of land. Data are taken from Tables 9.1 and 9.3

Production of bio-fuel on 'set-aside' land amounts to at least 75 GJ. This amount exceeds by far the total energy input needed for production of food and energy crops (22 GJ). The net energy gain through additional bio-fuel (53 GJ) is of the same order of magnitude as the net energy production through food in the organic system (64 GJ) on the same area (Fig. 9.2). Thus, instead of saving energy through low-input organic agriculture, conventional agriculture increases energy productivity.

9.6.2 *Ethanol Production*

In this example, the 'set-aside' area (see Fig. 9.2) is used to produce winter wheat for ethanol production in order to replace petrol for vehicles. We assumed a winter wheat yield of 5 Mg ha^{-1} and applied the following key data for the calculation: (i) the production of 1 Mg of ethanol requires 3.5 Mg of wheat to be fermented, which leaves 1.5 Mg dry matter in form of protein residues that can be used either as fodder or for combustion (Horne et al., 2003); (ii) the energy content of ethanol is 26.7 MJ kg^{-1} or 21.1 MJ L^{-1} and that of protein residues 15 MJ kg^{-1} dry matter; and (iii) the process energy required to ferment and distil 1 Mg ethanol amounts to 9.2 GJ.

According to these values, 5 Mg of grain produce 1.4 Mg of ethanol equivalent to an energy amount of 37 GJ. In addition to ethanol, 2.1 Mg protein residue dry matter containing 31 GJ are available for use as feed. Other more general figures derived from the calculation are that roughly 33% of the total energy in grain is transformed into ethanol energy, 43% remains in protein residues and 24% is required for the conversion and/or lost. The ratio between energy in ethanol and in protein residues is 44 to 56.

Based on this information, we calculated the number of cars that can be run on ethanol produced on 'set-aside' land if food is produced conventionally instead of organically. We assumed that an average Swedish car is driven 10,000 km per year using 0.1 L of petrol (32.6 MJ L^{-1}) per kilometre, which would be equivalent to 1000 L petrol per car and year. The equivalent consumption of ethanol would be 1545 L (21.1 MJ L^{-1}). Thus for each hectare of agricultural land converted to organic production, the additional land required to produce the same amount of food could provide fuel covering the annual consumption of an average Swedish car. As approximately 90,000 ha have been converted from conventional to organic cereal production (see Chapter 8; Andrén et al., 2008), fuel for 90,000 cars could be produced in addition to food if this area were converted back to conventional production.

9.7 Conclusions

The outcome of comparisons between organic and conventional systems depends on the original assumptions, the design of the study, the boundaries for the study, the use of data in space and time and, last but not least, the way of accounting for energy. Indeed, the same study can be used to demonstrate the superiority of either organic or conventional systems depending on how such factors are considered. In order to understand the energy characteristics of agricultural systems, it is necessary to calculate total or net energy production.

Production of agricultural crops, whether conventional or organic, results in a positive energy balance, with more solar energy bound than energy invested. The energy demand for N fertiliser production is the largest item in the energy budget of conventional systems, but the highly increased crop production when using N fertiliser results in a very positive energy balance, with at least a six-fold return on the energy invested for N fertiliser production. Growth of legumes for biological N fixation instead of artificial fertiliser production does not improve the energy budget of organic systems. It is a misunderstanding that exclusion of artificial N fertilisers means saving energy. The highly positive energy balance when using N fertilisers instead of biological N fixation needs to be fully recognised. Furthermore, the option exists to use renewable energy instead of fossil fuels for N fertiliser production.

As yields in conventional agriculture are about twice those in organic production, conventional methods allow both food and energy to be produced on the area of land required for organic food production alone. When bio-energy production is included in a comparison between conventional and organic systems, the more productive conventional systems are always preferable.

References

Andrén, O., Kätterer, T., and Kirchmann, H., 2008, How will conversion to organic cereal production affect carbon stocks in Swedish agricultural soils? in: *Organic Crop Production – Ambitions and Limitations*, H. Kirchmann and L. Bergström, eds., Springer, Dordrecht, The Netherlands.

Balwinder, P.S., and Fluck, R.C., 1993, Energy productivity of a production system: Analysis and measurement, *Agric. Syst.* **43**: 415–437.
Baumann, H., and Tillman, A.-M., 2004, *The Hitch-Hiker's Guide to LCA*, Studentlitteratur, Lund, Sweden, 543p.
Bergström, L., Bowman, B.T., and Sims, J.T., 2005, Definition of sustainable and unsustainable issues in nutrient management of modern agriculture, *Soil Use Manage.* **21**: 76–81.
Bertilsson, G., 1993, Environmental consequences of different farming systems using good agricultural practices, International Fertiliser Society, Proceedings No. 332, York, UK.
Brentrup, F., Küsters, J., Lammel, J., and Kuhlmann, H., 2004a, Environmental impact assessment of agricultural production systems using the Life Cycle Assessment (LCA) methodology. I. Theoretical concept of a LCA method tailored to crop production, *Eur. J. Agron.* **20**: 247–264.
Brentrup, F., Küsters, J., Lammel, J., Barraclough, P., and Kuhlmann, H., 2004b, Environmental impact assessment of agricultural production systems using the Life Cycle Assessment (LCA) methodology. II. The application to N fertiliser use in winter wheat production systems, *Eur. J. Agron.* **20**: 265–279.
Bonny, S., 1993, Is agriculture using more and more energy? A French case study, *Agric. Syst.* **43**: 51–66.
Carlgren, K., and Mattsson, L., 2001, Swedish soil fertility experiments, *Acta Agric. Scand.* (Section B) **51**: 29–78.
Casey, J.W., and Holden, N.M., 2006, Greenhouse gas emissions from conventional agri-environmental scheme and organic Irish suckler-beef units, *J. Environ. Qual.* **35**: 231–239.
Cederberg, C., and Mattsson, B., 2000, Life cycle assessment of milk production – a comparison of conventional and organic farming, *J. Clean. Prod.* **8**: 49–60.
Cleveland, C.J., 1995, The direct and indirect use of fossil fuels and electricity in USA agriculture, 1910–1990, *Agric. Ecosyst. Environ.* **55**: 111–121.
Connor, D., and Mingues, I., 2006, Looking at biofuels and bioenergy (Letters), *Science* **312**: 1743.
Corré, W., Schröder, J., and Verhagen, J., 2003, Energy use in conventional and organic farming systems, Proceedings No. 511, International Fertiliser Society, York, UK.
Dalgaard, T., Halberg, N., and Porter, J., 2001, A model for fossil energy use in Danish agriculture used to compare organic and conventional farming, *Agric. Ecosyst. Environ.* **87**: 51–65.
Eckert, H., Breitschuh, G., and Sauerbeck, D., 1999, Kriterien umweltverträglicher Landbewirtschaftung (KUL) – ein Verfahren zur ökologischen Bewertung von Landwirtschaftsbetrieben, *Agribiol. Res.* **52**: 57–76 (In German).
Finkbeiner, M., Inaba, A., Tan, R.B.H., Christiansen, K., and Klüppel, H.-G., 2006, The new international standards for life cycle assessment: ISO 14040 and ISO 14044, *Int. J. LCA* **11**: 80–85.
Fluck, R.C., 1992, *Energy in World Agriculture*, Elsevier, Amsterdam, The Netherlands, 367p.
Hirst, E., 1974, Food-related energy requirements, *Science* **184**: 134–138.
Horne, R.E., Mortimer, N.D., and Elsayed, M.A., 2003, Energy and carbon balances of biofuels production: Biodiesel and bioethanol, Proceedings No. 510. International Fertiliser Society, York, UK.
Hülsbergen, K.-J., and Kalk, W.-D., 2001, Energy balances in different agricultural systems – can they be improved? Proceedings No. 476, International Fertiliser Society, York, UK.
IFA, 2006, International Fertiliser Association, Statistics, www.fertiliser.org/ifa/statistics/indicators/ind_reserves.asp, Paris. Assessed July 2006.
ISO (International Organization for Standardization), 1997, Environmental management – life cycle assessment – Principles and framework, International Standard ISO 14040, ISO, Geneva, Switzerland.
Ivarson, J., and Gunnarsson, A., 2001, Försök med konventionella och ekologiska odlingsformer 1987–1998, Meddelande från Södra Jordbruksförsöksdistriktet. Nr.53, Swedish University of Agricultural Sciences, Uppsala, Swden, SJFD-M-53-SE, 165p (In Swedish).
Jenssen, T.K., and Kongshaug, G., 2003, Energy consumption and greenhouse gas emissions in fertiliser production, Proceedings No. 509, International Fertiliser Society, York, UK.
Jenssen, T.K., 2004, N_2O emissions trading – implications for the European fertiliser industry, Proceedings No. 538, International Fertiliser Society, York, UK.

Jørgensen, U., Dalgaard, T., and Kristensen, E.S., 2005, Biomass energy in organic farming – the potential role of short rotation coppice, *Biomass Bioenergy* **28**: 237–248.

Kirchmann, H., Bergström, L., Kätterer, T., Andrén, O., and Andersson, R., 2008, Can organic crop production feed the world? in: *Organic Crop Production – Ambitions and Limitations*, H. Kirchmann and L. Bergström, eds., Springer, Dordrecht, The Netherlands.

Lefroy, E., and Rydberg, T., 2003, Emergy evaluation of three cropping systems in southwestern Australia, *Ecol. Model.* **161**: 195–211.

Mäder, P., Fliesbach, A., Dubois, D., Gunst, L., Fried, P., and Niggli, U., 2002, Soil fertility and biodiversity in organic farming, *Science* **296**: 1694–1697.

Martin, J.F, Diemont, S.A.W., Powell, E., Stanton, M., and Levy-Tacher, S., 2006, Emergy evaluation of the performance and sustainability of three agricultural systems with different scales and management, *Agric. Ecosyst. Environ.* **115**: 128–140.

Odum, H.T., 1996, *Environmental Accounting: Emergy and Environmental Decision Making*, John Wiley and Sons, New York, USA, 370p.

Pimentel, D., Hurd, L.E., Belloti, A.C., Forster, M.J., Oka, I.N., Sholes, O.O., and Whitman, R.J., 1973, Food production and the energy crisis, *Science* **182**: 443–449.

Pimentel, D., 2006, Impacts of organic farming on the efficiency of energy use in agriculture, An organic center state of science review. www.organic-center.org/reportfiles/energy ssr.pdf. Assessed October 2006.

Ramirez, C.A., 2006, Monitoring energy efficiency in the food industry. SenterNovem, www.now.nl. Accessed June 2006.

Ratke, G.-W., Körschens, M., and Diepenbrock, W., 2002, Substance and energy balances in the "static fertilization experiment Bad Lauchstädt", *Archiv für Acker- und Pflanzenbau und Bodenkunde* **48**: 423–433.

Refsgaard, K., Halberg, N., and Kristensen, E.S., 1998, Energy utilization in crop and dairy production in organic and conventional livestock production systems, *Agric. Syst.* **57**: 599–630.

Rydberg, T., and Jansén, J., 2002, Comparison of horse and tractor using emergy analysis, *Ecol. Model.* **19**: 13–28.

Törner, L., 1999, Energibalans i ekologisk och anpassad-intgrerad växtodling, Internal Report, Odling i Balans, Sweden (In Swedish).

Uhlin, H.-E., 1998, Why energy productivity is increasing: an I-O analysis of Swedish agriculture, *Agric. Ecosyst. Environ.* **56**: 443–465.

Uhlin, H.-E., 1999, Energy productivity of technological agriculture-lessons from the transition of Swedish agriculture, *Agric. Ecosyst. Environ.* **73**: 63–81.

Van den Broek, R., Treffers, D-J., Meeusen, M., van Wijk, A., Nieuwlaar, E., and Turkenburg, W., 2001, Green energy or organic food. A life cycle analysis comparing two uses of set aside land, *J. Indust. Ecol.* **5**: 65–87.

Chapter 10
The Role of Arbuscular Mycorrhizas in Organic Farming

Megan H. Ryan and Mark Tibbett

Abstract Arbuscular mycorrhizal fungi (AMF) are ubiquitous in natural and agricultural ecosystems. AMF enhance uptake of nutrients by plants, particularly phosphorus (P), and may also improve plant drought avoidance and disease control. AMF may also be necessary for the long-term sustainability of ecosystems, particularly due to their role in the maintenance of soil structure, and plant community structure and diversity. In agricultural systems, high colonisation of roots by AMF is favoured by the absence of mineral fertilisers that supply readily soluble P, minimal soil disturbance, avoidance of non-host plants and bare fallows, and, perhaps, a high degree of plant diversity and minimal use of biocides. Colonisation by AMF is often higher on organic farms than conventional farms and there is some evidence of an increase in species diversity of AMF on organic farms. These differences appear primarily due to the lack of fertilisers containing readily soluble P on organic farms. Yet high colonisation by AMF is not an inevitable outcome of organic farm management and colonisation may be limited on some organic farms by high rates of tillage or residual high soil available P. There is some indication that organic farms can develop a community of AMF with an increased capacity to enhance plant P uptake. However, AMF do not substitute for fertiliser inputs as the nutrients taken up by the fungi primarily originate from the finite pool of soil available nutrients and their removal in farm products must be matched by inputs from off-farm sources. Indeed, high colonisation by AMF may be considered an indicator of low soil available P. As AMF depend on host photosynthate for energy, high levels of colonisation may reduce plant growth under some environmental or farm management conditions. For instance, monocultures, high soil fertility, and high rates of tillage may stimulate development of less beneficial communities of AMF. The need for inoculants containing AMF on organic farms is unknown, but they may prove beneficial if bare fallows or weed-free crops of non-hosts are regularly included in the rotation. Overall, the

M.H. Ryan (✉)
School of Plant Biology, University of Western Australia, 35 Stirling Hwy, Crawley, WA, Australia, 6009
e-mail: megan.ryan@uwa.edu.au

H. Kirchmann, L. Bergström (eds.), *Organic Crop Production – Ambitions and Limitations*, DOI 10.1007/978-1-4020-9316-6_10,

high abundance of AMF that often results from organic management suggests an important role for the fungi in the functioning of organic farming systems.

Keywords Arbuscular mycorrhizal fungi · Organic farming · Farming systems · Phosphorus · Sustainability

10.1 Introduction

Organic farming is often promoted as a superior alternative to conventional farming. The origins and current practices of organic farming are diverse (Kirchmann et al., 2008), but for the purposes of this review a number of alternative farming systems will be termed "organic" (biodynamic, organic, biological organic). Whilst the beliefs that determine the farm management practices permitted in organic farming may sometimes differ from the scientifically-based principles that underlie conventional farming (Kirchmann et al., 2008), in practice organic and conventional farming systems differ in two key aspects: (i) The replacement on organic farms of processed readily soluble fertilisers with leguminous crops and fertilisers which are generally less processed and less soluble (Watson et al., 2002; Kirchmann and Ryan, 2004). Fertiliser inputs also tend to be lower on organic farms and in some instances no fertiliser may be applied. (ii) The replacement on organic farms of the manufactured chemical biocides permitted on conventional farms with what are considered "natural" alternatives.

Arbuscular mycorrhizas are formed by a close association between roots and arbuscular mycorrhizal fungi (AMF), which are ubiquitous in natural and agricultural ecosystems. AMF have long fascinated agricultural researchers due to their ability to enhance host plant uptake of nutrients, especially phosphorus (P) (Koide and Mosse, 2004). Increasingly, AMF are being shown to influence host plants in other ways which include aiding with drought avoidance and providing a degree of disease control. There is now also evidence that AMF may be necessary for the long-term sustainability of agricultural systems due to their interactions with other components of the soil biological community, and their role in the maintenance of soil structure and, in permanent vegetation, plant community structure and diversity.

Thus AMF may be viewed as intermediaries between soil nutrients and host plants, and thereby as biological regulators of plant nutrient uptake. However, while host plant P uptake may be greatly enhanced by AMF (Smith and Read, 1997), when P is readily available to plants due to high soil available P or application of fertilisers containing readily soluble P, the occurrence of AMF may be greatly reduced (Smith and Read, 1997). This characteristic of AMF appears consistent with the view of proponents of organic farming that plants should not be "force-fed" readily soluble inorganic nutrients, instead acquiring nutrients in more desirable amounts and ratios by means of an active, nurtured, soil biological community (Kirchmann and Ryan, 2004).

In this chapter we attempt to characterise the role of AMF in organic farming by reviewing the relatively small literature that examines directly the role of AMF on organic farms as well as the much larger literature on the role of AMF in conventional farming systems. The chapter begins with a brief overview of the structure and function of AMF. The various methods for quantifying abundance, community diversity and structure, and function of AMF are then described. The effect on AMF of farm management (organic and conventional) is then reviewed, followed by consideration of the impacts on AMF of agronomic practices. The role of AMF on organic farms at scales ranging from individual plants through to the whole agricultural system is then discussed and the necessity for inoculation considered.

10.2 Arbuscular Mycorrhizas – Structure and Function

Mycorrhizas are currently classified into seven types based on morphology (Peterson et al., 2004). In this review we consider only one type – arbuscular mycorrhizas. Whilst other types of mycorrhizas are present in agricultural systems, notably ectomycorrhizas and ericoid mycorrhizas, the plants which form these mycorrhizas are generally much less common than those which form arbuscular mycorrhizas and there is almost no published work on the role of these other mycorrhizas in organic farming systems.

Arbuscular mycorrhizas are formed by fungi classified in the new phylum *Glomeromycota* (Schüβler et al., 2001). There are 198 described species of AMF (Schüβler, 2006), although this may be a gross underestimate owing to the small number of rigorous taxonomic studies. Eleven genera of AMF, placed into four orders, are currently recognised; *Glomus, Gigaspora, Scutellospora, Acaulospora, Kuklospora, Entrophospora, Pacispora, Diversispora, Paraglomus, Archaeospora* and *Intraspora* (Schüβler, 2006; Sieverding and Oehl, 2006). In addition, the genus *Glomus* is now considered nonmonophyletic and may soon be split into a number of new genera (Schwarzott et al., 2001). This on-going revision means most of the studies referred to in this chapter contain out-dated taxonomy. None-the-less, their findings remain useful when considering the role of AMF in agricultural systems.

AMF are obligate symbionts and cannot grow without being in association with a host root from which they attain all their energy (carbon) needs (Ho and Trappe, 1973; Pfeffer et al., 1999). This characteristic has resulted in an inability to utilise axenic culture for large scale inoculum production. AMF are believed to use between 4 and 20% of host plant photosynthate (Graham, 2000). When AMF colonise a root they penetrate through the epidermis and intraradical hyphae move from cell to cell or through intercellular spaces (Peterson et al., 2004). In some cells the hyphae form structures with fine branches known as arbuscules which allow exchange of carbon and nutrients with the host plant (Peterson et al., 2004). AMF may also form intraradical or extraradical lipid-filled vesicles, which seem to act as storage structures or, perhaps, propagules (Peterson et al., 2004). The fungi extend extraradical hyphae into surrounding soil and these absorb nutrients up to 11 cm away from the root (Jakobsen et al., 1992a; Leake et al., 2004). The distribution of the hyphae in soil, hyphal growth rate, the distance over which nutrients are transported and nutrient uptake per unit of mycorrhizal root can all differ with species of AMF (Jakobsen et al., 1992a,b). Spores are typically formed from the extraradical hyphae. AMF can constitute a significant portion of the soil microbial biomass (Olsson et al., 1999). For instance, Rillig et al. (2002) found that extraradical hyphae of AMF constituted greater than 50% of the total fungal hyphal length in annual grassland soil.

The extraradical hyphae of AMF, if undisturbed, form a common mycorrhizal network (CMN) of interconnected hyphae and host roots. CMNs may allow movement of nutrients between linked plants, although carbon flow is generally considered to be one way, from the host plant to the fungus (Pfeffer et al., 2004). While such processes are thought to be important in mediating the outcomes of plant competition, the extent to which interplant transfers occur in the field is not known (He et al., 2003; Leake et al., 2004).

Arbuscular mycorrhizas are formed in the majority of angiosperm families and are commonly found in most crop and pasture plants, in particular legumes and cereals (Newman and Reddell, 1987), as well as in many tree crops (Cuenca and Meneses, 1996; Graham et al., 1997; Mamatha et al., 2002). Hence AMF are ubiquitous in agricultural systems (Abbott and Robson, 1977; Talukdar and Germida, 1993; Sjöberg et al., 2004). A small number of crops are non-hosts for AMF including brassicas and some lupin species (Plenchette et al., 1983; Trinick, 1977).

AMF have traditionally been considered to be non-host specific as most species will colonise most host plants (e.g. Klironomos, 2003). However, the impact of AMF on host plant nutrient uptake, soil aggregation and even the relationship between plant species diversity and productivity can differ between species of AMF (Graham et al., 1997; Graham and Abbott, 2000; Klironomos et al., 2000; Piotrowski et al., 2004). The impact of a particular species of AMF on plant growth can also vary greatly between plant species (Klironomos, 2003) or crop cultivars and host plant species can affect AM fungal population growth rates and hence AM fungal community structure (Bever, 2002).

AMF are best known for their ability to enhance host plant P uptake, as well as uptake of other nutrients, although other benefits for the host plant may also occur. When benefits for the host plant outweigh carbon costs, host plant growth may be enhanced. Most studies of AMF have concentrated on their ability to enhance host plant P uptake under conditions of low P availability. Whilst agricultural systems in industrialised countries tend to present a high P soil environment, organic farms are often an exception (e.g. Derrick and Dumaresq, 1999; Ryan et al., 2000). Hence, AMF may be of particular importance in organic farming.

10.3 Assessing Abundance, Community Diversity and Structure, and Function of AMF

Crucial to advancing our understanding of the role of AMF in agricultural systems is the development of techniques for measuring the occurrence and function of AMF.

Abundance of AMF is most commonly assessed by estimating the percentage of root length colonised after roots have been cleared and stained (see Giovannetti and Mosse, 1980; Grace and Stribley, 1991; Brundrett et al., 1996). Strong correlations between changes in this measure and changes in plant growth or nutrient uptake have been observed in some studies (e.g. Ryan and Angus, 2003). However, when many studies are compared, the degree of change between plants or treatments in the percentage of root length colonised is not a particularly good indicator of host plant growth response, that is, mycorrhiza function (McGonigle, 1988; Jakobsen et al., 2001; Lekberg and Koide, 2005a). Factors responsible for this could include other limitations on plant growth or changes in root length confounding changes in the percentage of root length colonised. In addition, measuring the abundance of AMF as the percent of root length colonised treats mycorrhizas as a single entity taking no account of the diversity of species that can be present in the roots of one plant (Clapp et al., 1995), their spatial variation or physiological state. An attempt is sometimes made to elucidate the latter parameter by noting the occurrence of arbuscules, vesicles and hyphae (e.g. Johnson, 1993; Egerton-Warburton and Allen, 2000). However, in the absence of a more reliable, but equally simple, measure of mycorrhiza function, the percentage of root length colonised is often used as a surrogate.

The abundance of AMF is also commonly assessed by estimating the density of spores in soil. Spores are sieved from soil, identified, and counted under a microscope (e.g. Cuenca and Meneses, 1996). As it can be difficult to distinguish viable spores from dead spores, or determine the age of viable spores, this measure provides an assessment of the "accumulated sporulation history" (Hijri et al., 2006) which may not reflect the species currently colonising roots (Clapp et al., 1995). Also, spore counts may not reflect the relative abundance of species of AMF, as species can differ in rate and timing of sporulation (Oehl et al., 2003). Species which rarely sporulate may be absent and non-sporulating species will not be represented (Rosendahl and Stukenbrock, 2004).

Community diversity and structure of AMF is a particularly important parameter as it can be greatly changed by agricultural practices (Oehl et al., 2003) and species of AMF from the same soil can differ greatly in their impact on plant growth (Klironomos, 2003). Community diversity and structure has also traditionally been assessed by spore counts (Brundrett et al., 1996). While this method has the drawbacks mentioned above, communities of AMF can be usefully investigated (e.g. Oehl et al., 2004), with the exception of non-sporulating species. Establishing trap cultures can refine this method. Trap cultures involve growing plants in field soil in a glasshouse, with spores then being regularly sampled and identified (e.g. An et al., 1993). The longer the time before samples are taken the greater the number of species that will be confidently identified (e.g. Oehl et al., 2004). However, which of the identified species is actually colonising roots in the field at any point in time is unknown. While this issue could be investigated through careful description of colonisation morphology (Abbott, 1982; Rangeley et al., 1982), the development of molecular methods to identify individual species of AMF in roots now shows much promise (e.g. Hijri et al., 2006). However, not all molecular methods detect a wide range of AM fungal diversity (e.g. Helgason et al., 1998), and care must be taken with interpretation of results.

The impact of AMF on host plant growth and nutrition has been commonly assessed under glasshouse conditions using sterilised soil and single strain inoculants of AMF. The large growth benefits from AMF observed in these circumstances are often not found under field conditions (Rangeley et al., 1982; Fitter, 1985). In a meta-analysis of experiments examining the contribution of AMF to agricultural plants in non-sterile soils, Lekberg and Koide (2005a) found mycorrhizal benefit was substantially greater in the glasshouse (95% CI 84–304%) than the field (95% CI 35–83%). Possible reasons for this difference include the presence in field experiments of more variable and extreme environmental conditions (Gavito et al., 2005; Ryan et al., 2005), other beneficial or pathogenic soil organisms (Daniels Hetrick et al., 1988), or higher soil nutrient availability (Ryan and Graham, 2002). Unfortunately, assessment of mycorrhiza function in the field has many difficulties. In particular, the methods used to produce non-mycorrhizal controls, or treatments with reduced colonisation (e.g. fungicides, long bare fallow, tillage) also affect other soil organisms (Khasa et al., 1992; Ryan et al., 2002), soil nutrient availability (Thompson, 1990) and other parameters (Hulugalle et al., 1998). Thus, care must be taken with extrapolating results from glasshouse studies to field conditions and in the interpretation of results of field experiments.

10.4 Mycorrhizas and Organic Farming

A small number of studies have compared the abundance and diversity of AMF between organic (or low-input) and conventional farms or treatments. Across a broad range of agriculturally utilised host plants, agricultural systems and locations, the degree of colonisation of roots by AMF, spore density and species diversity of AMF are all generally reported as higher on the organic farms (Table 10.1; Limonard and Ruissen, 1989). For instance, Fig. 10.1 shows the results of a survey of dairy farms in SE Australia. Colonisation of clover (*Trifolium* spp.) by AMF was higher on the organic or biodynamic farm in 16 of the 19 pairs (Ryan, 1998; Ryan et al., 2000). Similarly in the UK, Eason et al. (1999) sampled grassland sites on organic and conventional farms and found colonisation by AMF of pasture roots and spore density were both higher, on average, for the organic sites (Table 10.2). For crops, the differences in colonisation reported between organic and conventional farms can be much larger than those reported for pastures (e.g. Sattelmacher et al., 1991; Ryan et al., 1994). Whilst colonisation by AMF is usually lower on conventional farms for both crops and pastures, it is rarely completely absent and above 10% of root length, and sometimes greater than 50% of root length, is usually colonised.

The higher abundance of AMF on organic farms is most commonly attributed to the absence of readily soluble P fertilisers and/or lower soil available P (especially in the Australian studies), minimal use of biocides and more diverse rotations (Table 10.1). However, of the papers referred to in Table 10.1, only the two sets

Table 10.1 Studies which compared the impact of organic and conventional farm management on variables related to AMF

Location	System	Plants present	Variables measured	Farm management under which variable was higher	Characteristics of organic system considered by authors as responsible for differences	Reference
Canada	Cropping	Flax	Root colonisation (%)	Organic	Lower soil P	Entz et al. (2004)
Pennsylvania, USA	Cropping	Maize, soy-bean	Community structure (measures including species density and Shannon index made using spores from field)	No difference	n.a.[a]	Franke-Snyder et al. (2001)
California, USA	Intensive hor-ticulture	Lettuce	Root colonisation (%)	Organic	No use of biocides and P and N fertilisers, more diverse rotation	Miller and Jackson (1998)
			Spore density	No difference		
California, USA	Orchard	Apple	Root colonisation (%)	Organic	More weeds, no pesticide use	Werner (1997)
California, USA	Horticulture	Straw-berries	Root colonisation (%)	Organic	No fumigation, low soil fertility	Werner et al. (1990)
Denmark	Cropping	Wheat	Clonal diversity and population genetic structure (PCR on single spores)	No difference	n.a.[a]	Stukenbrock and Rosendahl (2005)
Northern Germany	Cropping	Cereal rye	Root colonisation (%)	Organic	No use of fertilisers and agrochemicals, more diverse rotation	Sattelmacher et al. (1991)
Basel, Switzer-land (DOK experi-ment)	Cropping	Various	Root colonisation (%)	Organic	Lower soil P	Mäder et al. (2000)

Table 10.1 (continued)

Location	System	Plants present	Variables measured	Farm management under which variable was higher	Characteristics of organic system considered by authors as responsible for differences	Reference
Basel	Cropping	Grass, clover	Root colonisation (%)	Organic	Lower soil P	Oehl et al. (2004)
			Spore density	Organic	Lower soil P	
			Number of species (spores extracted from field and spores present in trap cultures)	Organic	Lower soil P	
			Community diversity (Shannon index using spores from field)	Organic	Lower soil P	
United Kingdom	Pasture	Pasture	Root colonisation (%)	Organic	Not known	Eason et al. (1999)
			Spore density	Organic	Not known	
Brazil	Orchard	Apple	Spore density	Conventional	Not known	Purin et al. (2006)
			Number of species (spores from field)	Organic	Not known	
			Community diversity (Shannon index using spores from field)	Conventional	Not known	
SE Australia	Crop-livestock	Wheat	Root colonisation (%)	Organic	No use of readily soluble P fertiliser	Dann et al. (1996), Ryan et al. (1994, 2004)
SE Australia	Permanent dairy pastures	White clover, ryegrass	Root colonisation (%)	Organic	No use of readily soluble P fertiliser	Ryan et al. (2000), Ryan and Ash (1999)

[a]n.a. = not available.

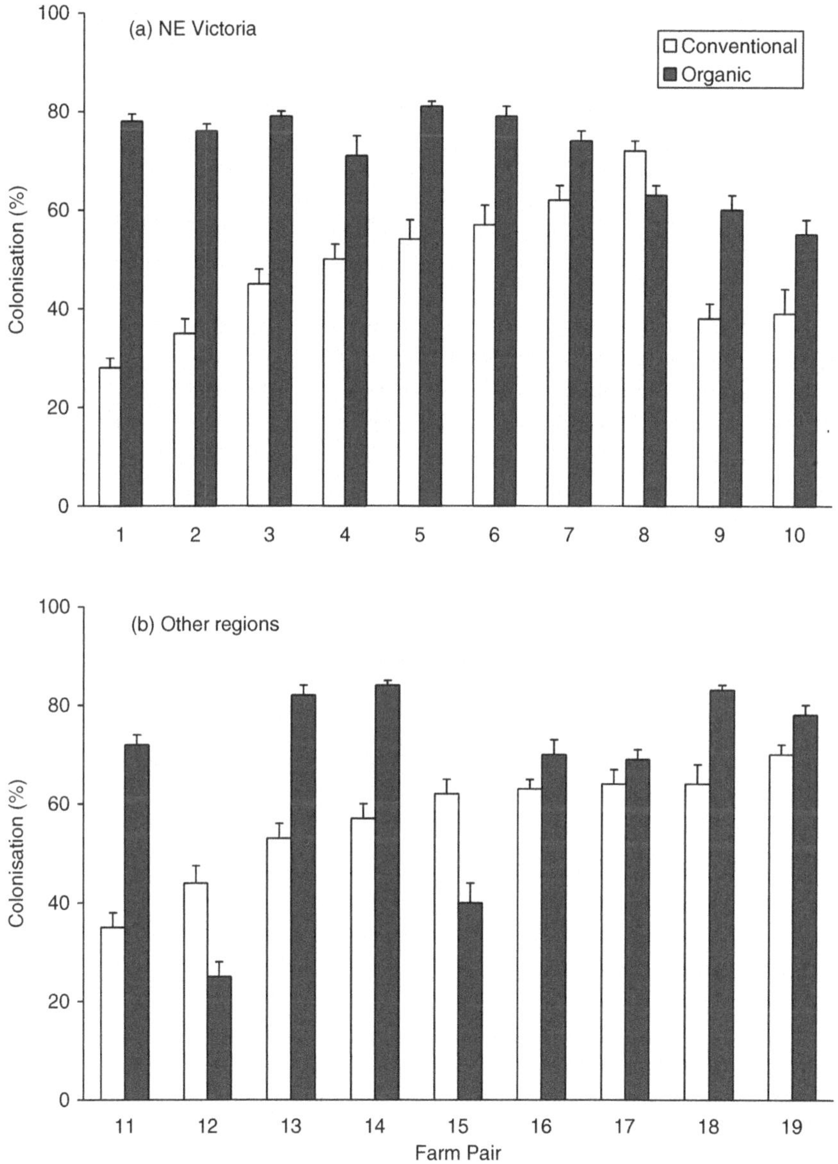

Fig. 10.1 The percentage of clover (*Trifolium* spp.) root length colonised by AMF on 19 biodynamic or organic dairy farms paired with a conventional neighbour located in (**a**) NE Victoria (Ryan et al., 2000) or (**b**) elsewhere in SE Australia (Ryan, 1998) (n = 20, mean ± SE)

Table 10.2 Soil extractable P, percentage of root length colonised by AMF and density of spores of AMF in soil from 13 conventional and 10 organic grassland sites in the United Kingdom (Eason et al., 1999)

Farm management	Soil extractable P (mg kg^{-1})	Colonisation by AMF (%)	Spore density (no. per gram dry soil)
Conventional	5.4	40.4	12.1
Organic	2.0	63.6	34.4
Significance	$P < 0.001$	$P < 0.001$	$P < 0.001$

of studies based in SE Australia included experimental investigation of the factors responsible for the differences in AMF. In addition, only a very few studies have investigated the functioning of the communities of AMF that develop on organic farms and their impact on plant nutrition or growth (Scullion et al., 1998; Eason et al., 1999; Muckle, 2003). We now examine the reasons behind the higher occurrence of AMF on organic farms through an examination of the effects of agricultural practices on AMF. The implications for plant and agricultural system function of the high occurrence of AMF on organic farms is then explored.

10.5 The Impacts of Agricultural Practices on AMF

The impacts of common agricultural practices on the abundance, community diversity and structure, and function of AMF have been examined in many review papers and books (e.g. Bethlenfalvay and Linderman, 1992; Smith and Read, 1997; Gosling et al., 2006). We will focus on areas relevant to understanding the role of AMF in organic farming and refer to other more detailed reviews as appropriate.

10.5.1 Mineral Fertilisers

Phosphorus, nitrogen (N), and potassium (K) are the nutrients most commonly applied as readily soluble fertilisers on conventional farms. On organic farms, biological N fixation by legumes is generally relied upon to supply N (e.g. Ryan et al., 2000; Ryan et al., 2004), although some N may also be supplied from composts or manures. Potassium is present in significant concentrations in manure and may also be applied on organic farms in other forms such as rock K (Watson et al., 2002). If P fertilisers are applied on organic farms they tend either to be relatively insoluble minerals such as rock phosphate or organic materials such as chicken manure or composts. Inputs of P on organic farms tend to be lower than those on conventional farms. Thus, over time, a decline in total and soil extractable P tends to occur on organic farms (Tables 10.2, 10.3; Penfold et al., 1995; Nguyen et al., 1995; Løes and Øgaard, 1997; Derrick and Dumaresq, 1999; Ryan et al., 2000; Oehl et al., 2002; van Diepeningen et al., 2006).

Table 10.3 Farming system, nutrient inputs, diversity of AMF and P budget for the DOK experiment. An identical seven year crop rotation was maintained in all farming systems

Treatment	Type of fertiliser or manure	Biocides	N	P	K	Species of AMF (field spores)[b]	Species of AMF (spores from 20 month trap cultures)[b]	P budget (1977–1988) (kg P ha^{-1} yr^{-1}) [c]
			Inputs (kg ha^{-1} yr^{-1}) [a]					
Mineral-conventional	Mineral	Synthetic	124	41	254	15	22	-5.0
Conventional	Manure, slurry plus mineral	Synthetic	154	39	258	15	24	3.8
Bio-organic	Manure, slurry	No synthetic	88	25	139	19	30	-5.7
Bio-dynamic	Manure, slurry, biodynamic preparations	No synthetic	91	22	159	17	26	-7.8
Control	No fertiliser		0	0	0	18	24	-20.9

[a] Mean nutrient inputs since 1985 (Oehl et al., 2004).
[b] Oehl et al. (2004).
[c] Oehl et al. (2002).

Readily soluble P fertiliser is often thought responsible for the lower occurrence of AMF on conventional farms (Table 10.1). While very low P supply will impede colonisation by AMF (e.g. Kelly et al., 2005), abundant P has been shown many times to reduce colonisation (Smith and Read, 1997), possibly due to a reduction in exudation of carbohydrates by roots (Graham et al., 1981). The reduction in colonisation limits the cost to the plant of supporting the AMF. Thus, readily soluble P fertiliser is most commonly reported to decrease colonisation by AMF and the density of spores in soil (Dann et al., 1996; Smith and Read, 1997; Miller and Jackson, 1998; Kahiluoto et al., 2000; 2001). The resulting increase in soil extractable P may lower colonisation levels for many years, even if no further fertiliser is added (Dekkers and van der Werff, 2001). If soil extractable P is very high, as may occur in conventional intensive horticultural systems, colonisation by AMF may be negligible (Ryan and Graham, 2002). Poorly soluble P fertilisers permitted on organic farms, such as rock phosphate, are generally reported to not reduce colonisation by AMF (Dann et al., 1996), although if applied at a rate high enough to increase soil available P presumably a negative impact on AMF would result.

In agricultural systems, negative correlations are often found between soil extractable P (or plant P status) and the percentage of root length colonised by AMF or spore density (e.g. Mårtensson and Carlgren, 1994; Cuenca and Meneses, 1996; Ryan et al., 2000). For instance, Fig. 10.2 shows high pasture foliar P strongly correlated to low colonisation by AMF of clover (*Trifolium* spp.) roots using data from 19 biodynamic or organic dairy farms paired with a conventional neighbour in SE Australia (Ryan, 1998; Ryan et al., 2000). However, even in low P soils, P-fertiliser does not always reduce colonisation by AMF (Xavier and Germida, 1997), especially if the soil is strongly P-fixing (Kabir and Koide, 2002) or the P is quickly immobilised by the microbial community. A decline in the community diversity of AMF with increasing soil extractable P, assessed by spore counts, was also reported by Cuenca and Meneses (1996), but was not found by Kahiluoto et al. (2001).

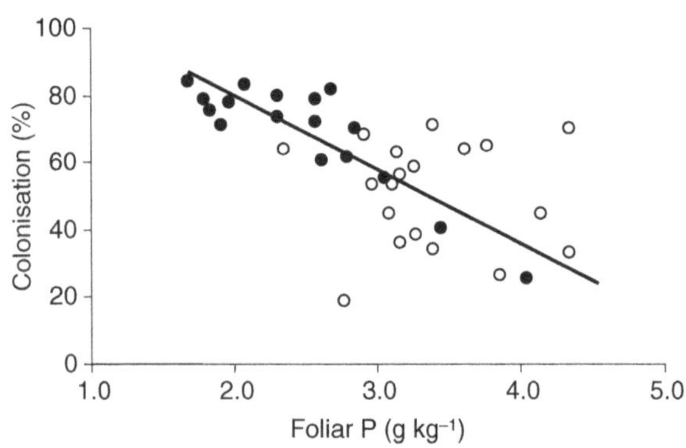

Fig. 10.2 The relationship between the percentage of clover (*Trifolium* spp.) root length colonised by AMF and whole pasture foliar P concentration for 19 biodynamic or organic dairy farms (•) paired with a conventional neighbour (○) located in SE Australia (Ryan, 1998; Ryan et al., 2000) (n = 20)

In SE Australia a series of studies clearly show readily soluble P fertiliser as the primary factor responsible for lowered colonisation by AMF on both irrigated permanent dairy pastures (Figs. 10.1 and 10.2; Ryan and Ash, 1999; Ryan et al., 2000) and conventional dryland livestock-cropping farms (Ryan et al., 1994; Dann et al., 1996). For instance, Ryan et al. (2000) compared colonisation by AMF of dairy pastures in two adjacent fields; the biodynamic field had received no P fertiliser for several decades while the conventional field had received regular applications of readily soluble P fertiliser. A strip in the conventional field had not received P fertiliser for 10 years and colonisation by AMF was higher than in the rest of the field and similar to the colonisation level in the adjacent biodynamic field. Thus the other practices of the conventional farmer, such as use of N fertiliser and biocides, and use of conventional livestock medicines, apparently had a minimal impact on colonisation level. Similarly, a glasshouse experiment undertaken by Ryan et al. (1994) suggested that readily soluble P fertiliser, but not N fertiliser, herbicides, seed fungicide dressings, wheat (*Triticum aestivum* L.) variety or soil pH, was responsible for lower colonisation by AMF on a conventional wheat farm compared with an organic neighbour. However, in these two examples, biocide inputs on the conventional farms were relatively small and for the cropping farms, rotations and tillage regimes were similar on the organic and conventional farms.

The impact on AMF of N fertiliser in agricultural systems is variable and generally does not match in magnitude the impacts of P fertiliser (e.g. Baltruschat and Dehne, 1989; Furlan and Bernier-Cardou, 1989; Ryan et al., 1994; Ryan and Ash, 1999). However, a number of recent studies in natural ecosystems have shown a shift in mycorrhizal community structure or decrease in the abundance of AMF in response to N fertiliser (e.g. Egerton-Warburton and Allen, 2000; Bradley et al., 2006). For instance, in coastal sage scrub vegetation in California, USA, Egerton-Warburton and Allen (2000) found high N was associated with displacement of the larger spored *Scutellospora* and *Gigaspora* species by small-spored *Glomus* species. The impact of K fertiliser on AMF has been little investigated, but has been reported to stimulate spore production (Furlan and Bernier-Cardou, 1989).

Fertiliser regime may be a major determinant of differences in AMF in the biodynamic, bio-organic and conventional (DOK) field experiment established near Basel, Switzerland, in 1978 (Mäder et al., 2000, 2002; Oehl et al., 2002, 2003, 2004; Hijri et al., 2006). At this site, two conventional farming treatments (mineral fertilisers only, mineral fertilisers plus farmyard manure) two organic farming treatments (bio-dynamic and bio-organic), and a non-fertilised control are subjected to identical seven year rotations, crop varieties and tillage. The treatments differ primarily in the type and amount of nutrients added and in the use of synthetic pesticides (Table 10.3). Fertiliser inputs of N, P and K have all been higher in the conventional treatments (Table 10.3). Colonisation by AMF was 30–60% higher in roots from the organic treatments than the conventional treatments (Mäder et al., 2000). Species richness of AMF, as assessed by spore abundance, was slightly higher in the organic treatments (Table 10.3) and spores of *Acaulospora* species and *Scutellospora* species were more abundant in the organic treatments (Oehl et al., 2004). While Oehl et al. (2004) concluded that higher P inputs and higher soil extractable

P in the conventional treatments (Oehl et al., 2002) were responsible for the lower colonisation and diversity of AMF, higher inputs of N and K, and lower inputs of organic matter, could also have played a role (Table 10.3).

Hijri et al. (2006) examined the diversity of AMF in five agricultural field sites, including the DOK experiment, using variable regions of ribosomal RNA genes to identify species of AMF within colonised roots. Whilst a high species diversity of AMF was found throughout the DOK experiment, low diversity and a complete absence of AMF in the genera *Acaulospora, Scutellospora, Gigaspora* and *Paraglomus* was found at two sites; a conventional intensively managed maize (*Zea mays* L.) monoculture and a leek (*Allium porrum* L.) field which had been managed organically for six years. Of the five field sites, the organic leek field had the highest soil P, due to a history of high P additions prior to conversion (soil N not given), while the maize site had the highest annual inputs of fertiliser P and N. Thus, it appears high availability of P, and perhaps N, may lower the diversity of AMF and over-ride any positive impacts from other organic management practices, including elimination of the biocides permitted in conventional farming systems.

There is some evidence that the communities of AMF that develop in frequently fertilised soils may be less beneficial to host plant growth. For instance, Johnson (1993) found use of fertiliser for eight years raised soil N from 1.6 to 8.6 mg kg^{-1} and extractable soil P from 26.5 to 62.2 mg kg^{-1} and caused the relative abundance of *Gigaspora* and *Scutellospora* species to decrease. The fertiliser addition also created a community of AMF less effective at promoting plant growth, as assessed by an indicator plant inoculated with soil from the different fertiliser treatments. These results suggest that organic farms may develop more effective communities of AMF than conventional farms due to presenting a lower P environment. This conclusion is supported by Scullion et al. (1998) who inoculated leek and white clover (*Trifolium repens* L.) growing in four soil types with spores of AMF isolated from pasture on organic or conventional farms. Shoot weight was greater with organic farm inocula in two of the four soils for leek (41 and 530%) and one of the four soils for clover (9%), although the higher colonisation levels produced by the organic inocula may have contributed to these results as well as differences in effectiveness. Overall, organic inocula were superior to conventional inocula in soil of lower P status. Similarly, Eason et al. (1999) compared growth in a low P soil of leek and white clover inoculated with spores from 13 conventional and 10 organic grassland sites (Table 10.4). The organic sites had, on average, lower soil extractable P (Table 10.2). The mean increase in yield was greater with organic inocula than the conventional inocula (Table 10.4), although some highly effective inocula were obtained from conventional farms and some relatively ineffective inocula from organic farms (Eason et al., 1999). Non-sporulating AMF and species that had not recently sporulated would have not been represented in the inocula in either study.

In summary, the prohibition of readily soluble P fertilisers in organic farming may be largely responsible for the higher levels of colonisation by AMF on organic farms and may also contribute towards a higher species diversity of AMF and development of communities of AMF with a greater ability to enhance host plant P-uptake. However, research which controls for the other management differences

Table 10.4 The response to inoculation with spores of AMF sieved from 13 conventional and 10 organic grassland sites in the United Kingdom for white clover (*Trifolium repens* L.) and leek (*Allium porrum* L.) grown in an irradiated low P soil (n = 3 per site, mean and range). Approximately 1000 spores were placed in each pot (Eason et al., 1999)

	Farm management	Total shoot weight (mg)	Shoot P content (mg)	Colonisation by AMF (%)
Clover	Conventional	9.4 (8.2–11.4)	7.6 (4.9–11.0)[a]	55 (0–68)
	Organic	10.3 (8.6–12.3)	8.2 (5.7–10.2)[a]	63 (44–73)
	Significance	$P < 0.05$	*ns*	*ns*
Leek	Conventional	0.39 (0.16–0.93)	0.97 (0.05–2.26)	57 (34–77)
	Organic	0.55 (0.16–1.22)	1.48 (0.37–3.30)	64 (53–76)
	Significance	$P < 0.05$	*ns*	*ns*

[a] From the third of three cuts.

between organic and conventional farms is required to confirm the impacts of fertiliser regime on diversity and function of AMF.

10.5.2 Organic Matter Amendments

Organic farmers are encouraged to apply organic amendments such as composts to the soil and plough in green manure crops and thus may apply greater amounts of organic carbon to soils (e.g. van Diepeningen et al., 2006). The impact of organic matter on colonisation by AMF seems to reflect the ease with which P is mineralised, which is highly dependent on the quality of the organic matter amendment (Smith et al., 2006). High levels of colonisation by AMF can occur after addition of high rates of compost if soil extractable P remains low (e.g. Gaur and Adholeya, 2000, 2002). Indeed an increase in extraradical hyphal length in response to organic matter addition is often noted (Gryndler et al., 2006). However, if the organic matter addition results in a rapid mineralisation of P it will act in the same manner as a readily soluble P fertiliser and have a negative impact on colonisation by AMF (Sáinz et al., 1998). Research is required to investigate the release of P and N from manure on AMF abundance and diversity.

10.5.3 Biocides

The effect of biocides, including fungicides, on AMF is complex and studies show impacts on abundance and function of AMF ranging from negative through to positive (see review by Johnson and Pfleger, 1992; and e.g. Sreenivasa and Bagyaraj, 1989; Plenchette and Perrin, 1992; Ryan et al., 1994; Udaiyan et al., 1999). For instance, Udaiyan et al. (1999) investigated the impact of six pesticide drenches at recommended rates on mycorrhizas of three cereals under field conditions. All pesticides reduced colonisation by AMF, but the impact of each pesticide often differed between host plants. The impact of the pesticides on spore density also differed between host plants with sporulation by AMF colonising *Panicum miliaceum* L. being stimulated by most of the pesticides. In contrast, Plenchette and Perrin (1992) found fungicides had minimal effect on the percentage of length of root colonised by AMF when applied after colonisation was already well established. The impact of biocides on community structure and diversity of AMF is essentially unknown, although one study does suggest that the decrease in host plant diversity that results from the use of herbicides can cause a decrease in the diversity of AMF (Feldmann and Boyle, 1999.

Thus it appears the prohibition on organic farms of many of the biocides permitted on conventional farms could contribute towards enhanced colonisation by AMF in some situations. However, biocides approved for organic farms may also be quite toxic (Edwards-Jones and Howells, 2001). For instance, Sreenivasa and Bagyaraj (1989) found copper oxychloride decreased colonisation by AMF, as well as spore density and, in particular, number of infective propagules. As noted by Gosling et al. (2006), for both conventional and organic farms there is a lack of studies, both short-term and long-term, which examine the impact of commercial biocide application regimes on AMF. Such regimes could involve several different biocides being added alone or in combination during the crop lifecycle. In a commercial context, the substantial yield benefits that can accrue from successful control of weeds, pests and diseases may outweigh any detrimental effects on AMF.

10.5.4 Tillage

Tillage or soil disturbance may reduce levels of colonisation by AMF, or reduce the effectiveness of colonisation, by disrupting the CMN (Miller, 2000). It is hypothesised that the ability to tap into an intact CMN allows fast colonisation of new roots and fast uptake of nutrients early in the season. As well as disrupting the CMN, excessive tillage or soil disturbance can greatly reduce subsequent colonisation by AMF (Mulligan et al., 1985) due to the destruction of inoculum (Jasper et al., 1991; Boddington and Dodd, 2000). However, in situations where inoculum concentrations are initially very high, sufficient inoculum may remain following tillage to ensure high levels of colonisation are quickly produced (Jasper et al., 1991). For instance, in the organic wheat crop examined by Ryan et al. (1994), 60% of root length was colonised 11 weeks after sowing even though the soil had been inverted with a mouldboard plough, left as a bare fallow for seven months, and then cultivated twice prior to sowing.

Tillage is known to have a significant effect on community structure of AMF. A number of recent studies have found that *Glomus* species predominate in cropping soils and it has been suggested that tillage may favour this genus (Land and Schönbeck, 1991; An et al., 1993; Helgason et al., 1998; Ezawa et al., 2000; Daniell et al., 2001; Jansa et al., 2002). For instance, Jansa et al. (2003) used PCR markers to examine roots of maize crops grown in an experiment established in 1987 and cropped continuously using a number of tillage treatments. Whilst tillage had little impact on *Gigaspora* species, colonisation by *Scutellospora* species was dramatically higher in no-tillage plots and there was an increased incidence of *Glomus* species in the tilled plots. Boddington and Dodd (2000) found soil disturbance caused the spore density of *Scutellospora* species to be greatly reduced; spore density of *Glomus* species was also reduced, while *Acaulospora* spore density was unaffected. Tillage may also reduce the species diversity of AMF (Menéndez et al., 2001). However, in Sweden, little difference was found in spore numbers or species diversity of AMF between semi-natural grassland and ploughed fields (Sjöberg et al., 2004).

Whilst a move towards low-tillage or no-tillage cropping has occurred in agricultural systems around the world, encouraged by benefits for soil structure and reduced soil erosion, such systems usually rely on herbicides for weed control, often making their adoption by organic farmers problematic. In addition, the restrictions on use of biocides on organic farms could result in increased tillage and soil disturbance in order to achieve adequate weed control. Although, in some circumstances, the greater use of pasture leys on organic farms could result in the net amount of tillage being lower on organic farms.

10.5.5 Rotations and Agricultural System

Perhaps the greatest impact of rotation on AMF can occur from a bare fallow. During a bare fallow AMF persist in the soil only as inactive inoculum in the form of spores, hyphae and colonised root segments (Bellgard, 1992; Jasper et al., 1993). Consequently the density of inoculum declines over time (Troeh and Loynachan, 2003) and colonisation is usually reduced in following crops (Thompson, 1987; Kabir and Koide, 2002; Ryan et al., 2002). In some instances this reduction in colonisation has corresponded to reduced crop uptake of P and Zn and a corresponding reduction in crop growth and yield (Thompson, 1987; Kabir and Koide, 2002). Non-host plants such as brassicas may act in the same manner as a bare fallow, providing there is no significant presence of weeds that host AMF (Ryan et al., 2002). The impact of bare fallows and non-hosts can be alleviated by the presence of plants that host AMF. For instance, in Pennsylvania, USA, the replacement of bare fallow with a rye (*Secale cereale* L.) or oats (*Avena sativa* L.) cover crop increased colonisation by AMF, shoot dry weight, shoot height and yield of marketable ears of following sweet corn (*Zea mays* L.) (Kabir and Koide, 2002). Bare fallows and non-host crops may not always have a negative impact on AMF. In western New South Wales, Australia, long fallow cotton (*Gossypium hirsutum* L.) (cotton, alternating with a bare fallow, is sown every other year) had similar colonisation by AMF to continuous cotton (Hulugalle et al., 1998). Low mean annual rainfall may have contributed towards a low rate of inoculum death.

An increase in plant diversity in a field or during the course of a rotation may increase AMF colonisation levels. For instance, in California, USA, Werner (1997) speculated that an increase in weeds was the cause of apple (*Malus domestica*

Borkh.) trees managed organically having higher colonisation than conventionally grown apples. In Indonesia, the density of AMF extraradical hyphae under maize grown in a monoculture was less than when maize was grown in an agroforestry system with a tropical tree legume (Boddington and Dodd, 2000).

Community structure and diversity of AMF can be quickly influenced by host crop and therefore by crop rotations (An et al., 1993; Daniell et al., 2001; Bever, 2002; Troeh and Loynachan, 2003). Thus monocultures of different crops may build up markedly different communities of AMF. For instance, in central Iowa, USA, after three years of cropping, *Glomus albidum* and *G. etunicatum* spores dominated under corn, while *G. constrictum* spores dominated under soybean (*Glycine max* (L. Merrill) (Troeh and Loynachan, 2003). Johnson et al. (1991) also observed distinct communities of AMF in plots with either corn or soybean cropping histories. Further experimentation suggested the communities were detrimental to yield of the crop they proliferated under, but beneficial to the other crop (Johnson et al., 1992).

There is some indication that an increase in plant diversity will lead to an increase in the species diversity of AMF. In a non-agricultural field experiment in Minnesota, USA, as plant diversity increased from one to 16 plant species per plot, sporulation and species numbers of AMF were increased and, in particular, sporulation by larger-spored *Gigaspora* and *Scutellospora* species was increased (Burrows and Pfleger, 2002). These relationships may have been mediated through factors such as midseason soil nitrate concentrations, plant density and specificity between host plants and species of AMF (Burrows and Pfleger, 2002). Feldmann and Boyle (1999) examined the impact on maize of removing all weeds, removing only weeds that host AMF or removing weeds that were non-hosts. The maize monoculture developed the least diverse community of AMF. When maize plants were later inoculated with each community of AMF, the community from the monoculture was the least effective at promoting growth. In western Kentucky, USA, An et al. (1993) found a higher species richness and higher diversity of AMF when crops were rotated than when soybean was continually grown, but only early in the season.

Eight sites in Europe with differing levels of plant diversity were examined by Oehl et al. (2003). Three sites were low-input very species rich grasslands, two sites were treatments in the DOK experiment which were farmed with a seven year rotation involving three years of grass-clover meadow as well as a range of crops, and three sites were high input continuous maize cropping. Spores of AMF were examined in soil samples and in trap cultures. The number of spores and species of AMF declined with decreasing plant diversity. However, strong conclusions can not be drawn from this experiment about the importance of plant diversity as lower plant diversity was correlated with increased farming intensity (i.e. increased inputs of P and N and tillage intensity).

In a farming system trial at the Rodale Institute Experimental Farm in Pennsylvania, USA, 15 years of more diverse rotations in the low-input treatments did not result in any substantial increase in species diversity of AMF (Franke-Snyder et al., 2001) with one species, *Gigapsora gigantean*, accounting for more than 60% of spore volume in most treatments. Franke-Snyder et al. (2001) speculated that the homogenised AMF diversity and failure to develop greater diversity in the low-input treatments reflected very high soil extractable P, 50–90 kg yr^{-1} of N inputs, and mouldboard ploughing of all treatments.

Organic farms do tend to support greater plant diversity than conventional farms (Gabriel et al., 2006) due to the importance of legumes for provision of N and the ensuing requirement for longer pasture leys than conventional farms, higher occurrence of weeds and a strong emphasis on crop rotation for control of weeds, pests and diseases (Watson et al., 2002). It seems high plant diversity could result in an increased species diversity of AMF and, perhaps, a community of AMF more beneficial to plant growth. Other characteristics of rotations likely to favour AMF, such as absence of non-host crops, may more reflect the nature of the agricultural system or preferences of individual farmers than whether a farm is under organic or conventional management.

10.5.6 Summary – The Combined Effect of Agricultural Practices on AMF

The abundance, community diversity and structure, and function of AMF are affected by many agricultural practices. While many studies have not examined farm management practices individually, it does appear that a reduction in abundance and community diversity of AMF is most likely to result from application of fertilisers which readily supply P to plants, as well as tillage and inclusion of bare fallows or non-host crops in rotations. The biocides used on conventional farms, and perhaps organic farms, may reduce abundance of AMF in some circumstances, although their impact on community diversity of AMF is unknown. There is some evidence that a more diverse community of AMF can develop if there is an increase in the diversity of plants within a rotation. Thus, the frequent finding of a higher abundance of AMF on organic farms than conventional farms – and the tendency towards a greater diversity of AMF on organic farms – most likely results from the elimination of readily soluble P fertilisers and the implementation of a more diverse rotation. Also, there is some evidence that less beneficial communities of AMF may develop under conditions of high P availability and in monocultures. However, organic farm management does not guarantee a high abundance of AMF and when factors such as high soil available P, frequent tillage or low crop diversity occur on organic farms the occurrence of AMF may be similar to, or lower than, on conventional farms.

10.6 Impacts of AMF in Agricultural Systems

In regards to agriculture, AMF are best known for their ability to enhance growth of plants due to their ability to increase plant P uptake. We will explore whether high levels of colonisation on organic farms automatically confers an important role for AMF in crop nutrition and growth and whether AMF can be considered as biological fertilisers. The non-nutritional impacts of AMF on host plants and on agricultural system sustainability are then examined.

10.6.1 Host Plant Growth

10.6.1.1 Biomass and Yield

AMF are generally regarded to have a positive effect on growth and yield of crop and pasture plants. A recent meta-analysis of mycorrhizal impact on agricultural plants grown in the field in non-sterilised soil found that inoculation with AMF increased yield by an average of 34% ± 9% (mean ± 95% CI) (Lekberg and Koide, 2005a). McGonigle (1988), using a different data set, found the average yield response of agricultural plants to inoculation in the field to be 37%.

The impact of the indigenous community of AMF on crop growth can be harder to quantify as eliminating or reducing colonisation for the purposes of comparison also changes other factors (e.g. Khasa et al., 1992). For instance, in the meta-analysis carried out by Lekberg and Koide (2005a), an increase in colonisation by AMF due to a shortened bare fallow had a very variable impact on yield with an average increase of 27±21% (mean ± 95% CI). A positive response to increased colonisation was most likely at low soil extractable P (<5mg kg^{-1}) and at sites where inoculum potential was low (<5% root length colonised) (Lekberg and Koide, 2005a). Unfortunately, such sites may be very much over-represented in the literature as studies are often carried out at sites where a response to inoculation is expected (e.g. eroded or reclaimed soils, or subsoils) (Lekberg and Koide, 2005a). The prevalence of such sites in established agricultural systems, especially on organic farms, may not be high as low soil extractable P is likely to correspond to high colonisation by AMF and high inoculum concentrations (e.g. Fig. 10.2). Thus while AMF are most commonly reported to have a beneficial effect on crop growth and yield, currently it is not possible to confidently predict the importance of AMF for a particular situation without experimentation. This situation reflects our limited understanding of how farm management and environmental variables affect the mycorrhizal symbiosis.

For instance, while arbuscular mycorrhizas are traditionally thought of as mutualisms, increasingly AMF are considered to function along a continuum from highly parasitic to highly beneficial (Johnson et al., 1997; Jones and Smith, 2004; Egger and Hibbett, 2004). The reliance of AMF on host photosynthate for all their energy requirements (Ho and Trappe, 1973) means a parasitic impact can occur if no benefits accrue to the plant (Johnson et al., 1997). Indeed, in their meta-analysis, Lekberg and Koide (2005a) found biomass of crops was significantly reduced in 2% of experiments where colonisation by AMF was increased. This could reflect plants not requiring the P supplied by the fungi, which could occur under conditions of high soil P or if other growth limiting factors were present (Jones and Hendrix, 1987; Graham and Eissenstat, 1998; Khaliq and Sanders, 2000; Kahiluoto et al., 2001). Alternatively, parasitic impacts could occur if plant carbon reserves were limited, such as under conditions of low light (Son and Smith, 1988), or if the fungi were unable to supply nutrients for a reason such as low temperatures (Cooper and Tinker, 1981). Under field conditions, seasonal fluctuations in climate and changes in the lifecycle stage of the host plant and AMF may result in the impact of AMF oscillating between beneficial and parasitic (Bethlenfalvay et al., 1982; Pearson and Schweiger, 1993; van der Heijden, 2001; Lerat et al., 2003; Ryan et al., 2005). This makes untangling the impact of AMF on crop growth complicated. For instance, if a crop finishes under dry conditions a mycorrhizal-induced increase or decrease in early biomass could greatly impact on available soil water during grain filling and thereby influence grain yield (e.g. Weber et al., 1993; Ryan et al., 2005).

10.6.1.2 Mycorrhizal Dependency

The benefit to agricultural crops from the presence of AMF can differ greatly between crop species (Plenchette et al., 1983; Khasa et al., 1992). This effect is termed "mycorrhizal dependency" and is defined as:

$$\frac{\text{weight of the colonised plant} - \text{weight of the uncolonised plant}}{\text{weight of mycorrhizal plant}} \times 100$$

In general, under P limiting conditions, plants with thick roots with few, short, thick root hairs will be most dependent on AMF and plants with finer roots and more, longer and thinner root hairs are likely to be less dependent on AMF as they can adequately access P without AMF (Crush, 1974; Baylis, 1975; Schweiger et al., 1995). Thus, grasses are often less dependent on AMF than legumes (Plenchette et al., 1983). Highly dependent crops include the onion family (*Allium* spp.), flax (*Linum usitatissimum* L.) and most legumes (Owusu-Bennoah and Mosse, 1979; Plenchette et al., 1983; Thompson, 1987; Schweiger et al., 1995; Kahiluoto et al., 2001). For legumes, the high P requirements for adequate nodulation and N-fixation also contribute towards high dependency (Crush, 1974).

Generalising about mycorrhizal dependency at a broad level can be problematic as dependency can vary with crop species and cultivar (Khalil et al., 1994; Hetrick et al., 1996; Xavier and Germida, 1997), species of colonising AMF (Graham and Abbott, 2000; Klironomos, 2003) and with environmental conditions such as the level of soil available P. For instance, mycorrhizal dependency in *Citrus* varies with host genotype (Graham et al., 1997), with high dependency linked to a relatively loose regulation of carbon expenditure on AMF (Jifon et al., 2002). Thus, highly dependent genotypes benefit from AMF at low P supply but are more likely to suffer a growth depression when P is plentiful (Jifon et al., 2002). The retention of an ability to form mycorrhizas in relatively poorly dependent plants such as grasses may indicate AMF influence host plants in areas unrelated to P nutrition, such as disease control (Newsham et al., 1995). On organic farms a reliance on legumes for N fixation, a tendency for more diverse rotations, and low soil available P may mean a greater number of highly dependent crops are grown than on conventional farms.

It has been suggested that modern crop breeding practices, especially the selection of crops under conditions where P is not limiting, may have resulted in modern cultivars generally possessing low mycorrhizal dependency and/or a poor capacity to become colonised by AMF (e.g. Hetrick et al., 1993). If this was the case, older or unimproved cultivars would be more suited to the low soil available P often found on organic farms, both for maximising yield benefits from AMF and obtaining other benefits such as disease control. Evidence for this hypothesis is sparse, with studies generally conducted under glasshouse conditions and often including only a few breeding lines. For instance, Khalil et al. (1994) examined three improved and three unimproved lines of corn and soybean (*Glycine soja* Siebold and Zucc. and *G. max* L.). Colonisation was uniformly high, but considerable variability in mycorrhizal dependency was evident across improved and unimproved lines. However, responsiveness to AMF does appear heritable (Azcón and Ocampo, 1981; Hetrick et al., 1992, 1993, 1996), and organic farmers may benefit from crop selection under low soil available P conditions for high mycorrhizal dependency and a high degree of mycorrhizal colonisation.

10.6.1.3 Impact of Location and Agricultural System

Both environmental and cultural characteristics of an agricultural system may determine the degree of reliance of crops on AMF for nutrient uptake and growth. Crops may be highly reliant on AMF in systems with one or more of the following characteristics: soil low in available nutrients particularly P; soil which causes fertiliser P to quickly become unavailable to crops; favourable conditions for rapid crop growth (i.e. high light levels, warm temperatures and adequate water) (e.g. irrigated summer crops); regular inclusion of dependent crops in rotations; and little addition of fertiliser (e.g. organic farms, developing countries). In particular, temperature may play an important role in delineating levels of colonisation and activity by AMF (Cooper and Tinker, 1981; Koske, 1987; Gavito et al., 2005). However, the impacts of temperature minimum or range are rarely examined in studies of AMF, perhaps due to methodological difficulties in both the field and the glasshouse (Tibbett and Cairney, 2007).

Many agricultural systems in the tropics and subtropics, which are often in developing countries, meet many of these criteria listed above (Khasa et al., 1992; Runge-Metzger, 1995; Lekberg and Koide, 2005b). For instance, in the subtropical NE cropping zone of Australia, low colonisation by AMF following long bare fallows can be associated with poor P and zinc (Zn) nutrition and low yields in following crops; a situation termed *long fallow disorder* (Thompson, 1987, 1994, 1996). Yet when crops are grown after bare fallows or non-host crops in the SE cropping zone of Australia, colonisation by AMF is reduced but growth and yield are often increased, suggesting AMF are generally parasitic in this region (Ryan et al., 2002, 2005; Ryan and Angus, 2003) and high colonisation may be contributing to low yields on organic farms (Kitchen et al., 2003; Ryan et al., 2004). The parasitic impact of AMF in SE Australia, which occurs primarily before anthesis for autumn-sown crops, could reflect soil temperatures below 5 °C reducing the ability of AMF hyphae to transport P, combined with short winter days and low light levels, at a time when AMF carbon demands are at a peak (Ryan et al., 2005).

10.6.2 Host Plant Nutrition

Increased plant uptake of P in the presence of AMF has been shown on many occasions (Smith and Read, 1997; Clark and Zeto, 2000) and is greatest when P is limiting (Schweiger et al., 1995; Smith and Read, 1997). The enhanced uptake is thought to result from the exploration of a greater volume of soil by colonised plants due to the growth of the extraradical hyphae of AMF past the nutrient depletion zones that form around roots (Jakobsen et al., 1992a,b). The enhancement of P-uptake by AMF has often been observed to increase nodulation by rhizobia and thereby indirectly enhance plant N nutrition (Lekberg and Koide, 2005b). Recent work also shows AMF are capable of transferring large amounts of N from the soil to plant roots (Govindarajulu et al., 2005) and there are some reports of AMF increasing host plant N uptake in both legumes and non-legumes (e.g. Gaur and Adholeya, 2002). Nitrogen may be transferred from legumes to non-legumes through the CMN (He et al., 2003).

AMF may also affect plant uptake of other nutrients including sulphur, boron, K, calcium, magnesium, sodium, Zn, copper, manganese, iron, aluminium, silicon and some other trace elements (Clark and Zeto, 2000). The impact of AMF on these elements can be positive, neutral or negative depending on soil type, host plant and other factors. A recent review by Clark and Zeto (2000) concluded that the nutrients most commonly enhanced in host plants by AMF are P, N, Zn and Cu, with K, Ca and Mg enhanced when plants are grown in acidic soils. Mn uptake is often reduced in mycorrhizal plants (Kothari et al., 1991). The tendency of AMF to decrease the root-shoot ratio of the host plant (e.g. Weber et al., 1993) may result in poor host uptake of nutrients in some instances (e.g. Ryan and Angus, 2003). AMF may also alleviate toxicities of some elements (Clark and Zeto, 2000).

Enhanced uptake of P by AMF has traditionally been thought to reflect increased access to soil available P (Bolan, 1991). However, the ability of AMF to access P from poorly soluble P sources is important when considering the role of AMF on organic farms as fertiliser P is often applied in either organic forms (e.g. composts, manure) or in poorly soluble mineral forms (e.g. rock phosphate). Moreover, an ability to increase weathering of P from insoluble forms in soil is the only means by which AMF can aid in the maintenance of a positive P balance for organic farms. If this does not occur, the ability of AMF to enhance plant P-uptake and crop yields also increases losses of P in farm products. AMF can enhance plant uptake of P from organic sources such as phytate due to the production of extracellular phosphatase by extraradical hyphae (Tarafdar and Marschner, 1994; Koide and Kabir, 2000). Recent evidence also suggests hyphae of AMF produce exudates such as citric acid that can aid with solubilisation of poorly soluble forms of inorganic P (Tawaraya et al., 2006). However, the impact of AMF on plant access to organic and insoluble inorganic P in an agricultural context is unknown and may not be large.

The contribution of AMF to crop P uptake has not been directly compared between organic and conventional farms. However, as soil extractable P is generally lower on organic farms than conventional farms and colonisation by AMF correspondingly higher, it should be safe to assume that AMF play a greater role in crop P nutrition on organic farms. A trend towards greater effectiveness in enhancing P uptake for AMF from organic farms was reported by Scullion et al. (1998) and Eason et al. (1999) (Table 10.4). Similarly, Muckle (2003) found P-uptake from root-free mycorrhizal compartments was around 80% higher in soil from organic wheat fields than conventional wheat fields.

Little information is available on the role of AMF on organic farms in the uptake of elements other than P. For organic wheat crops in SE Australia, Ryan et al. (2004) found concentrations in grain of Zn and copper (Cu) were higher, and concentrations of manganese (Mn) and P were lower, than in grain of conventional crops. Colonisation by AMF was 2–3 times higher in the organic crops. A series of related glasshouse and field experiments suggested that higher colonisation by AMF on the organic farms was responsible for the higher Zn and partly responsible for the lower Mn, with other agronomic factors affecting Cu, P and, partly, Mn (Dann et al., 1996; Ryan and Angus, 2003; Ryan et al., 2004). A further study found that the differences in grain Zn bioavailability between the organic and conventional grain were significant for human nutrition (Ryan et al., 2008). Graham et al. (2000) also found lower P and higher Zn concentrations in grain of organic crops of SE Australia in comparison to conventional crops.

Indeed, impacts of AMF on product quality generally seem related to enhanced grain nutrient concentrations (e.g. Kahiluoto et al., 2001), although Gupta et al. (2002) found inoculation with AMF enhanced oil content and essential oil yield of menthol mint (*Mentha arvensis* L.). For instance, in California, USA, Cavagnaro et al. (2006) compared a tomato (*Lycopersicon esculentum* Mill.) mycorrhizal defective mutant and its mycorrhizal wildtype and found AMF had little impact on yield, but enhanced fruit Zn concentration by around 24%. Although, as fruit P concentration was also enhanced by 41%, the additional Zn may have been bound in forms unavailable to human digestive systems such as phytate (Adeyeye et al., 2000).

10.6.3 Can AMF Substitute for Fertilisers?

AMF are sometimes considered as a biological substitute for P fertiliser. However, AMF can only aid crops in the depletion of a finite pool of P that will eventually need replenishment. In addition, AMF are unlikely to be as efficient as a readily soluble P fertiliser at enhancing plant P uptake as all their energy requirements come from host plant photosynthate (Ho and Trappe, 1973) and P release to the plant may not match the timing of plant P requirements. Comparing the impacts on plant growth of readily soluble P fertiliser and AMF in sterilised soil is essentially a comparison between plants supporting high colonisation in a low P environment (no fertiliser) and plants with low colonisation in a high P environment (plus fertiliser). The results of such comparisons vary (e.g. Rangeley et al., 1982), but the fertilised plants often grow best (Al-Karaki, 2002; Ortas, 2003; Ryan and Angus, 2003), even in soil from organic farms (Dann et al., 1996; Ryan and Ash, 1999).

Thus, the higher abundance and diversity of AMF on organic farms does not mean yields will match those of conventional farms with similar rotations, but higher soil available P and a lower abundance of AMF. For instance, in the DOK experiment, crops yields were 20% lower in the organic treatments due largely to low K supply and disease (Mäder et al., 2002). While colonisation by AMF was 40% higher in the organic treatments (Mäder et al., 2000), P-budgets were negative and yields were maintained using native soil reserves of P or residual P from earlier fertiliser applications (Oehl et al., 2002). In SE Australia, colonisation by AMF was higher in 9 of 10 paired biodynamic and conventional irrigated permanent dairy pastures, but soil extractable P and pasture shoot P were lower and P deficiency appeared responsible for lower milk yields on the biodynamic farms (Table 10.5; Burkitt et al., 2007a,b). Whilst AMF were presumably important for pasture P-uptake on the biodynamic farms, the fungi could not match the P supplied to conventional pastures in readily soluble fertiliser and, more importantly, could not compensate for P lost in farm products. Soil extractable P was declining over time for the biodynamic pastures (Table 10.5; Burkitt et al., 2007a).

Significant imbalances exist in global P management and many developing countries which have agricultural systems with low external inputs (Latin America, Sub-Saharan Africa and Asia) also have P as a constraining factor for increasing yield (Runge-Metzger, 1995) and soils with high P fixation capacities. AMF cannot overcome P shortages in such situations as they primarily provide P to plants from existing pools of soil available P. Thus, limiting inputs of readily soluble P in order to favour AMF may dangerously restrict yields. In both industrialised and developing countries returning some of the P lost from farms in produce, i.e. returning animal wastes, sewage and food wastes (Kirchmann and Ryan, 2004) is important to increase the sustainability of agricultural systems (although often not currently permitted by organic farming regulations). Doing so in a manner not detrimental to

Table 10.5 Nutrient inputs, soil and pasture nutrient concentrations, the percentage of root length colonised by AMF, milk yield and P-budget for 10 paired biodynamic and conventional dairy farms in SE Australia. All sampled fields were under permanent perennial irrigated pasture (Ryan et al., 2000; Burkitt et al., 2007a; b)

Characteristics	Conventional	Biodynamic	Significance
	Superphosphate, diammonium phosphate or urea	No P fertilisers	
N input (kg ha^{-1} yr^{-1})	17	none	–
P input (kg ha^{-1} yr^{-1})	27	none	–
Soil extractable P (mg kg^{-1})	18.0	8.4	$P < 0.002$
Pasture foliar P (g kg^{-1})	3.3	2.4	$P < 0.00001$
Pasture foliar N (g kg^{-1})	20.1	21.1	ns
Clover colonisation by AMF (%)	48	71	$P < 0.0004$
Grass colonisation by AMF (%)	38	48	$P < 0.02$
Milk yield (L yr^{-1} cow^{-1})	4585	3436	$P < 0.001$
P budget (kg P ha^{-1} yr^{-1})	15.9	−7.3	–

the abundance of AMF would allow the other benefits from AMF to be realised, but as this would involve restricting the solubility of P, it may not maximise yields.

10.6.4 Non-nutritional Impacts

The water balance of host plants, both adequately watered and droughted, can be influenced by AMF (Augé, 2001). In particular, AMF are frequently reported to improve drought avoidance, often due to improved P nutrition (Augé, 2001). While a recent review by Augé (2001) concluded that the effects of AMF on the water balance of host plants are often "subtle, transient and probably circumstance and symbiont specific", in terms of plant growth and reproduction, such small effects may still be very significant over time. For instance, Klironomos et al. (2001) showed that the impact of drought on colonisation by five species of AMF isolated from an old-field meadow in Canada differed from positive to negative and suggested that the association of particular plants with particular AMF could confer a competitive advantage at certain times. Perhaps the tendency for higher diversity of AMF on organic farms may confer more advantage in regards to plant water relations than the generally high levels of colonisation.

Many studies have also indicated that AMF can play a role in controlling plant diseases (Whipps, 2004). Most studies involve soil-borne fungal pathogens although bacterial pathogens have also been investigated (Whipps, 2004). For instance, Thompson and Wildermuth (1989) reported an inverse correlation between colonisation by AMF and infection by the fungal pathogen *Bipolaris sorokiniana*. Control of nematodes has also often been shown. While results of studies vary, a degree of control of these three groups of pathogens is generally reported. Foliar pathogens, conversely, are often stimulated by AMF presumably due to improved nutrition and greater physiological activity in mycorrhizal plants (Whipps, 2004). The mechanisms behind disease control by AMF are not well understood and the control is most likely the outcome of interactions between numerous mechanisms (Harrier and Watson, 2004). However, relatively few studies investigate the impact of AMF on pathogens under field conditions, partly due to methodological difficulties with such studies. Certainly, high colonisation by AMF is no guarantee of an agronomically relevant degree of pathogen control (Bødker et al., 2002; Ryan et al., 2002). Overall, the impacts of an enhanced community of AMF, as may occur on organic farms, on pathogen control is likely to be minor compared to its impact on plant nutrition and, perhaps, soil structure.

10.6.5 Agricultural System Sustainability

AMF can be considered as a bridge between plants and the soil ecosystem (Bethlenfalvay and Schüepp, 1994). The contribution by AMF of plant photosynthate to soil carbon pools is substantial. For example, Johnson et al. (2002) found that within 21 h of pulse-labelling a grassland sward, between 5 and 8% of the fixed ^{13}C lost by shoot respiration and translocation was passed through the extraradical hyphae of the colonising AMF through the soil to the atmosphere. AMF also directly impact on other soil organisms and vice versa through a variety of mechanisms and may act synergistically with other soil organisms to aid plant growth (see reviews by Douds and Johnson, 2003; Artursson et al., 2006). While AMF obviously have a large effect on the soil biological community, this is not discussed further in this chapter as the application of this knowledge to management of agricultural systems is currently not possible. Instead we briefly consider the impact of AMF on soil structure, and plant community structure and diversity.

10.6.5.1 Soil Structure

The extraradical hyphae of AMF aid in the maintenance of soil structure by assisting roots in entangling and enmeshing soil particles to form macroaggregates (Tisdall and Oades, 1979; Jastrow et al., 1998). For instance, in Pennsylvania, USA, Kabir and Koide (2002) reported a significant positive linear relationship between the density of AM fungal hyphae and aggregate stability under a sweet corn crop. In organic tomato production, Cavagnaro et al. (2006) found the presence of AMF improved aggregate stability in a poorly structured soil when N was also applied. Hyphal architecture can vary greatly between species of AMF (Jakobsen et al., 1992a) and the impact of AMF on soil aggregation can vary with AMF-host species combinations (Piotrowski et al., 2004). Piotrowski et al. (2004) found hyphal lengths of AMF and root biomass were not positively correlated with aggregation, which suggested that other physiological or architectural mechanisms were also important. One possible mechanism is glomalin.

Glomalin is a fungal protein (or class of protein) that is quantified from soil as glomalin-related soil protein (GRSP) (Wright and Upadhyaya, 1998; Rillig and Mummey, 2006). Glomalin is thought to act as a glue with hydrophobic properties, although actual biochemical evidence for this is not available (Rillig and Mummey, 2006). Several recent studies show GRSP to be positively correlated with water stabile soil aggregates in both natural and agricultural ecosystems (e.g. Rillig et al., 2002). For instance, Wright et al. (1999) found as the number of years of no-tillage increased, so too did aggregate stability and total glomalin, while Wright and Anderson (2000) found aggregate stability and glomalin varied together with crop rotation. Further research is required to better understand both the contribution of AMF to accumulation of GRSP and the impact of GRSP on aggregation and, perhaps, other soil processes. Increased production of GRSP is likely to be a benefit arising from high levels of colonisation by AMF on organic farms and may contribute to higher aggregate stability in soils on organic farms (e.g. Mäder et al., 2002).

10.6.5.2 Plant Community Structure and Diversity

In natural ecosystems AMF appear important in determining plant community structure (Hetrick et al., 1994; Klironomos et al., 2001; O'Connor et al., 2002). For instance, Klironomos (2003) inoculated a single local isolate of AMF onto 64 local plants with growth responses, compared to non-inoculated controls, ranging from −46 to +48%, suggesting no single "mycorrhizal effect", but a series of plant-fungal symbioses with differing effect. It was speculated that a high degree of variation in plant response to AMF may be a large contributor to plant species co-existence (Klironomos, 2003). There is also evidence to suggest that AMF may allow maximum productivity with fewer plant species, as the fungi allow greater access to soil resources (Klironomos et al., 2000). An increased diversity of species of AMF may have a strong positive effect on plant community diversity and productivity (van der Heijden et al., 1998). Thus in situations such as permanent pastures a diverse community of AMF may be more important for maximum yields to be obtained on nutrient-limited organic farms than on conventional farms and may also play a role in the maintenance of high plant diversity.

10.6.6 Summary – The Impact of AMF in Agricultural Systems

In agricultural systems where P is limiting plant growth, AMF may make an important contribution to host plant P nutrition, growth and yield. However, this contribution will vary between locations and agricultural systems for reasons that are not yet fully understood. Beneficial impacts of AMF on host plants resulting from enhanced uptake of other nutrients, and improvements in drought avoidance and disease control, may also occur in some instances. AMF may be necessary for maintaining the long-term stability and sustainability of agricultural systems through their influence on other components of the soil biological community, soil structure and, in permanent vegetation, plant community structure and diversity. As organic farms generally support high colonisation levels by AMF, they may harness more of the beneficial impacts of AMF than conventional farms and may benefit from crop breeding programs with a focus on increasing mycorrhizal dependency and the degree to which host plants become colonised. However, AMF do not substitute for external inputs of P and there may be a trade-off between supplying P in a form that maximises yield and the occurrence of AMF.

10.7 Do Agricultural Systems Select for "Weedy" AMF?

As discussed above, in natural ecosystems the host plant growth response to AMF may vary from highly parasitic to highly beneficial and this may be an important contributor to plant species co-existence (Klironomos, 2003). However, agricultural systems in industrialised countries are generally managed with the goal of maximising plant productivity, not diversity, and the long-term relationship between the communities of AMF and their host plants has been disrupted by regular applications of fertiliser and regular reseeding with crop and pasture species selected off-farm for maximum yield under well-fertilised conditions. Under these circumstances, the benefits to AMF from contributing to host plant growth and nutrition and thereby ensuring persistence of host species will be reduced and the fungi may act to maximise their own fitness at the expense of the host plant. Thus, Kiers et al. (2002) suggest that modern agricultural practices will select for less mutualistic, "weedy", mycorrhizal associations.

There is some evidence that agricultural practices do select for less beneficial AMF, particularly application of fertilisers, maintenance of monocultures and, perhaps, tillage (Johnson et al., 1991, 1992; Johnson, 1993; Scullion et al., 1998; Eason et al., 1999; Feldmann and Boyle, 1999; Menéndez et al., 2001; Kiers et al., 2002). Indeed, a positive plant growth response to inoculation with exotic AMF, even when colonisation by indigenous AMF is high (Owusu-Bennoah and Mosse, 1979; Plenchette et al., 1981; Rangeley et al., 1982; Ortas, 2003), could reflect the exotic AMF not originating under the farm management and host crops present in the target agricultural system.

Characteristics of less beneficial species of AMF could include a tendency to colonise roots slowly, but sporulate rapidly (Oehl et al., 2003), and a tendency for a greater proportion of colonisation to consist of fungal storage structures (vesicles) at the expense of hyphae and arbuscules (Johnson, 1993). Interestingly, Oehl et al. (2003) found more rapid spore formation from isolates of individual species of AMF when they originated from cropped soils in comparison to grassland soils. This suggests that changes in the function of the mycorrhizal community can occur not only through a shift in relative abundance of species, but also through development of ecotypes of individual species of AMF.

Whether organic farms would be susceptible to development of "weedy" communities of AMF is unclear, but certainly the prohibition of readily soluble P fertilisers and a tendency towards higher plant diversity could prove beneficial in this regard. The impact of a community of AMF with reduced benefit for crop growth on other ecosystem-level benefits of AMF, such as improved soil structure is unknown.

10.8 Inoculation with AMF

There is great interest in the use of inoculum of AMF in agricultural systems, based on the assumption that any introduction is likely to be beneficial for crop nutrition, crop yield and agricultural system sustainability (Gianinazzi and Vosátka, 2004). Indeed, there are greater than 33 companies currently engaged in the production of AM fungal inoculum for use in agriculture (Gianinazzi and Vosátka, 2004). However, only a tiny area of agricultural land has been inoculated with AMF. This reflects difficulties with the cheap production and application of large quantities of inoculum, as well as a lack of knowledge about the outcomes of inoculation, meaning reliable prediction of impacts on crop growth cannot be made without experimentation. Some of the issues associated with use of inoculum are briefly discussed below.

10.8.1 Exotic or Indigenous Inoculum?

Schwartz et al. (2006) suggest only local species of AMF be utilised as inoculum and list potential detrimental consequences of exotic inoculum as decreased yields, decreased survival of desirable plant species, increased fitness of noxious weeds, reduced diversity of native AMF and the risk of introduced AMF becoming invasive weeds. However, there is little evidence for most of these negative outcomes, but many reports of beneficial outcomes from inoculation, where the confounding factor of soil fumigation is not involved, which involved exotic strains being inoculated into soil where indigenous AMF were abundant (Owusu-Bennoah and Mosse, 1979; Plenchette et al., 1981; Rangeley et al., 1982; Ortas, 2003). For instance, Mamatha et al. (2002) inoculated 10 year old mulberry (*Morus alba* L.) plants and 1.5 year old papaya (*Carica papaya* L.) plants which had 20–30% of root length colonised by indigenous AMF. For mulberry, an additional 8% of root length becoming colonised corresponded with an increase in fresh leaf yield of 400 kg ha^{-1}, while for papaya an additional 15% of root length becoming colonised corresponded with an additional seven fruits per plant. Similarly, on a soil with low extractable P, inoculation of garlic (*Allium sativum* L.) increased root colonisation from 68 to 85% and increased fresh bulb yield by 23%. When 60 kg ha^{-1} of P was applied, colonisation increased from 25 to 30% and fresh bulb yield increased by 4% (Al-Karaki, 2002). These studies suggest that unless inoculum potential has been recently lowered by a specific circumstance, such as a bare fallow or a non-host crop, there may be little benefit to inoculating with indigenous AMF, as colonisation levely may be little increased. However, inoculation with exotic AMF that substitute for indigenous may provide benefits if the indigenous AMF community has responded to farm management practices with a decline in beneficial impact on host plants.

10.8.2 Possible Problems with Inoculation

The likely incidence of negative outcomes from inoculation is difficult to predict as, presumably, studies showing negative outcomes are less likely to be submitted and accepted for publication. However, it is clear that under some circumstances, such as high soil extractable P, inoculation may cause a decrease in host growth (e.g. Khaliq and Sanders, 2000). As discussed previously, there are also instances where indigenous AMF have either no impact or a negative impact on crop growth. A negative outcome from inoculation has financial implications for farmers.

Lekberg and Koide's (2005a) meta-analysis of studies involving inoculation with AMF showed that even when soil extractable P and inoculum potential were very low there was no guarantee of a beneficial response to inoculation. The reasons for the lack of response in some instances were not known, but perhaps very effective communities of indigenous AMF were present and hence low levels of colonisation did not correlate to low mycorrhizal activity and benefit (see van der Heijden, 2001). For instance, in field experiments, Rangeley et al. (1982) found inoculation increased colonisation but did not benefit growth of white clover on a peat soil with very low extractable P and low colonisation by indigenous AMF. On a brown soil with more abundant P and higher colonisation from indigenous AMF, the inoculated fungi partly replaced the indigenous fungi in roots but growth was increased only when P was also applied (Rangeley et al., 1982).

The length of time the benefits from inoculation persist is rarely investigated, but is important when considering the economic viability of inoculum addition, as is the time taken for a positive effect to occur. For instance, Harinikumar and Bagyaraj (1996) found *Glomus intraradices* persisted for only one season after inoculation and Rangeley et al. (1982) found that a positive effect on white clover growth did not occur until the second year after inoculation.

10.8.3 Inoculum Production

It may be easiest and cheapest to bulk-up inoculum, which consists of spores, hyphae and colonised root segments, on-farm (Douds et al., 2006). However, on-farm inocula are not readily processed for mechanical application and are best suited to labour intensive situations in developing countries or high value horticultural farms where inoculated seedlings are transplanted into the field (Douds et al., 2005) or, perhaps, inoculation of established tree crops (Mamatha et al., 2002).

It is worth noting that field-scale inoculation with AMF is always an introduction of more than just the target groups of AMF as the inoculum will contain a range of other bacteria and fungi that are typically not controlled for in experiments and may contribute to the beneficial or parasitic effects of AMF (Daniels Hetrick et al., 1988). While deliberate combined inoculation of AMF with other micro-organisms has been shown to yield beneficial results (Whipps, 2004; Artursson et al., 2006), the range of organisms contained in inoculum is rarely quantified. This could explain some of the variation in outcomes of inoculation. Introduction of disease organisms with inoculants must also be considered.

In conclusion, while many studies show advantages to inoculation with AMF, usually exotic isolates, and particularly when soil available P is low, a beneficial outcome for crop growth and yield from inoculation cannot be guaranteed without on-site testing. Research is required to determine whether organic farms derive the same benefit from inoculation as conventional farms. Perhaps organic farms may benefit most directly after conversion, when soil inoculum levels are likely to be lower. Inoculants may be best utilised when inoculum levels are deficient such as after long bare fallows or non-host crops or, for conventional farms, following soil fumigation. Comparison of economic outcomes between applying inoculum and a change in farming practices to favour AMF should be undertaken before inoculum production and application is commenced.

10.9 How Important are AMF in Organic Farming?

High abundance of AMF on organic farms probably provides many benefits for crop growth and agricultural system stability and sustainability, although these may vary between locations and agricultural systems. However, it is specific components of organic farm management that contribute to this high abundance, notably low inputs of readily soluble P fertilisers and high plant diversity. However, there may be trade-offs between maximising the occurrence and activities of AMF, maximising crop or pasture yields and maintaining a neutral or positive farm P budget. While conventional farms may have a tendency to develop weedy, less beneficial, communities of AMF and hence benefit from inoculation with exotic species of AMF, it is not known if this situation also occurs on organic farms. As stated by Douds and Johnson (2003), a much better understanding of the evolutionary mechanisms responsible for generating beneficial, neutral or parasitic impacts from AMF on host plants in agricultural systems is necessary before the fungi can be effectively managed to maximise desirable outcomes.

References

Abbott, L.K., and Robson, A.D., 1977, The distribution and abundance of vesicular arbuscular endophytes in some Western Australian soils, *Aust. J. Bot.* **25**: 515–522.

Abbott, L.K., 1982, Comparative anatomy of vesicular-arbuscular mycorrhizas formed on subterranean clover, *Aust. J. Bot.* **30**: 485–499.

Adeyeye, E.I., Arogundade, L.A., Akintayo, E.T., Aisida, O.A., and Alao, P.A., 2000, Calcium, zinc and phytate interrelationships in some foods of major consumption in Nigeria, *Food Chem.* **71**: 435–441.

Al-Karaki, G., 2002, Benefit, cost and phosphorus use efficiency of mycorrhizal field-grown garlic and different soil phosphorus levels, *J. Plant Nutr.* **25**: 1175–1184.

An, Z.-Q., Hendrix, J.W., Hershman, D.E., Ferriss, R.S., and Henson, G.T., 1993, The influence of crop rotation and soil fumigation on a mycorrhizal fungal community associated with soybean, *Mycorrhiza* **3**: 171–182.

Artursson, V., Finlay, R.D., and Jansson, J.K., 2006, Interactions between arbuscular mycorrhizal fungi and bacteria and their potential for stimulating plant growth, *Environ. Microbiol.* **8**: 1–10.

Augé, R.M., 2001, Water relations, drought and vesicular-arbuscular mycorrhizal symbiosis, *Mycorrhiza* **11**: 3–42.

Azcón, R., and Ocampo, J.A., 1981, Factors affecting the vesicular-arbuscular infection and mycorrhizal dependency of thirteen wheat cultivars, *New Phytol.* **87**: 677–685.

Baltruschat, H., and Dehne, H.W., 1989, The occurrence of vesicular-arbuscular mycorrhiza in agro-ecosystems. II. Influence of nitrogen fertilization and green manure in continuous monoculture and in crop rotation on the inoculum potential of winter barley, *Plant Soil* **113**: 251–256.

Baylis, G.T.S., 1975, The magnolioid mycorrhiza and mycotrophy in root systems derived from it, in: *Endomycorrhizas*, F.E. Sanders, B. Mosse, and P.B. Tinker, eds., Academic Press, London, pp. 373–389.

Bellgard, S.E., 1992, The propagules of vesicular-arbuscular mycorrhizal (VAM) fungi capable of initiating VAM infection after topsoil disturbance, *Mycorrhiza*, **1**: 147–152.

Bethlenfalvay, G.J., Brown, M.S., and Pacovsky, R.S., 1982, Parasitic and mutualistic associations between a mycorrhizal fungus and soybean: development of the host plant, *Ecol. Epidemiol.* **72**: 889–893.

Bethlenfalvay, G.J., and Linderman, R.G., eds., 1992, *Mycorrhizae in Sustainable Agriculture*, American Society of Agronomy, Madison, WI.

Bethlenfalvay, G.J., and Schüepp, H., 1994, Arbuscular mycorrhizas and agrosystem stability, in: *Impact of Arbuscular Mycorrhizas on Sustainable Agriculture and Natural Systems*, S. Gianinazzi, and H. Schüepp, eds., Birkhäuser Verlag, Basel, pp. 117–131.

Bever, J.D., 2002, Host-specificity of AM fungal population growth rates can generate feedback on plant growth, *Plant Soil* **244**: 281–290.

Boddington, C.L., and Dodd, J.C., 2000, The effect of agricultural practices on the development of indigenous arbuscular mycorrhizal fungi. I. Field studies in an Indonesian ultisol, *Plant Soil* **218**: 137–144.

Bødker, L., Kjøller, R., Kristensen, K., and Rosendahl, S., 2002, Interactions between indigenous arbuscular mycorrhizal fungi and *Aphanomyces euteiches* in field-grown pea, *Mycorrhiza* **12**: 7–12.

Bolan, N.S., 1991, A critical review of the role of mycorrhizal fungi in uptake of phosphorus by plants, *Plant Soil* **134**: 189–207.

Bradley, K., Drijber, R.A., and Knops, J., 2006, Increased N availability in grassland soils modifies their microbial communities and decreases the abundance of arbuscular mycorrhizal fungi, *Soil Biol. Biochem.* **38**: 1583–1595.

Brundrett, M., Bougher, N., Dell, B., Grove, T., and Malajczuk, N., 1996, *Working with Mycorrhizas in Forestry and Agriculture*, Australian Centre for International Agricultural Research Monograph 32, Pirie Printers, Canberra.

Burkitt, L.L., Small, D.R., McDonald, J.W., Wales, W.J., and Jenkin, M.L., 2007a, Comparing irrigated biodynamic and conventionally managed dairy farms. I. Soil and pasture properties, *Aust. J. Exp. Agric.* **47**: 479–488.

Burkitt, L.L., Wales, W.J., McDonald, J.W., Small, D.R., and Jenkin, M.L., 2007b, Comparing irrigated biodynamic and conventionally managed dairy farms. II. Milk production and composition, and animal health, *Aust. J. Exp. Agric.* **47**: 489–494.

Burrows, R.L., and Pfleger, F.L., 2002, Arbuscular mycorrhizal fungi respond to increasing plant diversity, *Can. J. Bot.* **80**: 120–130.

Cavagnaro, T.R., Jackson, L.E., Six, J., Ferris, H., Goyal, S., Asami, D., and Scow, K.M., 2006, Arbuscular mycorrhizas, microbial communities, nutrient availability and soil aggregates in organic tomato production, *Plant Soil* **282**: 209–225.

Clapp, J.P., Young, J.P.W., Merryweather, J.W., and Fitter, A.H., 1995, Diversity of fungal symbionts in arbuscular mycorrhizas from a natural community, *New Phytol.* **130**: 259–265.

Clark, R.B., and Zeto, S.K., 2000, Mineral acquisition by arbuscular mycorrhizal plants, *J. Plant Nutr.* **23**: 867–902.

Cooper, K.M., and Tinker, P.B., 1981, Translocation and transfer of nutrients in vesicular-arbuscular mycorrhizas IV. Effect of environmental variables on movement of phosphorus, *New Phytol.* **88**: 327–339.

Crush, J.R., 1974, Plant growth responses to vesicular-arbuscular mycorrhiza. VII. Growth and nodulation of some herbage legumes, *New Phytol.* **73**: 743–752.

Cuenca, G., and Meneses, E., 1996, Diversity patterns of arbuscular mycorrhizal fungi associated with cacao in Venezuela, *Plant Soil* **183**: 315–322.

Daniell, T.J., Husband, R., Fitter, A.H., and Young, J.P.W., 2001, Molecular diversity of arbuscular mycorrhizal fungi colonising arable crops, *FEMS Microbiol. Ecol.* **36**: 203–209.

Daniels Hetrick, B.A., Thompson Wilson, G., Gerschefske Kitt, D., and Schwab, A.P., 1988, Effects of soil microorganisms on mycorrhizal contribution to growth of big bluestem grass in non-sterile soil, *Soil Biol. Biochem.* **20**: 501–507.

Dann, P.R., Derrick, J.W., Dumaresq, D.C., and Ryan, M.H., 1996, The response of organic and conventionally grown wheat to superphosphate and reactive rock phosphate, *Aust. J. Exp. Agric.* **36**: 71–78.

Dekkers, T.B.M., and van der Werff, P.A., 2001, Mutualistic functioning of indigenous arbuscular mycorrhizae in spring barley and winter wheat after cessation of long-term phosphate fertilization, *Mycorrhiza* **10**: 195–201.

Derrick, J.W, and Dumaresq, D.C., 1999, Soil chemical properties under organic and conventional management in southern New South Wales, *Aust. J. Soil Res.* **37**: 1047–1055.

Douds, D.D., and Johnson, N.C., 2003, Contributions of arbuscular mycorrhizas to soil biological fertility, in: *Soil Biological Fertility – A Key to Sustainable Land Use in Agriculture*, L.K. Abbott, and D.V. Murphy, eds., Kluwer, Boston, MA, pp. 129–162.

Douds, D.D., Nagahashi, G., Pfeffer, P.E., Kayser, W.M., and Reider, C., 2005, On-farm production and utilisation of arbuscular mycorrhizal inoculum, *Can. J. Plant Sci.* **85**: 15–21.

Douds, D.D., Nagahashi, G., Pfeffer, P.E., Reider, C., and Kayser, W.M., 2006, On-farm production of AM fungus inoculum in mixtures of compost and vermiculite, *Bioresource Technol.* **97**: 809–818.

Eason, W.R., Scullion, J., and Scott, E.P., 1999, Soil parameters and plant responses associated with arbuscular mycorrhizas from contrasting grassland management regimes, *Agric. Ecosyst. Environ.* **73**: 245–255.

Edwards-Jones, G., and Howells, O., 2001, The origin and hazards of inputs to crop protection in organic farming systems: are they sustainable?, *Agric. Syst.* **67**: 31–47.

Egerton-Warburton, L.M., and Allen, E.B., 2000, Shifts in arbuscular mycorrhizal communities along an anthropogenic nitrogen deposition gradient, *Ecol. Appl.* **10**: 484–496.

Egger, K.N., and Hibbett, D.S., 2004, The evolutionary implications of exploitation in mycorrhizas, *Can. J. Bot.* **82**: 1110–1121.

Entz, M.H., Penner, K.R., Vessey, J.K., Zelmer, C.D. and Thiessen Martens, J.R., 2004, Mycorrhizal colonisation of flax under long-term organic and conventional management, *Can. J. Plant Sci.* **84**: 1097–1099.

Ezawa, T., Yamamoto, K., and Yoshida, S., 2000, Species composition and spore density of indigenous vesicular-arbuscular mycorrhizal fungi under different conditions of P-fertility as revealed by soybean trap culture, *Soil Sci. Plant Nutr.* **46**: 291–297.

Feldmann, F., and Boyle, C., 1999, Weed-mediated stability of arbuscular mycorrhizal effectiveness in maize monocultures, *J. Appl. Bot.* **73**: 1–5.

Fitter, A.H., 1985, Functioning of vesicular-arbuscular mycorrhizas under field conditions, *New Phytol.* **99**: 257–265.

Franke-Snyder, M., Douds Jr., D.D., Galvez, L., Phillips, J.G., Wagoner, P., Drinkwater, L., and Morton, J.B., 2001, Diversity of communities of arbuscular mycorrhizal (AM) fungi present in conventional versus low-input agricultural sites in eastern Pennsylvania, USA., *Appl. Soil Ecol.* **16**: 35–48.

Furlan, V., and Bernier-Cardou, M., 1989, Effects of N, P, and K on formation of vesicular-arbuscular mycorrhizae, growth and mineral content of onion, *Plant Soil* **113**: 167–174.

Gabriel, D., Roschewitz, I., Tscharntke, T., and Thies, C., 2006, Beta diversity at different spatial scales: plant communities in organic and conventional agriculture, *Ecol. Appl.* **16**: 2011–2021.

Gaur, A., and Adholeya, A., 2000, On-farm production of VAM inoculum and vegetable crops in marginal soil amended with organic matter, *Trop. Agric.* **77**: 21–25.

Gaur, A., and Adholeya, A., 2002, Arbuscular-mycorrhizal inoculation of five tropical fodder crops and inoculum production in marginal soil amended with organic matter, *Biol. Fertil. Soils* **35**: 214–218.

Gavito, M.E., Olsson, P.A., Rouhier, H., Medina-Penafiel, A., Jakobsen, I., Bago, A., and Azcón-Aguilar, C., 2005, Temperature constraints on the growth and functioning of root organ cultures with arbuscular-mycorrhizal fungi, *New Phytol.* **168**: 179–188.

Gianinazzi, S., and Vosátka, M., 2004, Inoculum of arbuscular mycorrhizal fungi for production systems: science meets business, *Can. J. Bot.* **82**: 1264–1271.

Giovannetti, M., and Mosse, B., 1980, An evaluation of techniques for measuring VAM infection in roots, *New Phytol.* **84**: 489–500.

Gosling, P., Hodge, A., Goodlass, G., and Bending, G.D., 2006, Arbuscular mycorrhizal fungi and organic farming, *Agric. Ecosyst. Environ.* **113**: 17–35.

Govindarajulu, M., Pfeffer, P.E., Jin, H., Abubaker, J., Douds, D.D., Allen, J.A., Bücking, H., Lammers, P.J., and Shachar-Hill, Y., 2005, Nitrogen transfer in the arbuscular mycorrhizal symbiosis, *Nat. Lett.* **435**: 819–823.

Grace, C., and Stribley, D.P., 1991, A safer procedure for routine staining of vesicular arbuscular mycorrhizal fungi, *Mycol. Res.* **95**: 1160–1162.
Graham, J.H., Leonard, R.T., and Menge, J.A., 1981, Membrane-mediated decrease in root exudation responsible for phosphorus inhibition of vesicular arbuscular mycorrhiza formation, *Plant Physiol.* **68**: 548–552.
Graham, J.H., Duncan, L.W., and Eissenstat, D.M., 1997, Carbohydrate allocation patterns in citrus genotypes as affected by phosphorus nutrition, mycorrhizal colonisation and mycorrhizal dependency, *New Phytol.* **135**: 335–343.
Graham, J.H., and Eissenstat, D.M., 1998, Field evidence for carbon cost of citrus mycorrhizas, *New Phytol.* **140**: 103–110.
Graham, J.H., 2000, Assessing costs of arbuscular mycorrhizal symbiosis in agroecosystems, in: *Current Advances in Mycorrhizae Research*, G.K. Podila, and D.D. Douds, eds., APS Press, St. Paul, pp. 111–126.
Graham, J.H., and Abbott, L.K., 2000, Wheat responses to aggressive and non-aggressive arbuscular mycorrhizal fungi, *Plant Soil* **220**: 207–218.
Graham, R.D., Humphries, J.M., and Kitchen J.L., 2000, Nutritionally enhanced cereals: a sustainable foundation for a balanced diet, *Asia Pac. J. Clin. Nutr.* **9**(Suppl.): S91–S96.
Gryndler, M., Larsen, J., Hršelová, H., Řezáčová, V., Gryndlerová, H., and Kubát, J., 2006, Organic and mineral fertilization, respectively, increase and decrease the development of external mycelium of arbuscular mycorrhizal fungi in a long-term field experiment, *Mycorrhiza* **16**: 159–166.
Gupta, M.L., Prasad, A., Ram, M., and Kumar, S., 2002, Effect of the vesicular-arbuscular mycorrhizal (VAM) fungus *Glomus fasciculatum* on the essential oil yield related characters and nutrient acquisition in the crops of different cultivars of menthol mint (*Mentha arvensis*) under field conditions, *Bioresource Technol.* **81**: 77–79.
Harinikumar, K.M., and Bagyaraj, D.J., 1996, Persistence of introduced *Glomus intraradices* in the field as influenced by repeated inoculation and cropping system, *Biol. Fertil. Soils* **21**: 184–188.
Harrier, L.A., and Watson, C.A., 2004, The potential role of arbuscular mycorrhizal (AM) fungi in the bioprotection of plants against soil-borne pathogens in organic and/or sustainable farming systems, *Pest Manage. Sci.* **60**: 149–157.
He, X.-H., Critchley, C., and Bledsoe, C., 2003, Nitrogen transfer within and between plants through common mycorrhizal networks (CMNs), *Crit. Rev. Plant Sci.* **22**: 531–567.
Helgason, T., Daniell, T.J., Husband, R., Fitter, A.H., and Young, J.P.W., 1998, Ploughing up the wood-wide web? *Nature* **394**: 431.
Hetrick, B.A.D., Wilson, G.W.T., and Cox, T.S., 1992, Mycorrhizal dependence of modern wheat varieties, landraces, and ancestors, *Can. J. Bot.* **70**: 2032–2040.
Hetrick, B.A.D., Wilson, G.W.T., and Cox, T.S., 1993, Mycorrhizal dependence of modern wheat cultivars and ancestors: a synthesis, *Can. J. Bot.* **71**: 512–518.
Hetrick, B.A.D., Hartnett, D.C., Wilson, G.W.T., and Gibson, D.J., 1994, Effects of mycorrhizae, phosphorus availability, and plant density on yield relationships among competing tallgrass prairie grasses, *Can. J. Bot.* **72**: 168–176.
Hetrick, B.A.D., Wilson, G.W.T., and Todd, T.C., 1996, Mycorrhizal responses in wheat cultivars: relationship to phosphorus, *Can. J. Bot.* **74**: 19–25.
Hijri, I., Sýkorová, Z., Oehl, F., Ineichen, K., Mäder, P., Weimken, A., and Redecker, D., 2006, Communities of arbuscular mycorrhizal fungi in arable soils are not necessarily low in diversity, *Mol. Ecol.* **15**: 2277–2289.
Ho, I., and Trappe, J.M., 1973, Translocation of ^{14}C from *Festuca* plants to their endomycorrhizal fungi, *Nat. New Biol.* **244**: 30–31.
Hulugalle, N.R., Antwistle, P.C., Cooper, J.L., Allen, S.J., and Nehl, D.B., 1998, Effect of long-fallow on soil quality and cotton lint yield in an irrigated, self-mulching, grey Vertosol in the central-west of New South Wales, *Aust. J. Soil Res.* **36**: 621–629.
Jakobsen, I., Abbott, L., and Robson, A.D., 1992a, External hyphae of vesicular-arbuscular mycorrhizal fungi associated with *Trifolium subterraneum* L. I. Spread of hyphae and phosphorus inflow into roots, *New Phytol.* **120**: 371–380.

Jakobsen, I., Abbott, L.K., and Robson, A.D., 1992b, External hyphae of vesicular-arbuscular mycorrhizal fungi associated with *Trifolium subterraneum* L. II. Hyphal transport of ^{32}P over defined distances, *New Phytol.* **120**: 509–516.

Jakobsen, I., Gazey, C., and Abbott, L.K., 2001, Phosphate transport by communities of arbuscular mycorrhizal fungi in intact cores, *New Phytol.* **149**: 95–103.

Jansa, J., Mozafar, A., Anken, T., Ruh, R., Sanders, I.R., and Frossard, E., 2002, Diversity and structure of AMF communities as affected by tillage in a temperate soil, *Mycorrhiza* **12**: 225–234.

Jansa, J., Mozafar, A., Kuhn, G., Anken, T., Ruh, R., Sanders, I.R., and Frossard, E., 2003, Soil tillage affects the community structure of mycorrhizal fungi in maize roots, *Ecol. Appl.* **13**: 1164–1176.

Jasper, D.A., Abbott, L.K., and Robson, A.D., 1991, The effect of soil disturbance on vesicular-arbuscular mycorrhizal fungi in soils from different vegetation types, *New Phytol.* **118**: 471–476.

Jasper, D.A., Abbott, L.K., and Robson, A.D., 1993, The survival of infective hyphae of vesicular-arbuscular mycorrhizal fungi in dry soil: an interaction with sporulation, *New Phytol.* **124**: 473–479.

Jastrow, J.D., Miller, R.M., and Lussenhop, J., 1998, Contributions of interacting biological mechanisms to soil aggregate stabilization in restored prairie, *Soil Biol. Biochem.* **30**: 905–916.

Jifon, J.L., Graham, J.H., Drouillard, D.L., and Syvertsen, J.P., 2002, Growth depression of mycorrhizal *Citrus* seedlings grown at high phosphorus supply is mitigated by elevated CO_2, *New Phytol.* **153**: 133–142.

Johnson, N.C., Pfleger, F.L., Crookston, R.K., Simmons, S.R., and Copeland, P.J., 1991, Vesicular-arbuscular mycorrhizas respond to corn and soybean cropping history, *New Phytol.* **117**: 657–663.

Johnson, N.C., and Pfleger, F.L., 1992, Vesicular-arbuscular mycorrhizae and cultural stresses, in: *Mycorrhizae in Sustainable Agriculture*, G.J. Bethlenfalvay, and R.G. Linderman, eds., American Society of Agronomy, Madison, WI, pp. 71–99.

Johnson, N.C., Copeland, P.J., Crookston, R.K., and Pfleger, F.L., 1992, Mycorrhizae: possible explanation for yield decline with continuous corn and soybean, *Agron. J.* **84**: 387–390.

Johnson, N.C., 1993, Can fertilisation of soil select less mutualistic mycorrhizae? *Ecol. Appl.* **3**: 749–757.

Johnson, N.C., Graham, J.H., and Smith, F.A., 1997, Functioning of mycorrhizal associations along the mutualism-parasitism continuum, *New Phytol.* **135**: 575–585.

Johnson, D., Leake, J.R., Ostle, N., Ineson, P., and Read, D.J., 2002, *In situ* $^{13}CO_2$ pulse-labelling of upland grassland demonstrates a rapid pathway of carbon flux from arbuscular mycorrhizal mycelia to the soil, *New Phytol.* **153**: 327–334.

Jones, K., and Hendrix, J.W., 1987, Inhibition of root extension in tobacco by the mycorrhizal fungus *Glomus macrocarpum* and its prevention by benomyl, *Soil Biol. Biochem.* **19**: 297–299.

Jones, M.D., and Smith, S.E., 2004, Exploring functional definitions of mycorrhizas: are mycorrhizas always mutualisms? *Can. J. Bot.* **82**: 1089–1109.

Kabir, Z., and Koide, R.T., 2002, Effect of autumn and winter mycorrhizal cover crops on soil properties, nutrient uptake and yield of sweet corn in Pennsylvania, USA, *Plant Soil* **238**: 205–215.

Kahiluoto, H., Ketoja, E., and Vestberg, M., 2000, Promotion of utilization of arbuscular mycorrhiza through reduced P fertilization 1. Bioassays in growth chamber, *Plant Soil* **227**: 191–206.

Kahiluoto, H., Ketoja, E., Vestberg, M., and Saarela, I., 2001, Promotion of AM utilization through reduced P fertilization 2. Field studies, *Plant Soil* **231**: 65–79.

Kelly, R.M., Edwards, D.G., Thompson, J.P., and Magarey, R.C., 2005, Growth responses of sugarcane to mycorrhizal spore density and phosphorus rate, *Aust. J. Agric. Res.* **56**: 9–12.

Khalil, S., Loynachan, T.E., and Tabatabai, M.A., 1994, Mycorrhizal dependency and nutrient uptake by improved and unimproved corn and soybean cultivars, *Agron. J.* **86**: 949–958.

Khaliq, A., and Sanders, F.E., 2000, Effects of vesicular-arbuscular mycorrhizal inoculation on the yield and phosphorus uptake of field-grown barley, *Soil Biol. Biochem.* **32**: 1691–1696.

Khasa, P., Furlan, V., and Fortin, J.A., 1992, Response of some tropical plant species to endomycorrhizal fungi under field conditions, *Trop. Agric.* **69**: 279–283.

Kiers, E.T., West, S.A., and Denison, R.F., 2002, Mediating mutualism: farm management practices and evolutionary changes in symbiont co-operation, *J. Appl. Ecol.* **39**: 745–754.

Kirchmann, H., and Ryan, M.H., 2004, Nutrients in organic farming – Are there advantages from the exclusive use of organic manures and untreated minerals? *New directions for a diverse planet*, Proceedings of the 4th International Crop Science Congress, Brisbane, Australia, 26 September–1 October 2004.

Kirchmann, H., Thorvaldsson, G., Bergström, L., Gerzabek, M., Andrén, O., Eriksson, L.-O., and Winninge, M., 2008, Fundamenta ls of organic agriculture – past and present, in: *Organic Crop Production – Ambitions and Limitations*, H. Kirchmann, and L. Bergström, eds., Springer, Dordrecht, The Netherlands.

Kitchen, J.L., McDonald, G.K., Shepherd, K.W., Lorimer, M.F., and Graham, R.D., 2003, Comparing wheat grown in South Australian organic and conventional farming systems. I. Growth and grain yield, *Aust. J. Agric. Res.* **54**: 889–901.

Klironomos, J.N., McCune, J., Hart, M., and Neville, J., 2000, The influence of arbuscular mycorrhizae on the relationship between plant diversity and productivity, *Ecol. Lett.* **3**: 137–141.

Klironomos, J.N., Hart, M.M., Gurney, J.E., and Moutoglis, P., 2001, Interspecific differences in the tolerance of arbuscular mycorrhizal fungi to freezing and drying, *Can. J. Bot.* **79**: 1161–1166.

Klironomos, J.N., 2003, Variation in plant response to native and exotic arbuscular mycorrhizal fungi, *Ecology* **84**: 2292–2301.

Koide, R.T., and Kabir, Z., 2000, Extraradical hyphae of the mycorrhizal fungus *Glomus intraradices* can hydrolyse organic phosphate, *New Phytol.* **148**: 511–517.

Koide, R.T., and Mosse, B., 2004, A history of research on arbuscular mycorrhiza, *Mycorrhiza* **14**: 145–163.

Koske, R.E., 1987, Distribution of VA mycorrhizal fungi along a latitudinal gradient, *Mycologia* **79**: 55–68.

Kothari, S.K., Marschner, H., and Römheld, V., 1991, Effect of a vesicular-arbuscular mycorrhizal fungus and rhizosphere microorganisms on manganese reduction in the rhizosphere and manganese concentrations in maize (*Zea mays* L.), *New Phytol.* **117**: 649–655.

Land, S., and Schönbeck, F., 1991, Influence of different soil types on abundance and seasonal dynamics of vesicular arbuscular mycorrhizal fungi in arable soils in North Germany, *Mycorrhiza* **1**: 39–44.

Leake, J., Johnson, D., Donnelly, D., Muckle, G., Boddy, L., and Read, D., 2004, Networks of power and influence: the role of mycorrhizal mycelium in controlling plant communities and agroecosystem functioning, *Can. J. Bot.* **82**: 1016–1045.

Lekberg, Y., and Koide, R.T., 2005a, Is plant performance limited by abundance of arbuscular mycorrhizal fungi? A meta-analysis of studies published between 1988 and 2003, *New Phytol.* **168**: 189–204.

Lekberg, Y., and Koide, R.T., 2005b, Arbuscular mycorrhizal fungi, rhizobia, available soil P and nodulation of groundnut (*Arachis hypogaea*) in Zimbabwe, *Agric. Ecosyst. Environ.* **110**: 143–148.

Lerat, S., Lapointe, L., Gutjahr, S., Piché, Y., and Vierheilig, H., 2003, Carbon partitioning in a split-root system of arbuscular mycorrhizal plants is fungal and plant species dependent, *New Phytol.* **157**: 589–595.

Limonard, T., and Ruissen, M.A., 1989, The significance of VA-mycorrhiza to future arable farming in the Netherlands, *Neth. J. Plant. Path.* **95** (Suppl. 1): 129–135.

Løes, A.K., and Øgaard, A.F., 1997, Changes in nutrient content of agricultural soil on conversion to organic farming in relation to farm-level nutrient balances and soil contents of clay and organic matter, *Acta Agric. Scand.* (Section B) **47**: 201–214.

Mäder, P., Edenhofer, S., Boller, T., Wiemken, A., and Niggli, U., 2000, Arbuscular mycorrhizae in a long-term field trial comparing low-input (organic, biological) and high-input (conventional) farming systems in a crop rotation, *Biol. Fertil. Soils* **31**: 150–156.

Mäder, P., Fliessbach, A., Dubois, D., Gunst, L., Fried, P., and Niggli, U., 2002, Soil fertility and biodiversity in organic farming, *Science* **296**: 1694–1697.

Mamatha, G., Bagyaraj, D.J., and Jaganath, S., 2002, Inoculation of field-established mulberry and papaya with arbuscular mycorrhizal fungi and a mycorrhiza helper bacterium, *Mycorrhiza* **12**: 313–316.

Mårtensson, A.M., and Carlgren, K., 1994, Impact of phosphorus fertilisation on VAM diaspores in two Swedish longterm field experiments, *Agric. Ecosyst. Environ.* **47**: 327–334.

McGonigle, T.P., 1988, A numerical analysis of published field trials with vesicular-arbuscular mycorrhizal fungi, *Funct. Ecol.* **2**: 473–478.

Menéndez, A.B., Scervino, J.M., and Godeas, A.M., 2001, Arbuscular mycorrhizal populations associated with natural and cultivated vegetation on a site of Buenos Aires province, Argentina, *Biol. Fertil. Soils* **33**: 373–381.

Miller, R.L., and Jackson, L.E., 1998, Survey of vesicular-arbuscular mycorrhizae in lettuce production in relation to management and soil factors, *J. Agric. Sci.* **130**: 173–182.

Miller, M.H., 2000, Arbuscular mycorrhizae and the phosphorus nutrition of maize: a review of Guelph studies, *Can. J. Plant Sci.* **80**: 47–52.

Muckle, G.E., 2003, The functioning of arbuscular mycorrhizal fungi in land under different agricultural management intensities, Ph. D. thesis, University of Sheffield, Department of Animal and Plant Sciences, Sheffield, UK.

Mulligan, M.F., Smucker, A.J.M., and Safir, G.F., 1985, Tillage modifications of dry edible bean root colonization by VAM fungi, *Agron. J.* **77**: 140–144.

Newman, E.I., and Reddell, P., 1987, The distribution of mycorrhizas among families of vascular plants, *New Phytol.* **106**: 745–751.

Newsham, K.K., Fitter, A.H., and Watkinson, A.R., 1995, Arbuscular mycorrhiza protect an annual grass from root pathogenic fungi in the field, *J. Ecol.* **83**: 991–1000.

Nguyen, M.L., Haynes, R.J., and Goh, K.M., 1995, Nutrient budgets and status in three pairs of conventional and alternative mixed cropping farms in Canterbury, New Zealand, *Agric. Ecosyst. Environ.* **52**: 149–162.

O'Connor, P.J., Smith, S.E., and Smith, F.A., 2002, Arbuscular mycorrhizas influence plant diversity and community structure in a semiarid herbland, *New Phytol.* **154**: 209–218.

Oehl, F., Oberson, A., Tagmann, H.U., Besson, J.M., Dubois, D., Mäder, P., Roth, H.-R., and Frossard, E., 2002, Phosphorus budget and phosphorus availability in soils under organic and conventional farming, *Nutr. Cycl. Agroecosys.* **62**: 25–35.

Oehl, F., Sieverding, E., Ineichen, K., Mäder, P., Boller, T., and Wiemken, A., 2003, Impact of land use intensity on the species diversity of arbuscular mycorrhizal fungi in agroecosystems of central Europe, *Appl. Environ. Microbiol.* **69**: 2816–2824.

Oehl, F., Sieverding, E., Mäder, P., Dubois, D., Ineichen, K., Boller, T., and Wiemken, A., 2004, Impact of long-term conventional and organic farming on the diversity of arbuscular mycorrhizal fungi, *Oecologia* **138**: 574–583.

Olsson, P.A., Thingstrup, I., Jakobsen, I., and Bååth, E., 1999, Estimation of the biomass of arbuscular mycorrhizal fungi in a linseed field, *Soil Biol. Biochem.* **31**: 1879–1887.

Ortas, I., 2003, Effect of selected mycorrhizal inoculation on phosphorus sustainability in sterile and non-sterile soils in the Harran Plain in South Anatolia, *J. Plant Nutr.* **26**: 1–17.

Owusu-Bennoah, E., and Mosse, B., 1979, Plant growth responses to vesicular-arbuscular mycorrhiza. XI. Field inoculation responses in barley, lucerne and onion, *New Phytol.* **83**: 671–679.

Pearson, J.N., and Schweiger, P., 1993, *Scutellospora calospora* (Nicol. & Gerd.) Walker and Sanders associated with subterranean clover: dynamics of colonization, sporulation and soluble carbohydrates, *New Phytol.* **124**: 215–219.

Penfold, C.M., Miyan, M.S., Reeves, T.G., and Grierson, I.T., 1995, Biological farming for sustainable agricultural production, *Aust. J. Exp. Agric.* **35**: 849–856.

Peterson, R.L., Massicotte, H.B., and Melville, L.H., 2004, *Mycorrhizas: Anatomy and Cell Biology*, NRC Research Press, Ottawa, Canada.

Pfeffer, P.E., Douds Jr., D.D., Becard, G., and Shachar-Hill, Y., 1999, Carbon uptake and the metabolism and transport of lipids in an arbuscular mycorrhiza, *Plant Physiol.* **120**: 587–598.

Pfeffer, P.E., Douds Jr., D.D., Bücking, H., Schwartz, D.P., and Shachar-Hill, Y., 2004, The fungus does not transfer carbon to or between roots in an arbuscular mycorrhizal symbiosis, *New Phytol.* **163**: 617–627.

Piotrowski, J.S., Denich, T., Klironomos, J.N., Graham, J.M., and Rillig, M.C., 2004, The effects of arbuscular mycorrhizas on soil aggregation depend on the interaction between plant and fungal species, *New Phytol.* **164**: 365–373.

Plenchette, C., Furlan, V., and Fortin, J.A., 1981, Growth stimulation of apple trees in unsterilized soil under field conditions with VA mycorrhiza inoculation, *Can. J. Bot.* **59**: 2003–2008.

Plenchette, C., Fortin, J.A., and Furlan, V., 1983, Growth responses of several plant species to mycorrhizae in a soil of moderate P-fertility I. Mycorrhizal dependency under field conditions, *Plant Soil* **70**: 199–209.

Plenchette, C., and Perrin, R., 1992, Evaluation in the greenhouse of the effects of fungicides on the development of mycorrhiza on leek and wheat, *Mycorrhiza* **1**: 59–62.

Purin, S., Filho, O.K., and Stürmer, S.L., 2006, Mycorrhizae activity and diversity in conventional and organic apple orchards from Brazil, *Soil Biol. Biochem.* **38**: 1831–1839.

Rangeley, A., Daft, M.J., and Newbould, P., 1982, The inoculation of white clover with mycorrhizal fungi in unsterile hill soils, *New Phytol.* **92**: 89–102.

Rillig, M.C., Wright, S.F., and Eviner, V.T., 2002, The role of arbuscular mycorrhizal fungi and glomalin in soil aggregation: comparing effects of five plant species, *Plant Soil* **238**: 325–333.

Rillig, M.C., and Mummey, D.L., 2006, Mycorrhizas and soil structure, *New Phytol.* **171**: 41–53.

Rosendahl, S., and Stukenbrock, E.H., 2004, Community structure of arbuscular mycorrhizal fungi in undisturbed vegetation revealed by analyses of LSU rDNA sequences, *Mol. Ecol.* **13**: 3179–3186.

Runge-Metzger, A., 1995, Closing the cycle: obstacles to efficient P management for improved global food security, in: *Phosphorus in the Global Environment*, H. Tiessen, ed., John Wiley, Chichester, UK, pp. 27–42.

Ryan, M.H., Chilvers, G.A., and Dumaresq, D.C., 1994, Colonisation of wheat by VA-mycorrhizal fungi was found to be higher on a farm managed in an organic manner than on a conventional neighbour, *Plant Soil* **160**: 33–40.

Ryan, M.H., 1998, The ecology of VAM fungi in contrasting Australian agricultural systems, Ph. D. thesis, The Australian National University, Department of Botany and Zoology, Canberra, Australia.

Ryan, M.H., and Ash, J.E., 1999, Effects of phosphorus and nitrogen on growth of pasture plants and VAM fungi in SE Australian soils with contrasting fertiliser histories (conventional and biodynamic), *Agric. Ecosyst. Environ.* **73**: 51–62.

Ryan, M.H., Small, D.R., and Ash, J E., 2000, Phosphorus controls the level of colonisation by arbuscular mycorrhizal fungi in conventional and biodynamic irrigated dairy pastures, *Aust. J. Exp. Agric.* **40**: 663–670.

Ryan, M.H., Norton, R.M., Kirkegaard, J.A., McCormick, K.M., Knights, S.E., and Angus, J.F., 2002, Increasing mycorrhizal colonisation does not improve growth and nutrition of wheat on Vertosols in south-eastern Australia, *Aust. J. Agric. Res.* **53**: 1173–1181.

Ryan, M.H., and Graham, J.H., 2002, Is there a role for arbuscular mycorrhizal fungi in production agriculture? *Plant Soil* **244**: 263–271.

Ryan, M.H., and Angus, J.F., 2003, Arbuscular mycorrhizae in wheat and field pea crops on a low P soil: increased Zn-uptake but no increase in P-uptake or yield, *Plant Soil* **250**: 225–239.

Ryan, M.H., Derrick, J.W., and Dann, P.R., 2004, Grain mineral concentrations and yield of wheat grown under organic and conventional management, *J. Sci. Food Agric.* **84**: 207–216.

Ryan, M.H., van Herwaarden, A.F., Angus, J.F., and Kirkegaard, J.A., 2005, Reduced growth of autumn-sown wheat in a low P soil is associated with high colonisation by arbuscular mycorrhizal fungi, *Plant Soil* **270**: 275–286.

Ryan, M.H., McInerney, J.K., Record, I.R., and Angus, J.F., 2008, Zinc bioavailability in wheat grain in relation to phosphorus fertiliser, crop sequence and mycorrhizal fungi, *J. Sci. Food Agric.* **88**: 1208–1216.

Sáinz, M.J., Taboada-Castro, M.T., and Vilarino, A., 1998, Growth, mineral nutrition and mycorrhizal colonization of red clover and cucumber plants grown in a soil amended with composted urban wastes, *Plant Soil* **205**: 85–92.

Sattelmacher, B., Reinhard, S., and Pomikalko, A., 1991, Differences in mycorrhizal colonisation of rye (*Secale cereale* L.) grown in conventional or organic (biological-dynamic) farming systems, *J. Agron. Crop Sci.* **167**: 350–355.

Schüβler, A., Schwarzott, D., and Walker, C., 2001, A new fungal phylum, the Glomeromycota: phylogeny and evolution, *Mycol. Res.* **105**: 1413–1421.

Schüβler, A., 2006, Darmstadt; http://www.AMF-phylogeny.com. Assessed January 2006.

Schwartz, M.W., Hoeksema, J.D., Gehring, C.A., Johnson, N.C., Klironomos, J.N., Abbott, L.K., and Pringle, A., 2006, The promise and the potential consequences of the global transport of mycorrhizal fungal inoculum, *Ecol. Lett.* **9**: 501–515.

Schwarzott, D., Walker, C., and Schüβler, A., 2001, *Glomus*, the largest genus of the arbuscular mycorrhizal fungi (Glomales), is nonmonophyletic, *Mol. Phylogenet. Evol.* **21**: 190–197.

Schweiger, P.F., Robson, A.D., and Barrow, N.J., 1995, Root hair length determines beneficial effect of a *Glomus* species on shoot growth of some pasture species, *New Phytol.* **131**: 247–254.

Scullion, J., Eason, W.R., and Scott, E.P., 1998, The effectivity of arbuscular mycorrhizal fungi from high input conventional and organic grassland and grass-arable rotations, *Plant Soil* **204**: 243–254.

Sieverding, E., and Oehl, F., 2006, Revision of *Entrophospora* and description of *Kuklospora* and *Intraspora*, two new genera in the arbuscular mycorrhizal Glomeromycetes, *J. Appl. Bot. Food Qual.* **80**: 69–81.

Sjöberg, J., Persson, P., Mårtensson, A., Mattsson, L., Adholeya, A., and Alström, S., 2004, Occurrence of *Glomeromycota* spores and some arbuscular mycorrhiza fungal species in arable fields in Sweden, *Acta Agric. Scand.* (Section B) **54**: 202–212.

Smith, S.E., and Read, D.J., 1997, *Mycorrhizal Symbiosis*, Academic Press, London.

Smith, M.T.E., Cade-Menun, B.J., and Tibbett, M., 2006, Soil phosphorus dynamics and phytoavailability from sewage sludge at different stages in a treatment stream, *Biol. Fertil. Soils* **42**: 186–197.

Son, C.L., and Smith, S.E., 1988, Mycorrhizal growth responses: interactions between photon irradiance and phosphorus nutrition, *New Phytol.* **108**: 305–314.

Sreenivasa, M.N., and Bagyaraj, D.J., 1989, Use of pesticides for mass production of vesicular-arbuscular mycorrhizal inoculum, *Plant Soil* **119**: 127–132.

Stukenbrock, E.H., and Rosendahl, S., 2005, Clonal diversity and population genetic structure of arbuscular mycorrhizal fungi (*Glomus* spp.) studied by multilocus genotyping of single spores, *Mol. Ecol.* **14**: 743–752.

Talukdar, N.C., and Germida, J.J., 1993, Occurrence and isolation of vesicular-arbuscular mycorrhizae in cropped field soils of Saskatchewan, Canada, *Can. J. Microbiol.* **39**: 567–575.

Tarafdar, J.C., and Marschner, H., 1994, Phosphatase activity in the rhizosphere and hyphosphere of VA mycorrhizal wheat supplied with inorganic and organic phosphorus, *Soil Biol. Biochem.* **26**: 387–395.

Tawaraya, K., Naito, M., and Wagatsuma, T., 2006, Solubilization of insoluble inorganic phosphate by hyphal exudates of arbuscular mycorrhizal fungi, *J. Plant Nutr.* **29**: 657–665.

Thompson, J.P., 1987, Decline of vesicular-arbuscular mycorrhizas in long fallow disorder of field crops and its expression in phosphorus deficiency in sunflower, *Aust. J. Agric. Res.* **38**: 847–867.

Thompson, J.P., and Wildermuth, G.B., 1989, Colonisation of crop and pasture species with vesicular-arbuscular mycorrhizal fungi and infection by *Bipolaris sorokiniana*, *Can. J. Bot.* **67**: 687–693.

Thompson, J.P., 1990, Soil sterilisation methods to show VA-mycorrhizae aid P and Zn nutrition of wheat in vertisols, *Soil Biol. Biochem.* **22**: 229–240.

Thompson, J.P., 1994, Inoculation with vesicular-arbuscular mycorrhizal fungi from cropped soil overcomes long-fallow disorder of linseed (*Linum usitatissimum* L.) by improving P and Zn uptake, *Soil Biol. Biochem.* **26**: 1133–1143.

Thompson, J.P., 1996, Correction of duel phosphorus and zinc deficiencies of linseed (*Linum usitatissimum* L.) with cultures of vesicular-arbuscular mycorrhizal fungi, *Soil Biol. Biochem.* **28**: 941–951.

Tibbett, M., and Cairney, J.W.G, 2007, The cooler side of mycorrhizas: their occurrence and functioning at low temperatures. *Can. J. Bot.* **85**: 51–62.

Tisdall, J.M., and Oades, J.M., 1979, Stabilization of soil aggregates by the root systems of ryegrass, *Aust. J. Soil Res.* **17**: 429–441.

Trinick, M.J., 1977, Vesicular-arbuscular infection and soil phosphorus utilisation in *Lupinus* spp., *New Phytol.* **78**: 297–304.

Troeh, Z.I., and Loynachan, T.E., 2003, Endomycorrhizal fungal survival in continuous corn, soybean and fallow, *Agron. J.* **95**: 224–230.

Udaiyan, K., Greep, S., Mithukumar T., and Chitra, A., 1999, Effect of fumigation and pesticide drenches on VAM status and growth in cereals, *J. Environ. Bot.* **20**: 167–175.

van der Heijden, M.G.A., Klironomos, J.N., Ursic, M., Moutoglis, P., Streitwolf-Engel, R., Boller, T., Wiemken, A., and Sanders, I.R., 1998, Mycorrhizal fungal diversity determines plant biodiversity, ecosystem variability and productivity, *Nature* **396**: 69–72.

van der Heijden, E.W., 2001, Differential benefits of arbuscular mycorrhizal and ectomycorrhizal infection of *Salix repens, Mycorrhiza* **10**: 185–193.

van Diepeningen, A.D., de Vos, O.J., Korthals, G.W., and van Bruggen, A.H.C., 2006, Effects of organic versus conventional management on chemical and biological parameters in agricultural soils, *Appl. Soil Ecol.* **31**: 120–135.

Watson, C.A., Atkinson, D., Gosling, P., Jackson, L.R., and Rayns, F.W., 2002, Managing soil fertility in organic farming systems, *Soil Use Manage.* **18**: 239–247.

Weber, E., Saxena, M.C., George, E., and Marschner, H., 1993, Effect of vesicular-arbuscular mycorrhiza on vegetative growth and harvest index of chickpea grown in northern Syria, *Field Crop Res.* **32**: 115–128.

Werner, M.R., Kluson, R.A., and Gliessman, S.R., 1990, Colonisation of strawberry roots by VA mycorrhizal fungi in agroecosystems under conventional and transitional organic management, *Biol. Agric. Hortic.* **7**: 139–151.

Werner, M., 1997, Soil quality characteristics during conversion to organic orchard management, *Appl. Soil Ecol.* **5**: 151–167.

Whipps, J.M., 2004, Prospects and limitations for mycorrhizas in biocontrol of root pathogens, *Can. J. Bot.* **82**: 1198–1227.

Wright, S.F., and Upadhyaya, A., 1998, A survey of soils for aggregate stability and glomalin, a glycoprotein produced by hyphae of arbuscular mycorrhizal fungi, *Plant Soil* **198**: 97–107.

Wright, S.F., Starr, J.L., and Paltineau, I.C., 1999, Changes in aggregate stability and concentration of glomalin during tillage management transition, *Soil Sci. Soc. Am. J.* **63**: 1825–1829.

Wright, S.F., and Anderson, R.L., 2000, Aggregate stability and glomalin in alternative crop rotations for the central Great Plains, *Biol. Fertil. Soils* **31**: 249–253.

Xavier, L.J.C., and Germida, J.J., 1997, Growth response of lentil and wheat to *Glomus clarum* NT4 over a range of P levels in a Saskatchewan soil containing indigenous AM fungi, *Mycorrhiza* **7**: 3–8.

Chapter 11
Organic Food Production and Its Influence on Naturally Occurring Toxins

Carl K. Winter

Abstract The levels of natural plant toxins and mycotoxins in foods may be influenced by the methods used (organic vs. conventional) for agricultural production. Research findings suggest that organic foods may possess higher levels of natural plant toxins than conventional foods based upon mechanistic similarities between natural plant toxin production and the production of plant secondary metabolites of nutritional interest. Specific field research confirming such differences has not yet been conducted. Mycotoxin levels in organic foods may also be higher as a few studies have demonstrated that synthetic fungicides and insecticides used in conventional production can reduce plant pathogen populations. Food product analysis, however, has not demonstrated consistent findings of higher levels of mycotoxins in organic foods as compared with conventional foods. In the event that subsequent research does conclusively demonstrate that differences exist in the levels of naturally occurring toxins in organic versus conventional foods, the toxicological significance of the differences, if any, still requires determination.

Keywords Natural toxins · Mycotoxins · Plant secondary metabolites · Plant stress · Pesticides

11.1 Introduction

The organic foods industry in the U.S. has grown dramatically in the past 20 years with annual sales reaching $13.8 billion in 2005 and growth averaging about 20 percent pear year (Organic Trade Association 2006). The adoption of U.S. national standards for organic food production in 2002 has likely led to increases in consumer confidence in organic products. Several surveys indicate that consumers purchase organic foods for a variety of reasons including the minimization of pesticide residues, perceived increases in nutritional value, and the avoidance of genetically

C.K. Winter (✉)
Department of Food Science and Technology, University of California Davis, CA 95616, USA
e-mail: ckwinter@ucdavis.edu

H. Kirchmann, L. Bergström (eds.), *Organic Crop Production – Ambitions and Limitations*, DOI 10.1007/978-1-4020-9316-6_11,

modified foods (Whole Foods Market, 2005). Worker safety and environmental concerns are also frequently cited. Another study indicated that over 90 percent of organic and conventional food consumers perceived a reduction in pesticide residue risk in organic foods while 50 percent perceived that organic foods had lower microbiological safety risks and 50 percent perceived that the risks from naturally occurring toxins in foods would be lower in organic foods (Williams and Hammitt, 2001).

The presence of naturally occurring toxins in food has been well established and many food toxins of plant and fungal origin have been identified (Ames et al., 1990; Beier and Nigg, 1994; Coulombe, 2000; Park et al., 2000). Comparative risk assessment studies focusing upon carcinogenic effects have indicated that the relative health risks posed by naturally occurring food toxins may exceed those from synthetic chemicals found in food (Ames et al., 1987, 1990; Ames and Gold, 1991). A National Academy of Sciences panel concluded that "natural components of the diet may prove to be of greater concern than synthetic components with respect to cancer risk, although additional evidence is required before this conclusion can be drawn with certainty" (NRC, 1996).

While considerable research has been conducted investigating the potential nutritional differences between specific organic and conventional foods (Winter and Davis, 2006), very little direct investigation has compared organic and conventional foods with respect to naturally occurring toxins. This chapter considers the potential differences in naturally occurring toxin levels in organic and conventional foods and discusses the potential influences of such differences on human health.

11.2 Naturally Occurring Toxins

11.2.1 Plant Toxins

Plants have been known to produce hundreds of different chemicals, known as plant secondary metabolites, that do not appear to be necessary for basic plant health (Beier and Nigg, 1994). While the roles of many of these secondary metabolites remain unknown and most have not been appropriately studied for their potential toxicological effects, several are considered to be of possible human health concern. Celery plants, for example, frequently produce elevated levels of linear furanocoumarins when grown under conditions of stress (Beier and Oertli, 1983; Dercks et al., 1990); these secondary metabolites are considered possible human carcinogens and are also associated with the ability to cause contact dermatitis in workers handling celery plants (Seligman et al., 1987). Plants such as potatoes and tomatoes are capable of synthesizing glycoalkaloids that serve as effective natural insecticides and inhibit the same types of enzymes in insects and in humans as do the synthetic organophosphate and carbamate insecticides (Friedman and McDonald, 1997).

The practices used to classify plant secondary metabolites as naturally occurring toxins are arbitrary and inconsistent. Ames et al. (1990) referred to all plant secondary metabolites as "nature's pesticides" while others (Winter, 1990; Coulombe, 2000) have more narrowly defined natural plant toxins as chemicals produced by plants for which studies have been conducted to demonstrate toxicological effects. Complicating the classification process further, it is clear that some of the naturally occurring toxins may also have potential health benefits for humans. Glycoalkaloids found in potatoes, for example, are of potential toxicological concern due to their ability to inhibit human cholinesterase enzymes and cause effects upon the nervous system (Friedman and McDonald, 1997); they have also been shown to possess antiallergic, antipyretic, and anti-inflammatory effects in humans, glycemic effects in rats, and antibiotic activities against pathogenic bacteria, viruses, protozoa, and fungi (Friedman, 2006).

Most studies comparing organic and conventional foods with respect to levels of plant secondary metabolites have focused upon secondary metabolites considered to be nutritionally beneficial rather than potentially hazardous. One review suggested that organic foods may possess elevated levels of vitamin C compared with conventional foods (Worthington, 2001) although other reviews (Woese et al., 1997; Bourn and Prescott, 2002) concluded that study designs and results were too variable to provide definitive conclusions concerning the influence of agricultural production methods (i.e. organic vs. conventional, synthetic vs. non-synthetic fertiliser use) upon nutrient and vitamin levels in the plants. It has been concluded that nitrate levels in organic foods are lower than nitrate levels in conventional foods, presumably because the use of synthetic fertilisers in conventional production provides more nitrogen capable of conversion into nitrates than is provided in organic fertilization practices (Woese et al., 1997; Worthington, 2001). The health significance of these differences in nitrate levels is debatable. While nitrates have been implicated as precursors to carcinogenic nitrosamines and can form methemoglobin following ingestion by humans, other research indicates that nitrates are converted to nitrites that provide protection in the oral cavity against infectious diseases (Duncan et al., 1997).

In the past decade, there has been a large increase in the number of controlled field studies that compare organic and conventional foods with respect to plant secondary metabolite nutrients such as organic acids, flavonols, carotenoids, vitamin C, and plant phenolics (Winter and Davis, 2006). Many of these plant secondary metabolites are considered to play a role as possible antioxidants in humans.

While some of these comparative studies failed to show differences between conventional and organic production methods with respect to plant secondary metabolites of nutritional interest (Hakkinen and Torronen, 2000; Mikkonen et al., 2001; Young et al., 2005), other studies have demonstrated elevated levels of the same plant secondary metabolites in organic products relative to conventional products. These include studies done by Asami et al. (2003) that demonstrated higher levels of phenolics and ascorbic acid in organic marionberries, corn, and strawberries. Peaches and pears grown organically had higher levels of total phenolics and polypholoxidase enzyme activity than their conventional counterparts (Carbonaro and Mattera, 2001). Veberic et al. (2005) demonstrated higher phenolic levels in organic apple pulp than conventional apple pulp, while another study showed that organic tomatoes had higher levels of vitamin C, carotenoids, and polyphenols than conventional tomatoes (Caris-Veyrat et al., 2004). Recently, a ten-year comparison of organic and conventional production systems in tomatoes showed that organic production systems resulted in the production of 79 percent higher levels of the flavonoid quercetin and 97 percent higher levels of the flavonoid kaempferol than those found in conventional systems (Mitchell et al., 2007).

Two plausible primary scientific explanations have been advanced to explain the relative increase in production of plant secondary metabolites of nutritional interest in organic foods compared with conventional foods (Winter and Davis, 2006). The first considers the impacts of different types of fertilisation practices upon plant secondary metabolite production. The synthetic fertilisers available to conventional producers make nitrogen more available to the plants than organic fertilisers; this may accelerate plant growth and allocate plant resources for growth functions at the expense of other functions such as the production of plant secondary metabolites including organic acids, flavonols, carotenoids, vitamin C, and plant phenolics.

The second hypothesis considers the mechanisms by which plants may respond to stressful environments that exist when plants are attacked by insects, weeds, and plant pathogens. Synthetic pesticides provide one tool for conventional food producers to use as a means to reduce pest pressures and stress on their plants. Reducing plant stresses through the use of synthetic pesticides is not an option for organic food producers and raises the possibility that such stresses, if not controlled using other methods, could result in increases in plant secondary metabolites. The increase in plant polyphenolics in organic foods in the study by Asami et al. (2003), for example, was attributed to plant defense characteristics.

It is logical to conclude that the same mechanisms that increase the production of plant secondary metabolites of nutritional interest under organic production practices (slower and more complex biochemical growth, plant defense) might also be expected to cause increases in the production of natural plant toxins. Unfortunately, no studies have yet been reported that simultaneously examined differences in secondary metabolite levels for both plant nutrients and for plant toxins when comparing organic and conventional production methods. Studies demonstrating a decrease in natural plant toxin production resulting from the use of synthetic pesticides have also not been published although one study documented increases in the production of plant secondary metabolites in broad beans, pinto beans, peas, celery, and cotton resulting from sublethal plant exposure to the herbicide aciflourfen (Komives and Casida, 1983).

11.2.2 Mycotoxins

Mycotoxins represent an additional class of naturally occurring toxins that present potential health risks. Mycotoxins are produced when fungi colonize food crops and synthesize their own toxins. Common classes of mycotoxins include aflatoxins, fumonisins, and tricothecenes (Murphy et al., 2006). Aflatoxins are frequently detected in many food products including peanuts, corn, and grains, and have been shown to be potent mutagens, carcinogens, and teratogens. Fumonisins have been linked with the development of human esophageal cancer and cause a number of other effects in animals such as liver damage in rats, pulmonary edema in pigs and leukoencephalomalacia in horses (Sydenham et al., 1990; Murphy et al., 2006). Tricothecenes are commonly found in grain products and low to moderate consumption of tricothecenes such as deoxynivalenol (DON) in laboratory animals may lead to immunological and gastrointestinal effects (Murphy et al., 2006).

Since conventional agricultural production methods allow for the use of synthetic pesticides such as fungicides to control plant pathogen growth or insecticides to prevent insect damage that might allow subsequent colonization of plants by plant pathogens, it seems reasonable to assume that mycotoxin levels might be higher in organic foods than in conventional foods. Some studies have, in fact, demonstrated that synthetic pesticide use can decrease mycotoxins levels. Aflatoxin B_1 levels from cultures of *Aspergillus flavus* were reduced in the presence of the fungicides chlorothalonil, dichloran, and mancozeb (Chourasia, 1992). The fungicides iprodione, propionic acid, and cuprosan have also been associated with reductions in mycotoxin production (Arino and Bullerman, 1993; Calori-Domingues and Fonseca, 1995). Application of the insecticides/nematicides fenamiphos, carbofuran, and aldicarb reduced the occurrence of *Fusarium* species in the roots and fruits of tomato plants while also inhibiting or reducing production of the mycotoxin zearalenone (El-Morshedy and Aziz, 1995).

Direct comparisons of food products produced organically with those produced conventionally have yielded mixed results with respect to the relative levels of mycotoxins detected. In one study, DON was found in more than 80 percent of both organic and conventional foods purchased from Italian supermarkets while fumonisin B_1 was detected in 20 percent of organic foods and in 31 percent of the conventional foods and fumonisin B_2 was found in more than 32 percent of both the organic and conventional foods. With respect to median concentration levels, DON was highest in conventional rice-based foods while fumonisin B_1 was highest in conventional maize-based foods and fumonisin B_2 was highest in organic wheat-based foods (Cirillo et al., 2003). Finamore et al. (2004) showed that organic wheat contained higher levels of the mycotoxins DON and ochratoxin A than did conventional wheat although bioassays demonstrated that rats fed the conventional wheat were at a higher risk of lymphocyte damage than rats fed the organic wheat. Median DON levels from wheat flour produced in southwest Germany were consistently higher from conventionally-produced wheat than from organically-produced wheat (Schollenberger et al., 2002) while organic winter wheat produced in Germany showed lower DON contamination and lower rates of *Fusarium* ear blight infection (Birzele et al., 2002). In contrast, an exposure simulation of French consumers indicated significantly greater exposure to DON from consumers of organic foods as compared to conventional food consumers (Leblanc et al., 2002).

A comparison of ochratoxin A levels in Polish cereals produced from conventional or organic farms showed considerably greater contamination of rye, wheat, and barley produced from organic farms (Czerwiecki et al., 2002), while Jorgensen and Jacobsen (2002) showed that organic rye and wheat samples contained higher levels of ochratoxin A than did conventional samples. Another study in red table wines showed reduced levels of ochratoxin A in organic wines relative to conventional wines (Miceli et al., 2003).

11.3 Summary and Conclusions

From a theoretical standpoint, it is logical to assume that levels of naturally occurring plant toxins and mycotoxins might be higher in organic foods than in conventional foods. Significant scientific evidence has been developed to demonstrate that plant secondary metabolites of nutritional interest are frequently produced in higher concentrations from food grown organically than from food grown conventionally. These differences between organic and conventional plant secondary metabolite levels are explained by two different but not mutually exclusive theories. Plants grown conventionally may be stimulated to grow through the use of synthetic fertilisers that make nitrogen more readily available and allocate more plant resources towards plant growth than to biosynthesis of more complex plant secondary metabolites. In addition, the use of synthetic pesticides in conventional food production may serve to alleviate plant stress due to the presence of pests such as insects, weeds, or plant diseases. Organic food production practices do not allow synthetic pesticides to be used to reduce plant stress, and, unless other production approaches are available to control pests, plants grown under organic conditions may increase their synthesis of plant secondary metabolites as a means of chemical defense. While studies have not specifically looked at the influence of production methods (organic vs. conventional) upon natural plant toxins, it is reasonable to assume that the same factors that increase the production of plant secondary metabolites of nutritional interest under organic production methods would lead to similar increases in the production of natural plant toxins.

It can also be argued that the use of synthetic pesticides such as insecticides and fungicides in conventional food production can reduce the potential for plant pathogen growth and subsequent production of mycotoxins, and some studies have demonstrated this effect. In other studies comparing mycotoxins levels found in organic and in conventional foods, however, the levels of mycotoxins were found to be lower in organic foods, and it has also been suggested that organic practices such as longer crop rotation, and substitution of animal or green manures for synthetic fertilisers my reduce the risk of plant infection by pathogens (Van Bruggen, 1995).

Scientifically, it is very challenging to be able to perform appropriate field tests to determine if there are differences in the chemical composition of organic foods and conventional foods. Natural variations in plant secondary metabolite levels exist, and adequate studies need to ensure that the organic and conventional foods are produced using similar conditions with respect to cultivar (variety), soil type and quality, and climate. Sample collection, handling, and analytical procedures need to be standardized for the two different production methods as well.

It is premature to state that either organic or conventional food production practices are safer with respect to naturally occurring toxins. Theoretical arguments suggest that naturally occurring toxins in organic foods may be higher, but these arguments have not been conclusively demonstrated to date from comparative studies in the field. Even in the event that one production method was associated with higher levels of naturally occurring toxins in foods than the other, any health impacts stemming from these differences would be based upon the differential doses of the toxins consumed. In general, the toxicological significance of the possible doses of naturally occurring toxins that humans may be exposed to from either conventional or organic foods has not yet been established.

References

Ames, B.N., Magaw, R., and Gold, L.S., 1987, Ranking possible carcinogenic hazards, *Science* **236**: 271–280.

Ames, B.N., Profet, M., and Gold, L.S., 1990, Dietary pesticides (99.99% all natural), *Proc. Natl. Acad. Sci. USA* **87**: 7777–7781.

Ames, B.N., and Gold, L.S., 1991, Risk assessment of pesticides, *Chem. Eng. News* **69**: 28–32, 48–49.

Arino, A.A., and Bullerman, L.B., 1993, Growth and aflatoxin production by *Aspergillus parasiticus* NRRL 2999 as affected by the fungicide iprodione, *J. Food Prot.* **56**: 718–721.

Asami, D.K., Hong, Y.J., Barrett, D.M., and Mitchell, A.E., 2003, Comparison of the total phenolic and ascorbic acid content of freeze-dried and air-dried marionberry, strawberry, and corn grown using conventional, organic, and sustainable agricultural practices, *J. Agric. Food Chem.* **51**: 1237–1241.

Beier, R.C., and Oertli, E.H., 1983, Psoralen and other linear furanocoumarins as phytoalexins in celery, *Phytochemistry* **22**: 2595–2597.

Beier, R.C., and Nigg, H.N., 1994, Toxicology of naturally occurring chemicals in food, in: *Foodborne Disease Handbook: Diseases Caused by Hazardous Substances*, Hui, Y.H., Gorham, J.R., Murrell, K.D., and Cliver, D.O., eds., Marcel Dekker, New York, pp. 1–186.

Birzele, B., Meier, A., Hindorf, H., Kramer, J., and Dehne, H.W., 2002, Epidemology of *Fusarium* infection and deoxynivalenol content in winter wheat in the Rhineland, Germany, *Eur. J. Plant Pathol.* **108**: 667–673.

Bourn, D., and Prescott, J., 2002, A comparison of the nutritional value, sensory qualities, and food safety of organically and conventionally produced foods, *Crit. Rev. Food Sci. Nutr.* **42**: 1–34.

Calori-Domingues, M.A., and Fonseca, H., 1995, Laboratory evaluation of chemical control of aflatoxin producton in unshelled peanuts (*Arachis hypogaea* L.), *Food Addit. Contam.* **12**: 347–350.

Carbonaro, M., and Mattera, M., 2001, Polyphenoloxidase activity and polyphenol levels in organically and conventionally grown peach (*Prunus persica* L., Regina bianca) and pear (*Pyrus communis* L., cv. Williams), *Food Chem.* **72**: 419–424.

Caris-Veyrat, C., Amiot, M.J., Tyssandier, V., Grasselly, D., Buret, M., Mikolajczak, M., Guilland, J.C., Bouteloup-Demange, C., and Borel, P., 2004, Influence of organic versus conventional agricultural practice on the antioxidant microconstituent content of tomatoes and derived purees; consequences on antioxidant plasma status in humans, *J. Agric. Food Chem.* **52**: 6503–6509.

Chourasia, H.K., 1992, Control of aflatoxin production with fungicides, *Nat. Acad. Sci. Letters (India)* **15**: 243–246.

Cirillo, T., Ritieni, A., Visone, M., and Cocchieri, R.A., 2003, Evaluation of conventional and organic Italian foodstuffs for deoxynivalenol and fumonisins B_1 and B_2, *J. Agric. Food Chem.* **51**: 8128–8131.

Coulombe, R.A., 2000, Natural toxins and chemopreventives in plants, in: *Food Toxicology*, Helferich, W., and Winter, C.K., eds., CRC Press, Boca Raton, FL, pp. 137–162.

Czerwiecki, L., Dzajkowska, D., and Witkowska-Gwiazdowska, A., 2002, On ochratoxin A and fungal flora in Polish cereals from conventional and ecological farms. Part 1: Occurrence of ochratoxin A and fungi in cereals in 1997, *Food Add. Contam.* **19**: 470–477.

Dercks, W., Trumble, J.T., and Winter, C.K., 1990, Impact of atmospheric pollution alters linear furanocoumarins content in celery, *J. Chem. Ecol.* **16**: 443–454.

Duncan, C., Li, H., Kykhuizen, R., Frazer, R., Johnston, P., MacKnight, G., Smith, L., Lamza, K., McKenzie, H., Batt, L., Kelly, D., Golden, M., Benjamin, N., and Leifert, C., 1997, Protection against oral and gastrointestinal diseases: Importance of dietary nitrate intake, oral nitrate reduction and enterosalivary nitrate circulation, *Comp. Biochem. Physiol.* **118A**: 939–948.

El-Morshedy, M.M.F., and Aziz, N.H., 1995, Effects of fenamiphos, carbofuran, and aldicarb on zearalenone production by toxigenic *Fusarium* spp. Contaminating roots and fruits of tomato, *Bull. Environ. Contam. Toxicol.* **54**: 514–518.

Finamore, A., Britti, M.S., Roselli, M., Bellovino, D., Gaetani, S., and Mengheri, E., 2004, Novel approach for food safety evaluation. Results of a pilot experiment to evaluate organic and conventional foods, *J. Agric. Food Chem.* **52**: 7425–7431.

Friedman, M., and McDonald, G.M., 1997, Potato glycoalkaloids: Chemistry, analysis, safety, and plant physiology, *Crit. Rev. Plant Sci.* **16**: 55–132.

Friedman, M., 2006, Potato glycoalkaloids and metabolites: Roles in the plant and in the diet, *J. Agric. Food Chem.* **54**: 8655–8681.

Hakkinen, S.H., and Torronen, A.R., 2000, Content of flavonols and selected phenolic acids in strawberries and *Vaccinium* species: Influence of cultivar, cultivation site and technique, *Food Res. Int.* **33**: 517–524.

Jorgensen, K., and Jacobsen, J.S., 2002, Occurrence of ochratoxin A in Danish wheat and rye, 1992–1999, *Food Add. Contam.* **19**: 1184–1189.

Komives, T., and Casida, J.E., 1983, Aciflourfen increases the leaf content of phytoalexins and stress metabolites in several crops, *J. Agric. Food Chem.* **31**: 751–755.

Leblanc, J.C., Malmauret, L., Delobel, D., and Verger, P., 2002, Simulation of the exposure to deoxynivalenol of French consumers of organic and conventional foodstuffs, *Reg. Toxicol. Pharmacol.* **36**: 149–154.

Miceli, A., Negro, C., Tommasi, L., and De Leo, P., 2003, Polyphenols, resveratrol, antioxidant activity and ochratoxin A contamination in red table wines, controlled denomination of origin (DOC) wines and wines obtained from organic farming, *J. Wine Res.* **14**: 115–120.

Mikkonen, T.P., Maatta, K.R., Hukkanen, A.T., Kokko, H.I., Torronen, A.R., Karenlampi, S.O., and Karjalainen, R.O., 2001, Flavonol content varies among black currant cultivars, *J. Agric. Food Chem.* **49**: 3274–3277.

Mitchell, A.E., Hong, Y.J., Koh, E., Barrett, D.M., Bryant, D.E., Denison, R.F., and Kaffka, S., 2007, Ten-year comparison of the influence of organic and conventional crop management practices on the content of flavonoids in tomatoes, *J. Agric. Food Chem.* **55**: 6154–6159.

Murphy, P.A., Hendrich, S., Landgren, C., and Bryant, C.M., 2006, Food mycotoxins: An update, *J. Food Sci.* **71**: R51–R65.

NRC, 1996, *Carcinogens and Anticarcinogens in the Human Diet*, National Research Council, National Academy Press, Washington, DC, 417p.

Organic Trade Association, 2006, *U.S. Organic Industry Overview. OTA's 2006 Manufacturer Survey*, Organic Trade Association, Greenfield, MA.

Park, D.L., Ayala, C.E., Guzman-Perez, S.E., Lopez-Garcia, R., and Trujillo, S., 2000, Microbial toxins in foods: Algal, fungal, and bacterial, in: *Food Toxicology*, Helferich, W., and Winter, C.K., eds., CRC Press, Boca Raton, FL, pp. 93–136.

Schollenberger, M., Jara, H.T., Suchy, S., Drochner, W., and Muller, H.M., 2002, *Fusarium* toxins in wheat flour collected in an area in southwest Germany, *Int. J. Food Microbiol.* **72**: 85–89.

Seligman, P.J., Mathias, C.G., O'Malley, M.A., Beier, R.C., Fehrs, L.H., Serrill, W.S., and Halperin, W.E., 1987, Phytophotodermatitis from celery among grocery store workers, *Arch. Dermatol.* **123**: 1478–1482.

Sydenham, E.W., Thiel, P.G., Marasas, W.F.O., Shephard, G.S., Van Schalkwyk, D.J., and Koch, K.R., 1990, Natural occurrence of some *Fusarium* mycotoxins in corn from low and high esophageal cancer prevalence areas of the Transkei, Southern Africa, *J. Agric. Food Chem.* **38**: 1900–1903.

Van Bruggen, A.H.C., 1995, Plant disease severity in high-input compared to reduced input and organic farming systems, *Plant Dis.* **79**: 976–982.

Veberic, R., Trobec, M., Herbinger, K., Hofer, M., Grill, D., and Stampar, F., 2005, Phenolic compounds in some apple (*Malus domestica* Borkh) cultivars of organic and integrated production, *J. Sci. Food Agric.* **85**: 1687–1694.

Whole Foods Market, 2005, *Whole Foods Market Organic Trend Tracker*, Whole Foods Market, Austin, TX.

Williams, P.R.D., and Hammitt, J.K., 2001, Perceived risks of conventional and organic produce: Pesticides, pathogens, and natural toxins, *Risk Anal.* **21**: 319–330.

Winter, C.K., 1990, Natural toxins in the human food chain, in: *Chemicals in the Human Food Chain*, Winter, C.K., Seiber, J.M., and Nuckton, C.F., eds., Van Nostrand Reinhold, New York, pp. 2218–2238.

Winter, C.K., and Davis, S.F., 2006, Organic foods, *J. Food Sci.* **71**: R117–R124.

Woese, K., Lange, D., Boess, C., and Bogl, K.W., 1997, A comparison of organically and conventionally grown foods – Results of a review of the relevant literature, *J. Sci. Food Agric.* **74**: 281–293.

Worthington, V., 2001, Nutritional quality of organic versus conventional fruits, vegetables, and grains *J. Alt. Complement. Med.* **7**: 161–173.

Young, J.E., Xhao, X., Carey, E.E., Welti, R., Yang, S.S., and Wang, W., 2005, Phytochemical phenolics in organically grown vegetables, *Mol. Nutr. Food Res.* **49**: 1136–1142.

Index

The manufacturer's authorised representative in the EU is Springer Nature Customer Service Centre GmbH, Europaplatz 3, 69115 Heidelberg, Germany. If you have any concerns regarding our products, please contact ProductSafety@springernature.com

Printed and bound by CPI Group (UK) Ltd, Croydon, CR0 4YY
07/09/2026
02211225-0002